# The Complete
# RANGER DIGEST
## VOLUMES VI - IX
### TIPS, TRICKS & INFO
### Revised

by
**RANGER RICK F. TSCHERNE**

***LOOSE CANNON ENTERPRISES***
*Paradise, CA*

No part of this book may be reproduced in any form whatsoever, or by any means without the prior written permission of the publisher, except for brief quotes used in connection with reviews written specifically for the inclusion in a magazine, news article, or website specializing in such reviews.

Copyright © 2017 Rick F. Tscherne
All Rights Reserved

*The Complete Ranger Digest: Vol. 6-9*
First edition - Revised
ISBN-13: 978-1-944476-33-5

**Produced and published by Loose Cannon Enterprises**
**www.loose-cannon.com**

This publication is also available as an ebook at Amazon.com, iTunes and other ebook sellers.

**LEGAL DISCLAIMER:** THIS BOOK IS NOT INTENDED, NOR DOES IT PURPORT TO PROVIDE FULLY SAFE OR ACCURATE ADVICE OR INFORMATION.
MANY OF THE ITEMS/CONCEPTS/IDEAS PRESENTED MAY BE HAZARDOUS, WRONG, AND OR DANGEROUS WITHOUT ADEQUATE TRAINING. THE MATERIAL IS INTENDED FOR INFORMATIONAL PURPOSES ONLY AND NOT TESTED OR CERTIFIED AS 'SAFE'. NO WARRANTY, NEITHER EXPRESSED NOR IMPLIED, IS CREATED WITH RESPECT TO YOUR USE OF THE INFORMATION CONTAINED IN THIS PUBLICATION. YOUR USE OF ANY OR ALL OF THE INFORMATION CONTAINED HEREIN IS AT YOUR OWN RISK AND WITH YOUR OWN JUDGMENT. THIS PUBLICATION IS SOLD AS IS.

# TABLE OF CONTENTS

| | |
|---|---|
| The Ranger Digest VI | 6 |
| The Ranger Digest VII | 109 |
| The Ranger Digest VIII | 203 |
| The Ranger Digest IX | 299 |
| About the Author | 412 |

## ABOUT THIS NEW EDITION

Welcome to Part II.

It has been a long time since Sergeant Rick Tscherne wrote the first RANGER DIGEST book back in 1988, since then he wrote a total of nine volumes of what Rick would call "No B.S. guides of hard-earned Tips, Tricks and Info for Soldiers and Outdoorsmen." The series has gone on to sell over 750,000 copies in the U.S. and overseas, and have been well received by American troops of all services and most ranks.

Since then the books have gone out of print for a while and were difficult to find. Seeing this as an opportunity to bring these classic tips and information to a new generation of soldiers and infantrymen, Loose Cannon partnered with the author to create a new Revised edition in digital eBook formats. Many items/tips may be a bit dated, particularly when referring to various equipment no longer currently used by the U.S. military or regarding products that may have been for sale when the guides were originally published. We have taken efforts to try to add updated info and or supplier links to various products/equipment mentioned in the books, but were not always able to recommend a suitable replacement. As such, note that some items or information may be out of date and thus was left in the guide to provide a semblance of what the original guide included within the context as a whole. The goal was to provide as much of the original structure and flavor of the original books in creating this new edition.

For more information about Ranger Rick, and his current projects see the Author Bio section at the end of the book.

    - S. Hutchins , Publisher - Loose Cannon Enterprises

- - -

### AUTHOR'S DISCLAIMER

The Ranger Digest is a series of training handbooks strictly designed for US military personnel and NOT for civilians. The author cannot be held liable for injuries or deaths caused by these tips & tricks. You are advised to "use them at your own risk!"

**BOOK SIX**

# FOREWORD

I know what your thinking, just when you thought you learned all the secret tips and tricks-of-the-trade from my previous *Ranger Digest* handbooks, along comes another. Right?

Well, that's what the Ranger Digest series is all about, learning new ways in how to survive in a military outdoor environment. And if this is your first Ranger Digest handbook, well, you ain't seen nothing yet. There's a Ranger Digest I, II, III, IV, & V handbook. (See enclosed book advertisement)

I want to thank all of you readers out there who took the time to write in either sharing a tip or trick or two with me or just to say "Keep Up The Good Work Ranger Rick!" I appreciate your kind letters, compliments, and recognition even if the United States Army never appreciated my field tips and tricks.

Be advised that over the years I have tried my best to get the Department of the Army to accept my many training tips, tricks, and ideas. But each time I submitted them through the proper military channels, they always came back disapproved stating, "Sorry, not feasible for combat related field conditions." But it's obvious they were wrong, other wise there wouldn't be a Ranger Digest I, II, III, IV, V, & VI today, right?.

Well, that's about it for now, guys and gals. As I always say in all my Ranger Digest books, if you got the time to drop me a line or two, I'd appreciate hearing from you. Whether it's to tell me about a tip or trick, say hello, or just to tell me I'm doing a good job or a lousy one. Because your letters, comments, and input DO COUNT. Take care for now.

PS: If someone should ask where you learned these tips and tricks from, you just tell 'em... *"From my buddy Ranger Rick."*

*Rick F. Tscherne*

# SPECIAL THANKS

This page is dedicated to those who took the time to write in sharing their favorite field tips and tricks with the rest of us. And if it wasn't for these caring soldiers, readers, and leaders there wouldn't be a *Ranger Digest VI* today. Hurraah!

## RANGER DIGEST VI CONTRIBUTORS

CPT J. D. THOMPSON   SFC EDGAR W. DAHL   SSG CRAIG MARTS

SSG SEAM P. GILDAY   SGT MICHAEL PILSNER   SGT RICH REITZ

SGT ROB ROBINSON   SGT LANCE HEFINGTON   SGT PAUL R. HEADEN

SGT LEROY WOLPERT   SGT MATTHEW O'BOYLE   SGT MARTIN DUDEL

CPL GREG ESTTLAKE   SPC DAVID E. BROWN   SPC ROBERT WARMACK

SPC DAVID YIM   PFC RICH FONGEALLAZ   PFC CHRISTINE HUTMAN

PFC RYAN C. SHIPLEY   CSGT R. PRIMERANO   CSGT JOSHUA BERRIER

C. J. KUCHINSKAS   LARRY R. STANTON   DAVID J. WHITE

RICHARD M. DOBSON   SPC SPURLOCK   ALAN FOSTER

*The Complete Ranger Digest: Vol. VI*

# RANGER DIGEST UPDATE

HOME-MADE FIELD AMMO - (Ref: Ranger Digest I) A convenient place to store your "Home-Made Field Ammo" to insure you'll always have it when you need it the most, is in the butt stock of your weapon. NOTE: Used strictly for hunting and survival purposes only.

FILLING CANTEENS - (Ref: Ranger Digest II) *MR. JAMES FERGUSON* sent me a tip on how he was able to fill a canteen from a 5 gallon water container without spilling a drop. Which is simply by using a 6 foot long garden hose and keeping it stored always inside the container itself for future uses.

Well, I'd like to take his idea a little bit further, why not keep a few six foot hoses inside a water container so that you can fill "several" canteens simultaneously. Why stand around or wait in a slow moving line for each soldier to fill their canteens when you can speed up the process, right?

Another useful tip and idea for units, is to purchase several of these cheap D-battery operated water pumps, they work great inside a plastic or metal water container. They'll save you time in filling and spilling precious water. They take 2 X D size (BA30) batteries and can fill 4 X Quarts a minute.

```
       HANOVER HOUSE
       P.O BOX # 2
       HANOVER,   PA.
       17333-0002
```

Price: $ 19.95 each

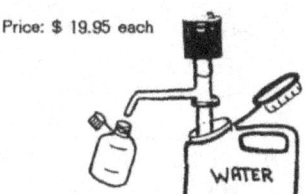

WATER BOTTLES & TUBING - (Ref: Ranger Digest IV) *Larry R. Stanton* says, "Rather than using a drinking tube with a military canteen, just carry a few plastic water bottles and modify one of the plastic bottle caps and just change bottles. The tubing can be purchased at any pet or fish aquarium supply store. When making a hole in the cap, be careful as to not to make it too large, or the tubing and water will both come out unexpectedly. Add a little bit of silicone (NOT glue) to insure it doesn't come apart.

*Rick F. Tscherne*

UN JOB OPPORTUNITIES - (Ref: Ranger Digest IV) Though I mentioned there are many United Nation job opportunities available. I failed to mention that there are also many private companies and contractors who "work for" the UN too. Here's a short list (below) of a few companies who hire only english speaking people.

If your looking for adventure and you want to make a lot of tax-free US dollars, these companies just might have some interesting job opportunities for you. (No, not mercenary work.) When writing for information about what kind of jobs are available, send copies of your military record, licenses, certificates, etc.

HOT: If you really want to improve your chances of landing a job, then you should hire a professional resume writer. By paying a professional resume writer $50-$100 to make you look good on paper, it will dramatically improve your chances of finding and landing a job.

```
DEFENSE SYSTEM LIMITED 7th Floor, Egginton House 25-28
Buckingham Gate. London SW1E 6LD England

PARC TECHNICAL SERVICES 24 Adam and Eve Mews London W8
6UJ England

ALLMAKES LIMITED 176 Milton Park Abingdon, Oxfordshire
OX14 4SW  England

DYNA CORPORATION 1st Floor 43 Queens Rd. Aldershot
Hampshire GU 113 JE

BROWN & ROOT, Fax  tt 713-676-5111 (Note: USA company.)
```

UPDATE: A good online source for finding Private military contractor job openings: www.Feraljundi.com

CELLULAR PHONES - (Ref: Ranger Digest V) I mentioned how easy it was to purchase and share the cost of a cellular telephone with unit members. Well, according to the Wall Street Journal, AT&T is pushing ahead with a plan to start providing wireless communication services to their customers.

They say AT&T plans on selling cellular telephones for only $1 each. That's right, only one dollar! Provided.... you agree to stick to their cellular service plan for a certain period of time. Thought this sounds like a terrific deal for field soldiers, I'd strongly recommend that you shop around and compare costs before signing up for one.

*The Complete Ranger Digest: Vol. VI*

# ALWAYS CARRY A PRE-MADE EMERGENCY TOURNIQUET

When an arm or leg is penetrated by a rifle or grenade projectile, chances are it will probably rupture an artery. If not stopped in time, a soldier could bleed to death within minutes.

Instead of waiting, struggling, or looking around tor something to make a tourniquet when the wound occurs, carry a "pre-made tourniquet" inside your first aid pouch. A piece ot wood attached to some 550 parachute cord makes an excellent torniquet.

Soldiers need to practice how to emplace a tourniquet on themselves using one hand (not both) just in case a medic or buddy isn't available to assist them due to combat conditions,

By practicing and knowing how to emplace a torniquet on yourself ONE-HANDED, it will prepare you for any type of a life threatening wound and increase your chances of survival.

SPECIAL NOTE: To insure casualties are quickly evacuated from a battle field, individual soldiers should carry their own casualty card inside their first aid pouches. Along with all the pertinent information filled in (name, unit, blood type, etc) except the type of wound they sustained.

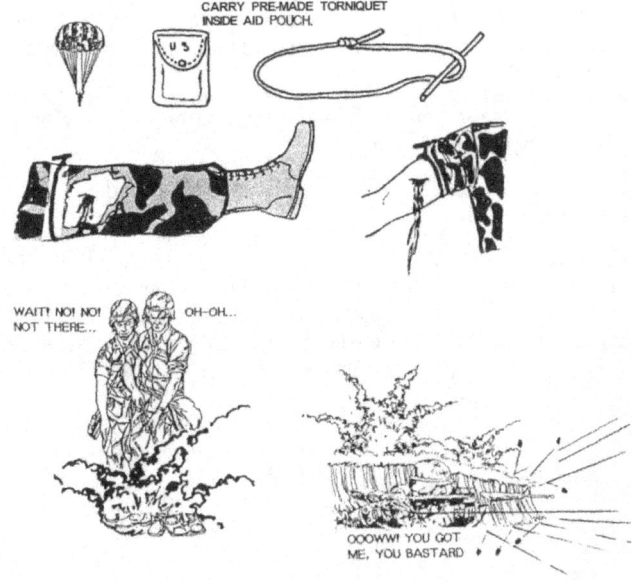

*Rick F. Tscherne*

UPDATE: Recommended commercial tourniquets
C-A-T (Combat Application Torniquet: http://www.combattourniquet.com/
SOF-TT http://www.remotemedical.com/SOF-Tactical-Tourniquet

# PRE-MADE RAPID RAPPEL SEAT

*Submitted By: C.J Kuchinskas* (A New York City Medic)

Dear Ranger Rick...

"I recently bought all five copies of your Ranger Digest handbooks and would like to submit to you a few of my own tips, tricks, and ideas for your next book."

In reference to your Belgium Commando rapid rappel seat, I use an entirely different and much easier method. Unlike YOUR rappel seat, mine is a lot quicker to put on and I can still make moderate leaps & bounds off cliffs. (See drawing below.)

When making or using this type of a rappel seat....

1. Use either civilian mountaineering rope or military "assault rope." It's much more flexible and won't coil or bunch up like the standard issued military rappel rope.

NOTE TO READERS: Civilian mountaineering and military assault rope is "braided cores of strands," while standard issue military rappel rope is "twisted rope." (See drawing)

2. Measure the rope to your body and a little bit more than what you need after the knots have been tied in place.

3. Use the civilian made "locking" snap links and NOT the military issued "spring loaded" snap links.

**WARNING**: Rappeling is extremely dangerous and could be very hazardous to your health. If you have never rappelled before or if you are not experienced in the art of military rappeling, DO NOT ATTEMPT to try out this new rappel seat. Only those who are well trained or fully qualified in the art of military rappeling may use it.

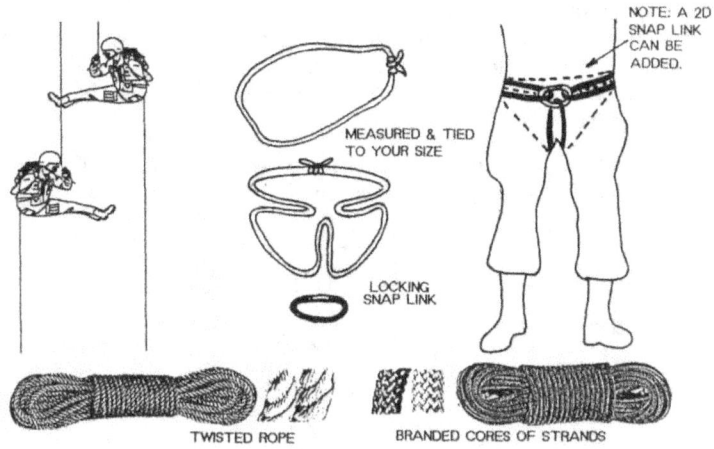

## BOOT LIGHTS

Moving tactically at night through thick vegetation can be tricky and difficult, especially when unit members must follow one another in a single file.

Well, have you ever seen those new type of tennis shoes with the little "motion lights" built into the heels? You know what I'm talking about, they're mini electric lights that blink on & off as one runs or walks.

Well, I made something similar to this except I used a pair of combat boots and a couple of mini lightsticks. And guess what? Yep, it worked terrific! Here's what I did..

STEP 1: Buy a package of those 1 1/2 inch mini lightsticks that are sold in almost every military supply store. Take one out and measure the width of it. Now take a drill bit "slightly smaller" than what you actually need and attach it to your drill gun. Now drill a hole completely through the center portion of the boot heel.

STEP 2: Take the mini lightstick, bend, snap, shake, and try to place it inside the hole. If it doesn't go in with a little bit of force, it's too small and needs to be drilled larger. If it slides in too easy, you screwed up and drilled the hole too large.

The lightstick should go into the hole with only a little bit of force with the aid of an M16 cleaning rod. To remove the lightstick, push it all the way through until it pops out the other end of the hole.

You can control the amount of light simply by moving the lightstick further in or out of the boot heel with the M16 cleaning rod. If you shove the lightstick all the way forward to the sole, then the light can only be seen when the foot is raised up off the ground. Experiment and choose the amount of glow light you need.

SPECIAL NOTE: Only one boot heel is needed and not all unit members need a boot heel light, only selected personnel.

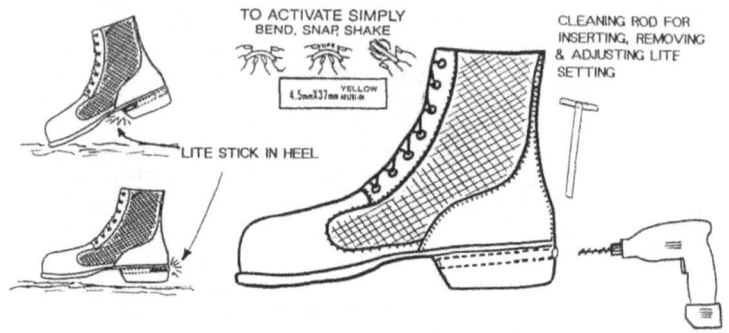

## SPACE BLANKET MODIFICATIONS

Lately, I've been getting a lot of mail from readers claiming they much prefer to use the OD green or camouflage space blankets rather than the poncho liner & poncho. They're lightweight, tough, durable, and reflect up to 80% of your body heat back to you. Plus you can use it as both, a shelter and a blanket.

Well, if you've been reading my other Ranger Digest handbooks, you know that I'm not all that crazy about using space blankets in the field. They're OK for emergency situations, but I still prefer to use a poncho liner & poncho or the ol' military sleeping bag.

Well, I'm not gonna argue with you any longer or try to convince you which is better. But for those of you who prefer to use space blankets in the field, you may want to make a few modification like what I did to mine recently.

A_. Install a sleeping bag zipper along the edges so that it can be used either as an emergency sleeping bag or sleeping bag cold weather/waterproof ground cover.

B_. Install a 12-18 inch zipper in the center so that you can wear it over your head as a wet weather or cold weather emergency poncho.

C_. Attach some 550 para-cord so that you will always have it readily available in case you should ever need to use it as an emergency shelter.

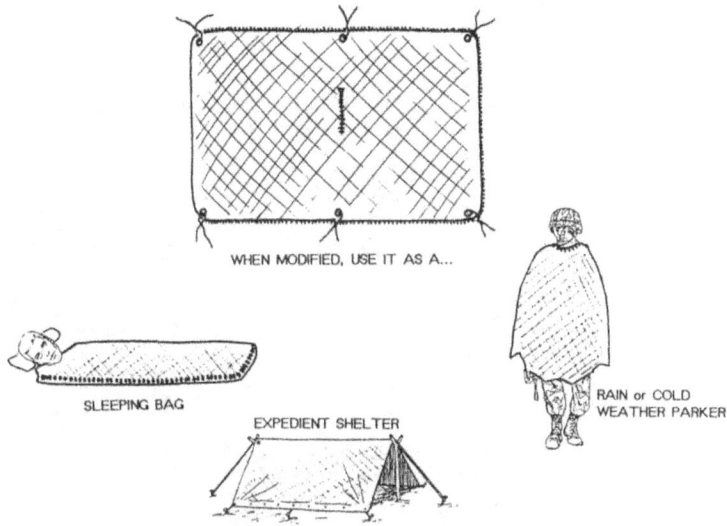

## M80 CLAYMORE MINE SIMULATOR

All soldiers hate to carry worthless things on an FTX. In particular, a weapon system that doesn't go BOOM, BANG or POW! I don't blame'em, if your going to "simulate" firing a weapon, you may as well "simulate" carrying or taking one to the field. Right?

Well, when I was back in the Ranger Battalion (1978-80) I came up with a few training devices that made some of these weapon systems go "BOOM!" And boy, did the troops love'em.

One particular device that they really enjoyed was my M80 Claymore Mine Simulator. Here's what you need to make one:

> 1) 50 feet of rubber coated "double stranded" wire. (Claymore training wire or "stereo speaker wire" will do nicely.)
> 2) 2 X nails (or screws).
> 3) Tape (100 mph, duck, or electrical tape).
> 4) 4 X "C" or "D" batteries.
> 5) An electric switch or "home-made" firing device.
> 6) Male and female electrical plugs.
> 7) 2X4 piece of wood about 8 inches in length.
> 8) 1 X large drill bit, 1 X small drill bit, and a drill.
> 9) Some M80 TOW Blast Simulators. (You can get these from any Anti-Tank Sect/Pit, though you may have to do a little "wheeling & dealing" to get 'em.)

NOW HERE'S WHAT YOU GOTTA DO: Take your large drill bit and drill two holes 2-3 inches deep into one narrow side of the wood, (see drawing). These holes must be wide and long enough to place 2 (tree or bush) sticks inside for legs.

On the wide side of the wood, hammer in 2 nails about 1 inch apart and about 3/4 of the way down in so that they remain exposed and protruding out. On the left and right side of these nails, drill two "smaller holes" through the wood.

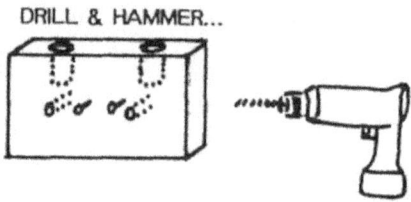

Go to the end of your wire, separate'em about 6 inches down and strip 1-2 inches of rubber coating off the ends. Twist together the "hair thin" strands of wire so that they are not loose and run them through each of the wooden holes and secure to the protruding nails.

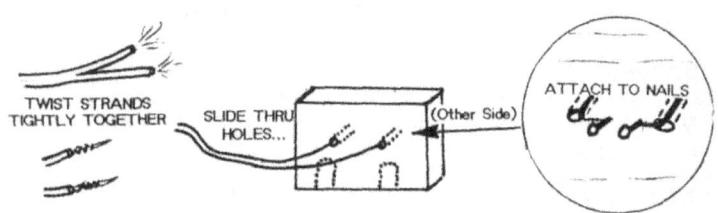

Take 4 X batteries, place them end-to-end (positive-to-negative) and secure them together with some tape. To insure the batteries remain straight and touching one another, tape along the sides of the batteries some pieces of wood.

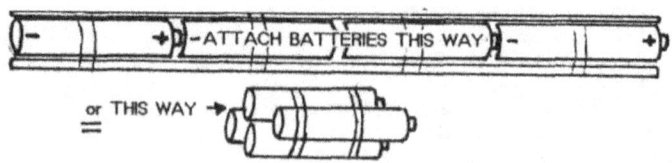

Cut off a "single" 24 inch piece of wire from the double stranded wire. Now remove about 1 inch of rubber coating from the ends of the wire (all 4 ends).

Connect the "short end" of the double wire to the negative (-) end of the battery with a thick rubber band. Then connect the one end of the "single" cut-off piece of wire to the positive ( + ) end of the battery with the same thick rubber band.

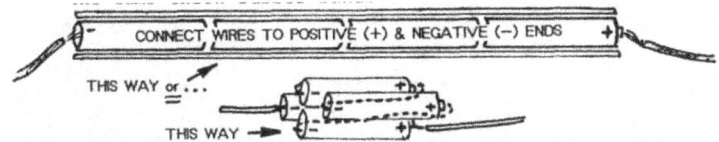

Take your electrical switch or home-made "firing device" and connect it to the other end of the wires that lead away from the batteries. (Note: If need, later on you may shorten these two wires to make them even or shorter in length).

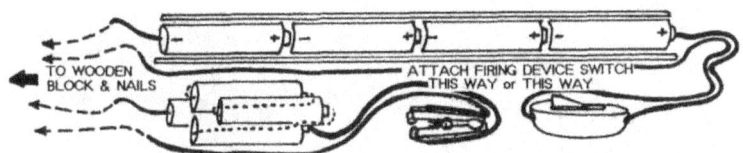

To insure you have a good electrical hook up, place your moist "tongue" to the two nails and press down on the firing device. You should receive a "mild" electric shock. If you didn't, then there's a problem either with the...

A. Batteries not properly touching one another end-to-end.

Solution: Re-position, re-tape, & add more rubber bands B. Batteries are dead or weak. Solution: Replace batteries. C. Wiring is not properly connected to the batteries or to

the firing device. Solution: Re-position and re-tape. D. Wire is defective. Solution: Replace wire.

Once you are sure that you have a good electrical circuit, measure 12 inches below the batteries and cut through both wires and install the male/female plugs onto both of the ends.

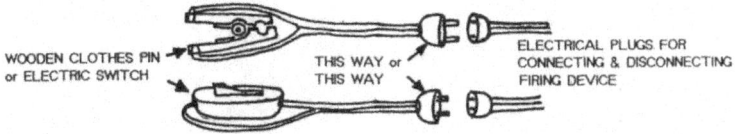

After connecting the male & female plugs together, check your electrical circuit one more time by placing your tongue to the two nails and pressing down on the firing device. Again, you should have received a "mild" electric shock.

The reason for installing these plugs into the wire, is so that you can disconnect and carry the "firing device" with you like the real thing. Plus it will also prevent some "idiot" from playing around with the firing device while your down range installing the M80. (Makes sense, don't it?)

If everything checks out OK, then your ready to test fire your first M80 TOW blast simulator. Select an area free of personnel and carefully set up your M80 Claymore Mine. Hook up your firing device, look down range, press down on the firing device and you should hear a "BOOM!"

**WARNING**: Use extreme CARE and CAUTION when using M80s, they can be DANGEROUS if they are not used or handled correctly. USE COMMON SENSE and NEVER place this training device where it can possibly cause harm or injuries to a fellow soldier.

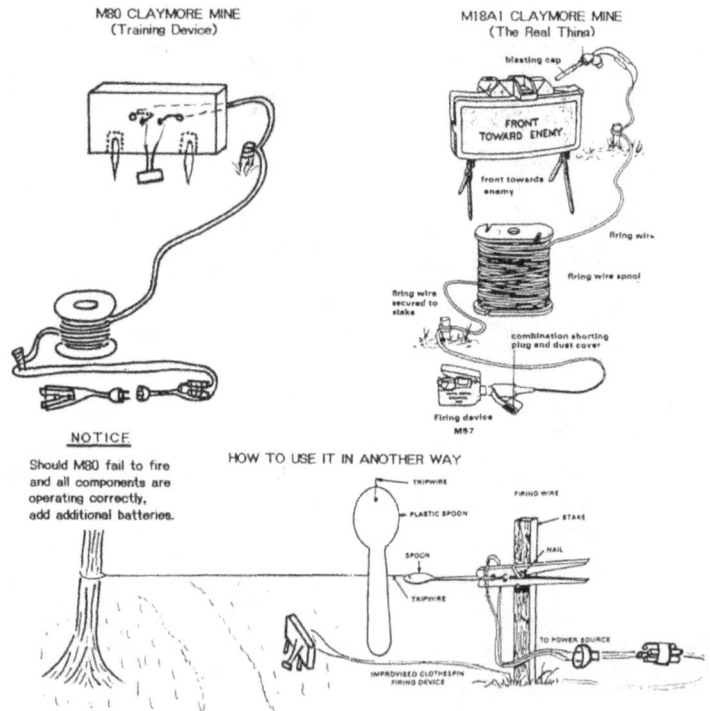

*The Complete Ranger Digest: Vol. VI*

# MILITARY JOKES

Three Army colonels were arguing over who had the toughest and bravest soldiers in the entire United States Army. They decided to settle the argument with a test of courage.

The first colonel, from the 82d Airborne Div., ordered one of his men to board a plane and to jump from it at an altitude of 500 feet. The soldier yelled "Airborne, sir!" Then boarded the plane and successfully jumped from it at 500 feet.

The Airborne colonel turns to the other two and says, "Did you see that, now that took a lot of courage to do."

"That's nothing," says the colonel from the 10th Special Forces Group. He then orders one of his men to board the plane and to jump from it at 300 feet with only one parachute on his back. The soldier shouts "Huaah, Can do, sir," then boarded the plane and successfully jumped from it at 300 feet while only wearing one parachute, (no reserve). The Special Forces colonel turns to the other two and says, "Now that's what I call bravery."

"Big deal!" said the colonel from 75th Rangers. He then commanded one of his Rangers to board the plane and to jump from it at an altitude of 300 feet WITHOUT wearing a parachute. The Ranger turns to the colonel and yells, "Sir, you gotta be outta your f—ken mind!" The colonel then turns to the other two and says, "Now that took a lot of balls!"

An 82d Airborne Division NCO was tasked to teach a class of West Point cadets on how to drive a military vehicle at night while wearing night vision goggles.

The NCO warned the cadets that when wearing night vision goggles at night, it's a little difficult driving and keeping the vehicle on the road at the same time. One puzzled cadet raised his hand and asks, "But sergeant, if it's that's difficult, how do we avoid going off the road?

"Simple," replied the NCO, "when your driving down a road and you start hearing or feeling thumping sounds on one side of the vehicle. That means you need to start steering the vehicle onto the other side of the road away from the thumping sounds."

Sir, the only difference between us and the Boy Scouts is that the Boy Scouts is the Boy Scouts have adult supervision.

## CLEANING MILITARY GEAR

I received some tips from a few readers on the best way to clean TA-50 gear after returning back from the field. And to be honest with you, most of them were pretty common sense, except one.

This one fella claims he finds it a lot easier to clean his gear if (now get this...) he puts it in the "TOILET BOWEL" first. He says he just keeps flushing the toilet until all the heavy dirt is removed, then he finishes the job by wearing it into shower. (Peeeyew!)

Well, if you think this idea "stinks,"then try some of these;

A. When on stand-down in the field, try to find a nearby creek, stream, or pond to rinse off the thick dirt from your gear. This will make it much more easier and faster to clean when you get home and more free time for drinking, partying, and "SEX."

B. If your unit has a vehicle washing point, lay or hang your gear up on a vehicle and blast the shit off with the "JET HOSE."

C. If your in a light infantry unit and don't have a vehicle washing point, no problem. Get a few of your buddies together and head on down to your local self-service car wash. You know, the ones with the hand held jet hoses? Place the switch first to soap, then to rinse, but DON'T PUT IT ON WAX. A garden hose will work pretty well too provided you have the right nosel.

D. You can also clean your gear pretty good if you either wear it into the shower or hang it so that the shower head is effectively spraying and hitting the gear. Or just fill the tub with hot water and soap, let it soak, scrub, and rinse.

E. Of course, washing machines work great too. But make sure you take apart all the LBE pieces (first aid, ammo, & canteen pouches) before placing it inside or you may damage the washer. And don't add the same amount of detergent or soap like your washing clothes, too much liquid and soap and you'll eventually discolor/ruin your gear over a period of time.

And of course, the best and easiest way to dry your gear is by either hanging it up outside and leting the air or sun dry it naturally, or by using a clothes dryer. **WARNING:** If you intend to use a clothes dryer, use the big heavy duty coin-operated ones at the laundromat and not your own.

## CS TEARGAS POWDER/CRYSTALS

GAS! GAS! GAS! Man, does this next trick bring back old memories. In 1975, I was deployed to England with my unit (1/509th Airborne Battalion Combat Team) on a NATO Training Exercise. Before we went, our platoon leader ordered a case of CS teargas spray from a store in the US and gave everyone a can. It was supposed to be used just in case we needed it, such as to prevent from being captured. Only one small problem with this, he or we didn't know it was totally illegal and against the law to have in England.

Well, to make a long story short...We rented some vehicles, dressed like civilians, and then drove through the entire exercise area behind enemy lines collecting intelligence. When the designated bad guys (the Brits) finally caught onto our scheme, they tried to capture one of our vehicles at a check point. The recon team was not going to give up so easy, so they just let'em have it at close range. *Bam!* Right smack in their faces at about 12 inches away.

One Brit was rushed to the hospital with complications in breathing and the other one was terribly ill. And our nov famous Recon Platoon leader was relieved of the platoon because of this incident. And not only that, our entire platoon was dismissed from the NATO exercise and sent immediately back home to Italy before we could all be prosecuted. Yep, those were the good old days....

Another CS incident that comes to mind happened at Fort Bragg back in 1985. During a division exercise, our platoon (3/325th) was given a container of "powdered CS crystals" to play around with. Man, is this shit really nasty. And to be honest, none of us really knew how to use it correctly. But we managed to rig up an interesting device that seemed to work out pretty well.

We took a few empty smoke grenade cardboard shipping containers and punctured a hole in one of the ends just big enough for a grenade simulator's fuze to fit in. Then the} were filled about 1/3 full of CS powder, along with the simulator inside, and sealed securely closed with tape.

OK, now your wondering just how in the hell did we fill these containers without choking and coughing ourselves to death, right? Simple, one guy put on his NBC mask and suit. When he was finished filling and sealing them, the outside portion oi the containers were washed off with water to insure nc crystals escaped.

Then all you gotta do is pull the fuze, throw it hard & far, and run like hell. *CAUTION: Danger, Handle With Care!*

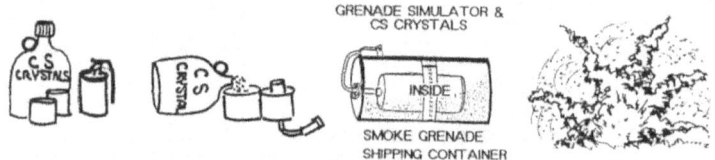

## ALWAYS CARRY YOUR OWN CASUALTY CARD

No one really wants to be a combat casualty, but sometimes shit just happens, you know what I mean? To increase your chances of being quickly evacuated off the battle field, always carry your own casualty card inside your first aid pouch. You should be able to get one from either your supply or battalion aid station.

When you get one, fill in all the pertinent information (Name, Rank, SS#, Blood Type, etc) except the type of wound. You might also want to "scotch tape" a small pencil to the card too. This way if your wounded, conscious or unconscious, the medic will be able to record your condition and vital signs and quickly evacuate you off to a field hospital.

SPECIAL TIP FOR LEADERS: Make it SOP for everyone in your unit to carry a casualty tag and pencil in their first aid pouch.

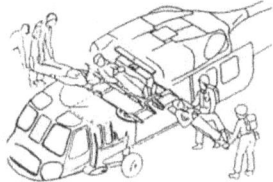

## WALLET SIZE MILITARY RECORDS

Do you want to know how to keep track of all the military information that's on your DA 201/2-1? Easy! Try reducing these documents down to wallet size with a xerox machine. Yea, they may be a little hard to read, but you'll still be able to see the information. So the next time you need to refer to it, just pull out your wallet size copy.

SPECIAL TIPS FOR LEADERS: Make all your men carry reduced copies of their records in their wallets, and or at least keep a copy of these documents in their individual unit files.

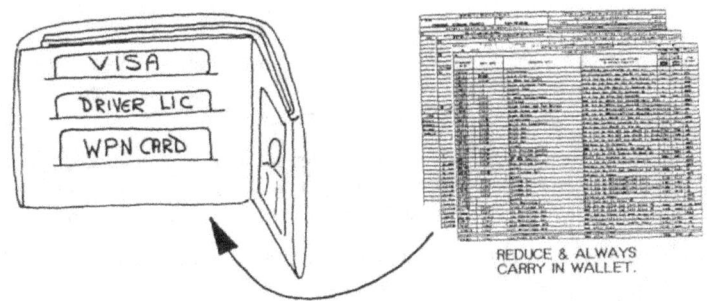

REDUCE & ALWAYS CARRY IN WALLET.

## DO-IT-YOURSELF BUG REPELLENT

*Submitted By: Pfc Christine Hutman*

Pfc Hutman writes, "Most of the tips that I know I think everyone else does too, except maybe this one. 1 learned it when I was in the Marines back at Parris Island."

To help keep mosquito and sand-fleas away, mix some Avon "Skin-So-Smooth" cream with a little bit of alcohol. It will not only keep them away, but it smells pretty good too.

## HAND LOTION & DENTAL FLOSS

*Submitted By: Spc. David Yim*

Hand lotion cream is not just for hands, you can use it instead of shaving cream for the field. It's a lot lighter and easier to pack in a ruck sack.

Another useful item for the field is dental floss. It comes in it's own lightweight compact container and can be used for repairing damaged web gear or for making animal snares.

## MORE M258 USES

*Submitted By: Cpt. Jonathan D. Thompson*

Captain Thompson writes, "In your Ranger Digest handbooks, you talked about the many uses a M258 NBC decon container has. But I was surprised you didn't mentioned the most obvious use, as a crush-proof water-proof cigarette container .

## KEEP THOSE CIG BUTTS DRY

*Submitted By: Sgt. Matthew O'Boyle*

A trick that I learned from a friend of mine was how to keep a pack of cigarettes dry, not in a container, but inside the plastic pack itself. But this trick only works with "soft packs," NOT the hard packs.

Whenever your carrying a pack of cigarettes in your shirt pocket, don't tear off the entire top of the wrapper. Instead, take your Bic lighter and melt a small hole in one of the corners and tear a small hole in the aluminum paper.

Now if your in the field and it should rain, most of your cigarettes in your shirt pocket will stay dry. To insure they stay dryer, place the pack in your pocket up-side-down.

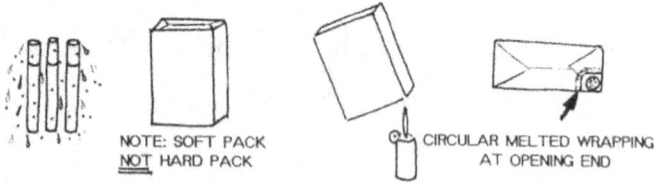

*The Complete Ranger Digest: Vol. VI*

# IMPROVISED RAFT/BOAT

Your not going to believe this next trick. First, take a good hard look at this photo and guess what this boat is made from

"Ding!" Times up, your not going to believe it, but it's made from a "Truck Canvas Cover." No, I'm not BSing you, it's true. It's one of the many tricks that I learned when I went through Belgium Commando School. Those instructors are experts when it comes to adapting and improvising for the field. Oh by the way, that young "dork"on the far left is ME.

Well, I bet you wanta know how to make one of these, right? First of all, you gotta find a 2 1/2 or 5 ton truck canvas cover that doesn't have any holes. Good luck finding one, I haven't seen very many truck canvas covers without any holes. But, if you do find a hole-less one, lay it open on the ground.

Now you'll need to find 12 strong tree poles about half the length of the truck canvas. If the tree poles are too thick -the boat will be too heavy to float. They should be no thicker than the width of 3 fingers. You will also need 8-10 smaller poles about 2 feet tall (same thickness) and a whole bunch of tie-down (commo wire, 550 cord, rope, etc).

Now carefully follow the drawings and my *Do's & Don'ts.*

1. DO tie the tree poles securely together with plenty of tie-down. The more tie-down, the more secure it will be.

2. DO check to insure that the ends of the tree poles are not sharp. If need, cushion ends with ponchos /clothes.

3. DO test the boat in shallow water for flotation, stability and sea worthiness before using it in deep water.

4. DON'T ever enter the boat wearing web gear & boots, place everything on the canvas floor. (Absolutely no rucks.)

5. DON'T ever allow personnel to just jump in, they must enter 2 at a time to help keep the weight distributed / balanced

6. DON'T ever overload the boat with excess personnel and equipment, carry only what it can handle safely.

WARNING: Never allow "non-swimmers" in the boat unless they have some sort of a flotation device. Example: Fluffed up sleeping bag inside a waterproof bag, etc. Also, assign strong swimmers to weak swimmers just in case the boat should come apart and they need help.

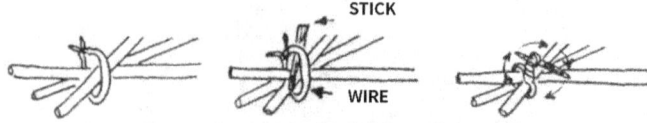

HOW TO WIRE & TIGHTEN POLES TOGETHER

NOTICE TO ALL: This field expedient canvas boat DOES NOT meet Coast Guard Approved Safety Standards. Use at your own risk.

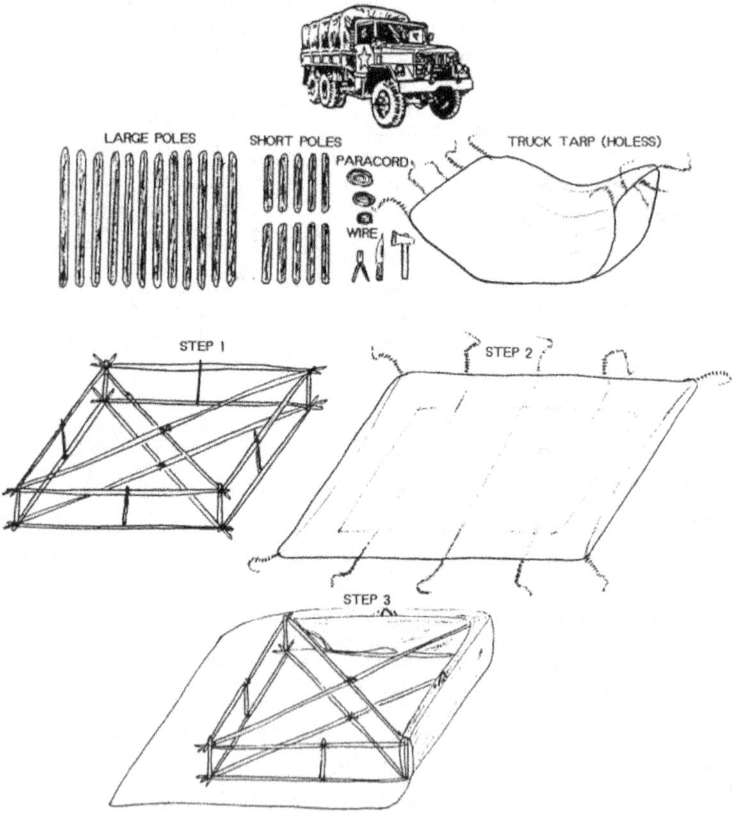

## PACKING FOR A REAL WORLD DEPLOYMENT

*Submitted By: Cpt. Jonathan D. Thompson*

In my tens years as an infantry officer, I've noticed that most of us EMs, NCOs & Officers have two sets of uniforms and boots. One set for the field and one set for garrison.

The field uniforms are usually worn out, frayed, and or torn & repaired in several places. The boots are usually showing wear & tear in the soles and heels, discolored, and have a crack or two somewhere in the leather.

I'd be willing to bet that most of us GRUNTS are guilty of this practice, I know I am. After all, who wants to trash and dirty a new set of BDUs in the field and waste money buying a new set, right?

A TRUE STORY: Right before the invasion of Panama (Operation Just Cause), many soldiers either wore their field uniforms or packed them away in their deployment bags. Most of them thought it was going to be a short "FTX" or battle.

Unfortunately, they failed to take into consideration Panama's hot tropical climate and humidity. And due to this, many soldiers saw their "field uniforms" quickly deteriorate and in some cases disinagrate right before their eyes.

What did the troops learn from this little deployment?

LESSON #1 - Never pack or wear worn out field uniforms & boots for a real world mission. They're OK for FTXs, but they won't last very long in a tropical, desert, or cold weather environment.

LESSON #2 - Leaders (NCOs & Officers) should be held fully responsible for inspecting and insuring all their soldiers wear, pack, and deploy with the best serviceable uniforms and NOT the worn out field BDUs.

Captain Thompson ended his letter by saying, "I hope you will pass along my tips to your readers, even if it helps only one soldier I will feel it was worth mentioning."

# MILITARY HUMOR

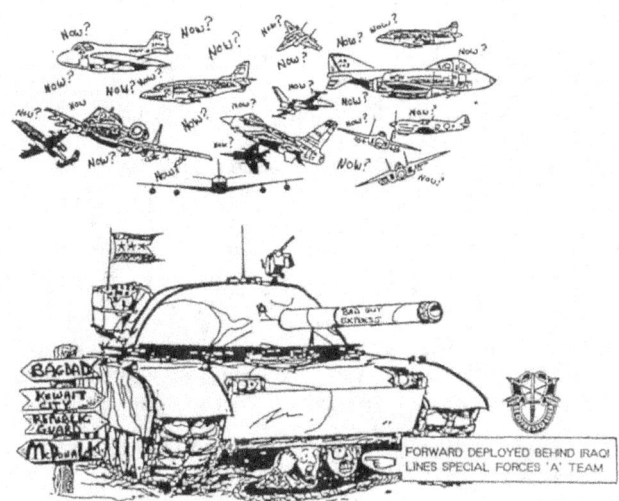

*WAIT! NO! NOT YET GUYS!* Be Patient, I'll Let You Know When ...

# SUPPLY TIPS, TRICKS & NSN's

*Submitted By: An Anonymous Supply Clerk*

Dear Ranger Rick,
When I first went to supply school, the most important thing they emphasize is to "Take Care Of The Soldier." Personally, I've always tried my best to get everything my unit asked me for. Evidently, I must be doing something right, as I have a lot of soldiers & leaders from other units asking me tor help

You can usually tell what kind of supply clerks the other units have by simply asking the soldiers a few simple questions about their unit supply room. But I have to admit, there are many supply clerks who either don't know their jobs or simply "Don't-Give-A-Shit."

If you want to know if your supply sergeant is up to snuff...

(1) Ask him if you can look through two of his books called the "ARMYLOG" and "FEDLOG." If he doesn't have one or the other or neither of these books, then he's missing the most important books needed in a supply room to order things.

(2) When you request to order supplies, if the clerk doesn't give you a copy of the receipt or tell you what the document control number is, ASK FOR IT! This way you can check on the status of your things yourself by visiting the unit S-4. Don't let your supply clerks BS you. If they ordered the items, then

they should produce a copy of the receipt or tell you what your control number is.

(J) Wait a few weeks before checking on the status of your things and raising hell. If the clerk claims that he can't find out the status...Bull Shit! If he ordered the items and forwarded it to the S-4/G-4, then all he has to do is punch in the document numbers into his computer log (LOGISTIC INFORMATION FILE) and the status should pop up on his screen. If he can't or won't do that, then check on the status of your things by either calling or visiting the unit S-4/G-4 yourself. If it's not in their LIF, then your supply clerk(s) either failed to order it or screwed-up on the paperwork.

To those of you who have trouble getting things from your unit supply room, I have enclosed a list of all the BASIC COMMON SUPPLY ITEMS that troops always seem to want to order on a regular basis. Just write down the NSN#, the QUANTITY you want, and hand it over to your favorite supply sergeant. He shouldn't have any reason for not ordering these things.

By the way Ranger Rick.... Please don't publish my name, somebody might not like me giving away this information. It's sort of a supply secret, if you know what I mean? Thanks.

The Anonymous Supply Clerk

| ITEM | NSN # | ITEM | NSN # |
|---|---|---|---|
| MRE HEATERS | 8970-01-321-9153 | HEAT TABS........ | 9110-00-263-9865 |
| BABY POWDER | 8105-00-817-0295 | FOOT POWDER...... | 6505-01-008-3054 |
| 4X4 ZIP LOCK BAG | 8105-00-837-7753 | 6X6 ZIP LOCK BAG | 8105-00-837-7754 |
| 8X8 ZIP LOCK BAG | 8105-00-837-7755 | 10X10 ZIP LOCK | 8105-00-837-7756 |
| 12X12 ZIP LOCK... | 8105-00-837-7757 | 2X2 PRESS DRESS | 6510-00-200-3075 |
| 4X4 PRESS DRESS.. | 6510-00-200-3080 | SKIN CLOSURE.... | 6510-00-054-7255 |
| BANDAGE/BANDANA | 6510-00-201-1755 | BAND-AIDS | 6510-00-913-7909 |
| "9 VOLT" BATTERY. | 6135-00-900-2139 | GLUE, SUPER | 8040-01-024-6988 |
| "AAA" BATTERIES | 6135-00-826-4798 | "AA" BATTERY..... | 6135-00-985-7845 |
| "C" CELL BATTERY. | 6135-00-985-7846 | "D" BATTERY...... | 6135-00-835-7210 |
| 10Z GLASS BOTL | 8125-00-933-4414 | 4 OZ. PLASTIC BOTL | 8125-00-174-0855 |
| 12OZ SPRAY BOT... | 8125-00-488-7952 | PLASTIC BOX...... | 8115-00-761-8912 |
| FORM.409 CLEANER | 7930-00-926-5280 | ARMORALL PROT | 8030-01-103-2858 |
| EARPLUGS (FOAM) | 6515-00-137-6345 | DISPOSABLE LITE | 6230-00-125-5528 |

PLIERS..GERBER  5110-01-346-5339      CRN LENS,FL-LITE  6230-00-504-8341
WOODEN MATCHES  9920-00-221-0613     BRASSO POLISH    7930-00-266-7136
RAZOR, SURGICAL  6515-01-363-1212     RUBBER STRAP,10" 5340-00-340-0980
RUBBER STRAP,15" 5340-01-029-9084     TAPE, ENGINEER...8315-00-260-0341
BREAKFREE LIQ  9150-01-054-6453       FLSHLGHT BLUB...6240-00-155-8675
AMBER LENS, LITE.6230-00-504-8342     RED LENS,FL-LITE. 6230-00-111-0190
Q-TIPS (6")      6515-01-234-6838     CHAMOIS CLOTH....8330-00-823-7545
AMMONIA INHALANT 6505-00-106-0875    ASPIRIN..........    6505-00-118-1948
TOOTH BRUSH, WPN 1005-00-494-6602    HANDI-WIPES......  8520-00-782-3554
PARA-CORD 440    4020-00-935-5761     PARA-CORD 550....4020-00-246-0688
BLACK WATCH CAP..8405-01-006-1074     CHEM-LITE STICK..6260-00-106-7478
TAPE,100MPH  7510-00-074-4960        WATER PURI-TABLET 6850-00-985-7166
SURVIVAL FISHNET 8465-00-300-2138 SURVIVAL PKT-SAW 5110-00-570-6896

## "JOE COOL" BACKWARD HATS

When I was a squad leader and platoon sergeant, one of the annoying things that I hated to see soldiers do was wear their hats backwards in the field. Man, that aggravated the f---en hell out of me. But you know what, I learned something from these guys. I learned that if you wear your hat backwards in a hot desert or tropical climate, your head will stay much cooler. Hey, I'm not BSing you! I'm dead serious. Here's how I figure it works.

Heat rises, right? So as you sweat and produce body heat underneath your clothes, it has no where else to go except up and out around the opening of the neck. Do ya follow me?

Well, as this body heat escapes from around the neck, the sun visor of your BDU hat somehow catches or "reflects" this heat back into your face, thus making your head and face hotter. But, if you wear your hat backwards with the sun visor to the rear, the body heat won't get caught or reflected back to you. Does this make sense to you?

If it doesn't, then the next time your in a hot environment, try wearing your hat backwards and see if it makes a difference to you. I'll bet many of you will say, "Son of bitch, that damn Ranger Rick was right!"

*Rick F. Tscherne*

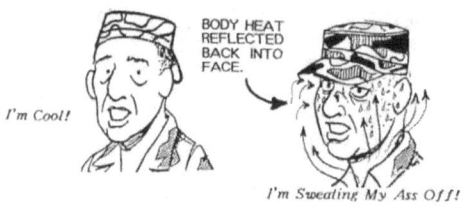

## RAMBO BANDANNAS

OK, OK, OK, we all know they're here to stay, but let's not get carried away with it. The real purpose of wearing a bandana around your forehead (like Rambo) or neck (like Chuck) is to keep the sweat from running down over your face and chest. But if your wearing one just to look like a bad ass, your only fooling yourself, buster. Wise up and stop imitating movie stars, don't wear'em unless you need'em.

## REMOVABLE BOOT TONGUES

*Submitted By: SPC David E. Brown*

Dear Ranger Rick,
When I used to go on a road march the leather tongues inside my boots use to bother me, the would keep rubbing against my shins and ankles. But I solved the problem, I cut them out out and replaced them with a pair of boot blousers. Here's what I did...

STEP 1: Remove the leather tongues by cutting them off with a pair of scissors or knife.

STEP 2: Take a pair of boot blousers and cut the velcro part in half.

STEP 3: Take this piece of cut velcro and sew it into the lower portion of the boot where the leather tongue use to be.

STEP 4: Now take the boot blouser and connect it to the velcro part and "**PRESTO**," you now have a removable boot tongue.

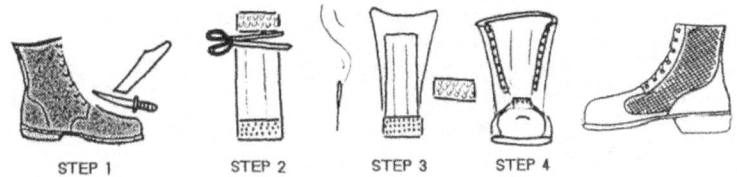

## FIELD PILLOWS

I hear many of you field troopers are getting pretty tired of using your NEC mask, poncho, and kevlar for a field pillow. OK, then here's what else you can do....

1) Get yourself an old pair of pants and cut off either one or both parts of the legs. 2) Take the cut end and sew it completely closed all the way across. 3) If the other end of the pant leg still has the cuff strap, then leave it in place. If it doesn't, either run some 550 para-cord through it or sew in a zipper about the same width as the pant leg. 4) Now you can either stuff it with cloth, foam padding, or place a small pillow inside.

NOTE: When it needs cleaning, remember to always remove the inside portion before placing it in the washing machine.

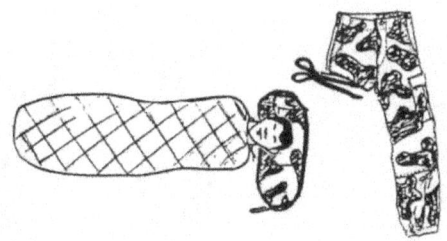

## UNIFORM TIPS

*Submitted By: SSG. CRAIG MARTS*

Ssg. Craig Marts writes, "You know, Ranger Rick, I thought everyone knew how to modify and mold their berets. But I found quite a few who didn't, both, the newbies & older timers. To mold & fold your beret to make it look much sharper:

> STEP 1 - Cut out and remove only the beret's liner, do not remove the cardboard stiffener.
> STEP 2 - Wet thoroughly the entire beret.
> STEP 3 - Place the beret on your head, grab the front portion and fold it down forward over the forehead.
> STEP 4 - Reshape and mold the rest of the beret to the way you want it and carefully remove it from your head.
> STEP 5 - Let the beret dry gradually on it's own or use a hair blow dryer to speed up the process.

Note: When wetting, folding, and molding, the "unit flash" should already be attached to the beret. Insure the flash is sew through both, the material and the cardboard stiffener, and NOT just the material.

When removing the beret from your head, be very careful not to disturb the fold and shaping. If you have a basketball or volley ball around the house, place it on top of the ball.

For those of you who are high-speed super troopers with a few tabs (Airborne, Ranger, & SF), you can save yourself some bucks by sewing the tabs and unit patches together before taking them to a sewing shop. The standard rate they charge for sewing patches on a uniform is about $2 a patch.

If you got two tabs (Airborne & Ranger) and the unit patch, that adds up to $6 per uniform. If you sew the tabs and patches together yourself before turning it into a sewing shop, it will only cost you $2 per uniform. Save $$$.

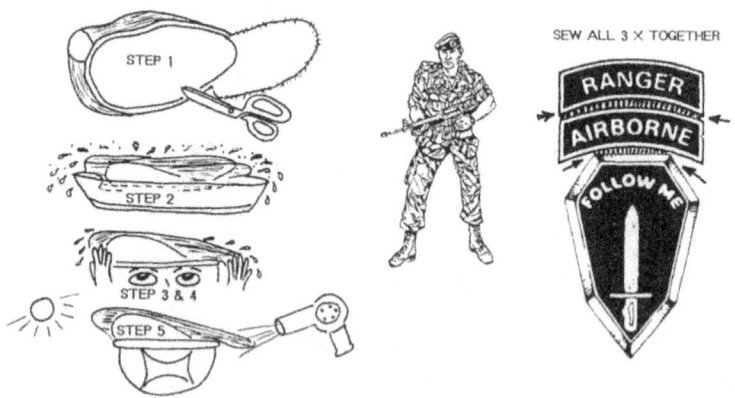

## DID YOU KNOW THAT BDUs...

SHOULD NEVER BE IRONED. It's true, they're strictly "wash and wear" uniforms. Which means your only required to wash'em, dry'em and put'em on. Though they do look a lot nicer when pressed with an iron, they're NOT suppose to be ironed. Hot irons will only scorch and or wear out the camouflage coloring much faster.

If you want to be smart about it and look sharp, keep a separate set of uniforms strictly for. garrison and another set for the field. Then only iron the garrison uniforms, NOT the field uniforms.

CAN BE REINFORCED. It's true, if you turn the pants inside out and cut through one layer of the material over the knee area. You can add some foam padding, cardboard, or other material inside these knee "pockets" to protect them from getting bruised and banged up. Thus providing a bit more protection from the rugged terrain and or the cold ground.

CAN BE MODIFIED & IMPROVED. It's true, you can remove and replace the leg cuff strap with a commercial elastic band. Simply tie the elastic band to one end of the pants cuff strap and pull it entirely out until the elastic band pops through. Then remove the cuff straps, tie the two elastic ends together and you now have a set of built-in blousing rubbers.

Or you might want to do what SGT MICHAEL J. PILSNER did with his cuff straps, he calls it his "Blousing Stirrup System." Take the pants cuff straps, overlap the ends and sew them together. Put on the pants, place the "stirrup strap" under the heel of the foot and then put on your boots. Now the pant leg cuffs will remain inside the boot and won't come out.

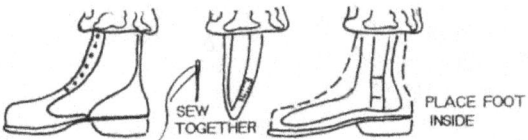

RANGER RICK'S COMMENTS: Got some lint hairs or fuzz balls on your uniform that you need to get rid of? Wrap a piece of masking tape around your hand and then either "pat" or "roll" it across your uniform. The fuzz balls and lint hairs will come right off.

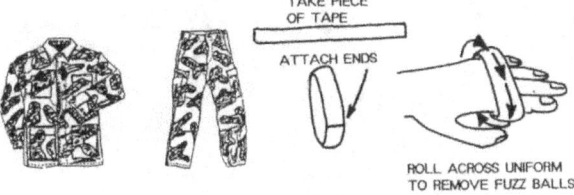

*Rick F. Tscherne*

# MRE HEATER GRENADE SIMULATORS?

*Submitted By: Sgt. Rich Reitz*

Sgt. Reitz tells me, "Hey Ranger Rick, here's a neat little trick that we used to surprise some OPFOR during a tactical field training exercise. They didn't think anyone was issued any grenade or artillery simulators, but boy, did we surprise them."

Here's what you do...
Take 2 X MRE Heaters, open up the green square element and take out all the "gray shit" that's inside. Crumble it up real good and then drop it inside a "dry" plastic coke bottle (NOT glass). Add a small amount of water, crank on the bottle cap real tight, shake it a little bit and throw it quickly in the desired direction.

Here's what happens....
The gas inside the bottle quickly builds up and rupture the bottle almost as loud as a real grenade simulator. Try it, it works great!

RANGER RICK'S COMMENT: Warning to Readers! This is the first time that I've ever heard of this, I have never tested nor tried this trick. WARNING: Use At Your Own Risk With Caution.

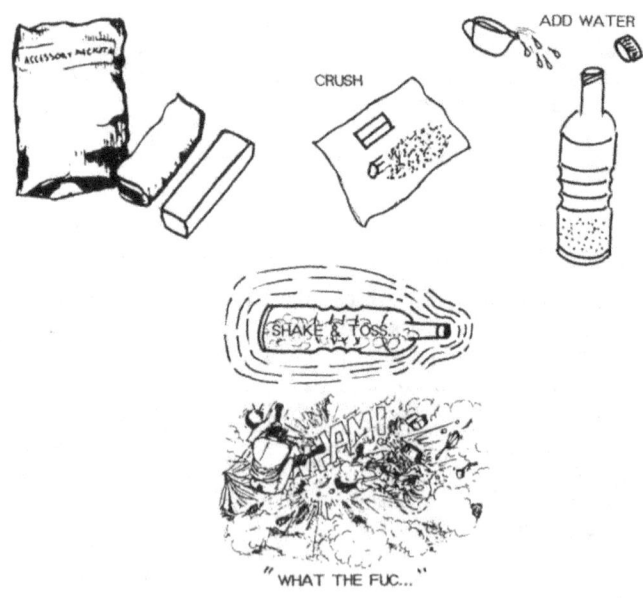

## HOW WELL DO YOU KNOW YOUR MILITARY LINGO?

So, ya think you know your military lingo pretty good, huh? Well, let's see how ya do matching these up correctly. (Answers on bottom of page.)

1. TWO DIGIT MIDGET    A. A truck that sells assorted snacks
2. G.I. SHITS    B. Fuckin New Guy (Gal).
3. GAS ATTACK    C. When serious things start to happen.
4. PEONS    D. Latrine Toilet Paper.
5. F.N.G/NEWBY    E. Uncontrollable, son shit, diarrhea.
6. GRUNT/LINE PUPS    F. Rear Echelon Mother Fucker, Spt. Troops
7. CLERKS & JERKS    G. Civilian life outside the military.
8. CHICKEN SHIT    H. To be killed, death of a soldier.
9. F.T.A    I. Medics or hospital personnel.
10. FUBAR    J. A 2d Lieutenant.
11. DILLIGAF    K. Meaning "your ass is in trouble."
12. COMFU    L. Same Old Shit (Same Old Stuff).
13. SPOONS    M. EMs who think they know the rules/rags.
14. REMF    N. An Infantry soldier.
15. PUZZLE PALACE    O. Unit PT or organized unit sports.
16. SHIT HIT THE FAN    P. A series of nasty smelling farts.
17. WHITEWALLS    Q. A slow minded/thinking soldier.
18. THE WORLD    R. Non-combat arms, rear support soldiers.
19. SANDPAPER/TP    S Lowest ranking soldier, E-1 to E-2.
20. YA ASS IS GRASS    T. Means-Does It Look Like I Give A Fuck.
21. ROACH COACH    U. One who has 10-99 days left (PCS/ETS).
22. PRONE POSITION    V. Comfortable intercourse or firing pos.
23. PECKER HEAD    W. Artillerymen.
24. TO BUY IT/BOUGHT IT    X. Wimp, scared, gutless, no balls.
25. KNOCK OUT SOME Zs    Y. A sleeping bag.

26. TO BLOW SMOKE      Z. Situation's Normal - All Fucked Up!
27. M.F.I.C.           AA. To successfully accomplish a mission.
28. JUG FUCK           BB. Keep It Simple Stupid.
29. JESUS FREAK        CC. One who always lies or exaggerates.
30. BEDPAN-HANDLERS    DD. Fun, Travel, &Adventure(Fuck-The-Army).
31. TURTLE             EE. Bare sided haircut that Rangers wear.
32. CHERRY             FF. Candy, snacks, junk food.
33. K.I.S.S.           GG. When things are fucked up/disorganized
34. POGIE BAIT         HH. Low ranking new soldiers, a virgin.
35. PUCKER FACTOR      II. Tight ass muscles, scared shitless.
36. BULL SHITTER       JJ. To deceive, mislead, false information.
37. KICKED ASS         KK. Mother-Fucker-ln-Charge.
38. ORGANIZED GRAB-ASS LL A unit headquarters (Bn, Bde, etc).
39. FART SACK          MM. To go to sleep.
40. SNAFU              NN. A very religious God fearing soldier.
41. CANNON ROCKERS     OO. A dumb, thick headed soldier or leader.
42. SHIT HOUSE LAWYER  PP. Cooks or pers. who work in the mess hall.
43. SOS                QQ. Completely Fucked Up.
44. BUTTER BAR         RR. Fucked Up Beyond All Recognition.

<u>ANSWERS</u>

| 1-U | 6-N | 11-T | 16-C | 21-A | 26-JJ | 31-Q | 36-CC. |
| --- | --- | --- | --- | --- | --- | --- | --- |
| 2-E | 7-R | 12-QQ | 17-EE | 22-V | 27-KK | 32-HH | 37-AA |
| 3-P | 8-X | 13-PP | 18-G | 23-OO | 28-GG | 33-BB | 38-0 |
| 4-S | 9-DD | 14-F | 19-D | 24-H | 29-NN | 34-FF | 39-Y |
| 5-B | 10-RR | 15-LL | 20-K | 25-MM | 30-I | 35-II | 40-Z |
| 41-W | 42-M | 43-L | 44-J | | | | |

# RANGER RICK'S FAVORITE TRAPS & SNARES

Before I came in the Army (1972), I was a very active camper, fisherman, hunter, and trapper. Coming from a small town and farm community (Berwick, PA), there were only two types of sportsmen. Those who played ball and those who preferred to hunt & fish.

Well, to be honest, I wasn't very good in basketball nor football,(I always got picked last).But I was very good when it came to hunting, fishing, and trapping. I usually shot, caught, or trapped whatever I was going after, plus more.

When 1 came in the military, I put my skills to good use by always qualifying expert rifleman, knowing how to read and follow a map & compass, build a fire, sneak through the woods without being detected or seen, and much, much more.

But one area of military training that I've always disagreed with, was how to trap and snare animals for food. Now if you look through the survival section of the Ranger or Special Forces Handbook, you will see nice pictures of how traps and snare are supposed to be set up. It's funny how all these traps and snares are rigged with a piece of food for bait.

Well, if I'm starving or I'm in a life or death survival situation, I'm not going to give up my last few pieces of food to some lousy animal that might not come along.

And another thing, most of these traps & snares are so damn complicated and hard to remember how to make, you need to always keep a survival book in your pocket to remember how.

That is, all except one. It's a "hanging snare" that requires no food for bait and can securely hold onto an animal when it's caught in the trap. (See drawing) When setting this "hanging snare"....

1. Use a strong but flexible piece of wire or cord such as military boobytrap trip wire or the nylon strands from the inside of portion of some 550 parachute cord.
2. When making a knot in the cord/wire, keep it loose.
3. When using a tree or bush branch as the "spring," use one that is alive and firmly planted in place.
4. Use a strong but dry piece of wood as the "trigger/hook."
5. Use ashes or plant roots to cover up any human scent you left on the snare wire or cord while you Were rigging it.
6. Place this hanging snare preferably along animal trails or near their burrows. Note: An active animal trail has fresh droppings, packed down grass or weeds, dug up or chewed up vegetation around the trail and immediate area.

Yep, this hanging snare works pretty well, and it's very, very effective. So if your like me and can't remember how to make all these different types of deadfalls, snares, and traps. Then just remember one, THIS ONE!

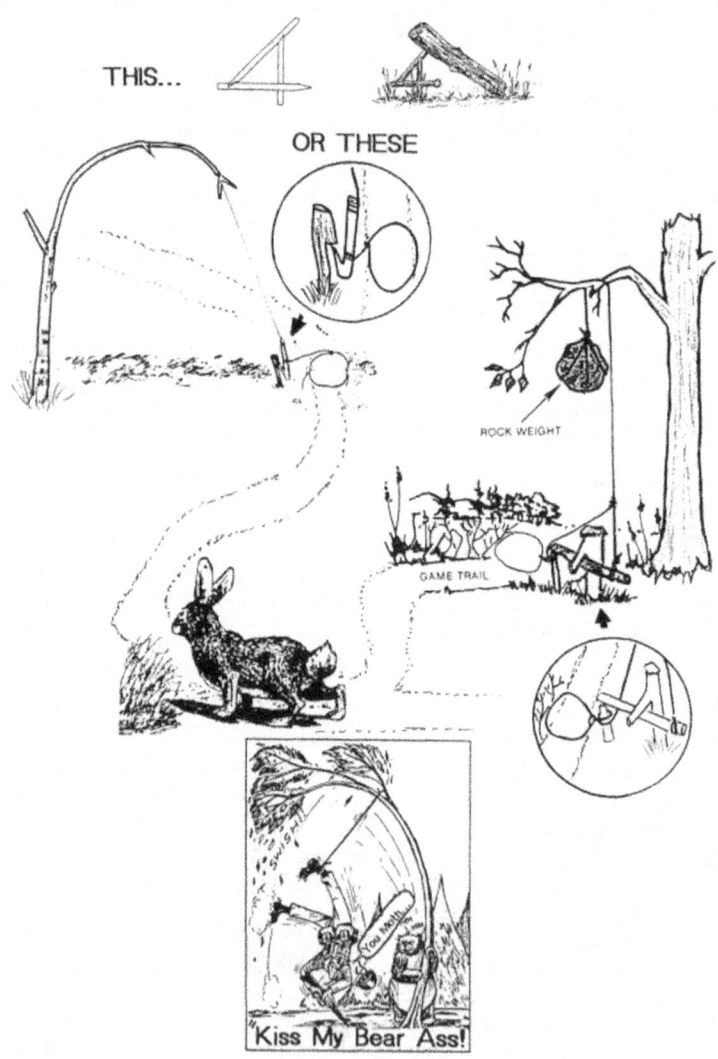

## HOW TO BREAK DOWN AN M16 LOWER (TRIGGER) RECEIVER

Do you always have problems cleaning the inside portion of an M16 lower receiver with Q-tips and pipe cleaners? How many times has your squad leader or unit armor inspected the lower receiver and said, "Hey bozo, it's still dirty, do it over!" A whole bunch of times, right?

Well, ya wanta know an easier way to clean it? Just break it down. Oh, I know it's not supposed to be taken apart unless your a qualified armor, but who cares. What's important is that you know how to break it down and put it back together again in the proper sequence.

The key to remembering how to correctly assemble and disassemble the lower receiver, is to look closely and remember how each piece was positioned before you remove it. Then, as you remove each piece one-by-one, study how it came out and how it's suppose to go back inside the receiver. As you remove the pieces, lay them down on the floor or table in the sequence that they were removed, DO NOT MIX THE PIECES.

NOTE: Where you see written (L>R) or (R>L), this means you must remove or install the pins from either Left to Right or Right to Left of the receiver. It's important that you remove the pins in a certain manner, DO NOT DO IT ANY OLD WAY.

TO DISASSEMBLE: First, insure the selector lever is set on "BURST" or "FIRE." Then with the use of a nail, punch out (L>R) the automatic sear pin (1) and remove the automatic sear (2) and selector lever (3). Now take the nail and punch out (R>L) the hammer pin (4) and remove the hammer (5). Then take the nail and punch out (R>L) the trigger pin (6) and remove the trigger assembly & components (7).

And that's it, now your ready to clean the inside portion of the lower receiver and the individual parts. Now tell me, was that so hard to do?

TO REASSEMBLE: Follow the same steps except in reverse sequence. First install the trigger assembly & components (7), trigger pin (6), the hammer (5) and the hammer pin (4). Then install the selector lever (3), the automatic sear (2), and then the automatic sear pin (1).

Once you have reassembled the lower receiver, check to insure all the pins are resting "flush" against the surface. Then do a function check to insure all the pieces were put back together correctly.

WARNING: Never break down the lower receiver unless your doing it over a table or poncho, one lost piece and your weapon will become totally KAPUT!

*Rick F. Tscherne*

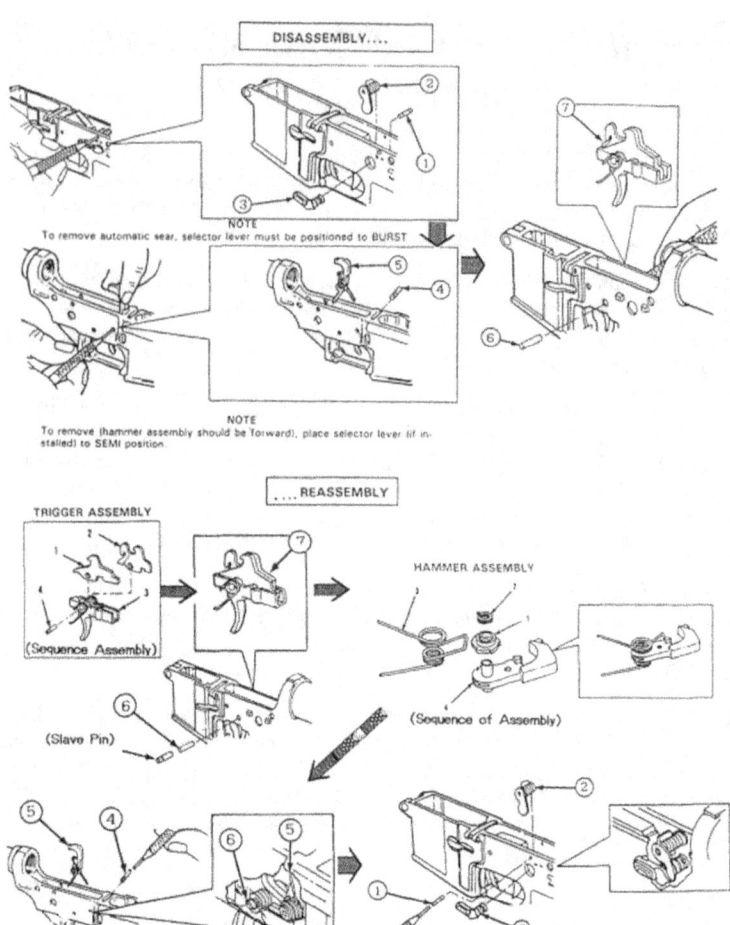

**SPECIAL NOTE:** *Should you screw this up and need your unit armorer's assistance, tell him it's Ranger Rick's fault. I'll take the blame even though it was YOUR screw up.*

*The Complete Ranger Digest: Vol. VI*

# USEFUL HAND & ARM SIGNALS

Hand & arm signals are used for communicating and controlling unit's while on the move. Though there are many hand & arm signals, here's a few useful ones that you might not have seen,

PERSONNEL COUNT - Patting top of head with palm of hand. MEANING: Head count of personnel to insure no one is missing.

LEADER NEEDED - Hand to upper part of arm displaying fingers for which leader. MEANING: Need to talk with a certain leader.

LISTEN/I HEAR - Cupping of one hand to ear and then pointing in the direction you want someone to listen. MEANING: Listen up.

LOOK/I SEE - Two fingers to eyes and then pointing in the direction you want someone to look. MEANING: To look or observe

SPLIT UP/BREAK OFF - Palm extended, fingers locked straight together and placing it between the eyes in an "up & down" motion. MEANING: Split up / break off from the unit.

QUIET - One finger held vertically to lips. MEANING: Shut-up.

DANGER AREA - Motion of one finger across throat. MEANING: An open area of land or a road/trail where you are vulnerable.

OBJECTIVE/TARGET - A closed fist to the heart and then pointing MEANING: Location of the objective or target.

READY/I AM READY - A fist with the thumb held in an upward position. MEANING: Are you ready? Or I am ready! (A display of fists and thumbs to one another mean both parties are ready).

PACE COUNT - Patting of the leg/foot. MEANING: Distance traveled

ENEMY - Pointing of index finger w/thumb down. MEANING: Enemy.

FRIENDLY - Pointing of index finger w/thumb up. MEANING: Friendly

DISREGARD - Crossing of hands across the face. MEANING: Disregard previous hand & arm signal or previous instructions.

CALL /YOU HAVE A RADIO CALL - Thumb and pinky finger to ear & mouth simulating a radio handset. MEANING: You have a call.

BREAK TIME: - Two fists side-by-side, simulating breaking a stick MEANING: Take a break, rest. (Fingers indicate how long.)

SCREW YOU: - Making a fist and then raising the middle finger. MEANING: To disapprove or disagree with someone.

KORNUTOE: - Making a fist and raising only the index and pinky finger together. MEANING: (In Italy) To hope someone is sexually enjoying the hell out of your spouse or lover while your away. (This is used to really, really piss someone off.)

| PERSONNEL COUNT |  LEADER NEEDED |  LISTEN/I HEAR |
| --- | --- | --- |
|  LOOK or I SEE |  SPLIT/BREAK OFF |  QUIET |
|  DANGER AREA |  OBJECTIVE/TGT |  READY |
|  PACE COUNT |  ENEMY |  FRIENDLY |
|  DISREGUARD |  YOU HAVE A CALL |  BREAK TIME |

SCREW YOU

KORNUTOE:

*The Complete Ranger Digest: Vol. VI*

# PISS TREES & BOTTLES

There's no doubt about it, we male GIs are pretty nasty SOBs when it comes to urinating. I still don't understand why some "boys" can't lift the seat before they piss. Man, do I hate finding piss drops on the seat when I have to take a shit.

When I was a squad leader and platoon sergeant, I always made sure either I or one of my NCOs selected a special spot for the troops to shit and piss. The location had to be very convenient for everyone, if it wasn't, then I knew the troops would shit & piss wherever they wanted. Especially at night.

So what I did every time we set up a patrol base or defensive position, was select several "piss trees" where the troops were only allowed to piss, NOT SHIT. This way all the piss would be located in one or two places and not all over the place.

But when it came to living outta GP tents, well, this was another story. It seems the colder and darker it was outside, the closer to the tent the troops would piss. Instead of pissing in the nearby port-a-potties or at the designated piss tree, they started pissing closer and closer to the tent. Pretty soon it started to smell pretty nasty outside.

That is until I got hold of a bunch of empty plastic water bottles and made everyone keep one near their bunk. If anyone had the urge to go in the middle of the night and couldn't hold it beyond the tent door, they were told to use their piss bottle. If they didn't, and they got caught pissing outside the door, I would make them pull guard duty around the tent.

Well after a few more warnings and a few soldiers getting caught and put on "dick watch," the pissing problem came to a complete halt. If it worked for me, it'll work for you too.

```
ATTENTION TO ALL SERVICE MEMBERS
         USING THIS LATRINE

SOLDIERS with short rifles please stand within firing range.
SAILORS  with small anchores, please drop it near the pier.
AIRMEN   with little engines, please taxi up to the terminal.
MARINES  with tiny pistols, please test fire it somewhere else.
LADIES   please remain seated during the entire performance
```

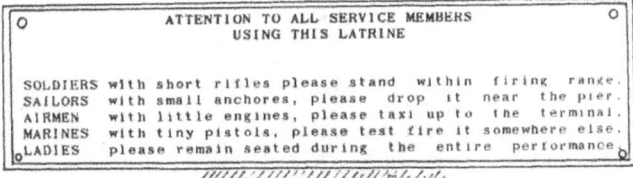

*What weird weather we're having, that's the 4th time it started to rain.*

*Rick F. Tscherne*

## THE TOP 10 PROBLEM CHILDREN THAT ARE IN EVERY UNIT

**THE KNOW-IT-ALL**
(or Think They Do)
Doesn't like to be
wrong or corrected.

**THE COMPLAINER**
Always has something
to say or bitch about
never, ever satisfied.

**THE WHINER**
Always feeling sorry
for themselves, think
world's against them

**THE KISS ASS**
Always sucking up
and agreeing with
someone with power.

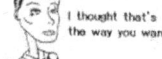

**THE DUMMY**
Always screwing up,
can't do or get any
thing right.

**THE INSTIGATOR**
(or Double Face)
Mislead/twist things
to start trouble.

**THE BULL SHITTER**
Always bragging or
exaggerating to win
someone's attention.

**THE BACK STABBER**
A sneeky, revengeful
jealous person, enjoys
putting others down.

**THE HOT HEAD**
Gets angry fast for
no reason, always
looking for trouble

**THE FOLLOWER**
Which ever way the
wind is blowing, they
go along with the
"mood of the crowd."

*The Complete Ranger Digest: Vol. VI*

# HOW TO CONVERT A MINI-HAMMOCK INTO A TACTICAL CAMMIE NET

There's a product on the market called a "Mini-Hammock," you can find it in almost any outdoor camping or military supply store. On the side of the box it states that it can be used as a hammock, emergency stretcher, rope, fish net, etc. But the manufacturer missed one other use, it makes an excellent CAMOUFLAGE NET.

To use it in this manner, you will need to permanently remove the two metal O-rings attached to the ends and replace them with some 550 parachute cord. The easiest way to do this is by cutting them off with a hacksaw and opening them up with a pair of pliers.

You will then need 2 X lengths of OD green 550 parachute cord at least 4 meters (12 feet) long. Before running the parachute cord through the 4 nylon net loops, melt the ends of the cord with a match or lighter and then mold them into a point. This will make it easier for the 550 paracord to go through the nylon net loops.

CAUTION: When removing the 4 X nylon strands/loops from the metal O-rings, remove them one-at-a-time and immediately run the parachute cord through the loop. This will prevent any mix up or unravelings in the strands.

When both of the rings have been removed and replaced with 550 paracord, lay the hammock on the ground and inspect each of the loops to insure they are not twisted, worn, or broken. Then move all the strands to the middle portion of the parachute cord and tie a figure "8" knot into the cord.

When you want to use it as a hammock, just leave the figure "8" knots in place and tie the running ends of the cord to a couple of sturdy trees. When you want to use it as a camouflage net, just untie the figure 8 knot, stretch out the cord and space the nylon loops out evenly along the paracord.

NOTE: When using it in the camouflage net mode, you may need some separate pieces of parachute cord (or bungy cord) to keep it fully stretched out over your position. Then add some light foliage (grass, leaves, etc) on top of it from the surrounding terrain.

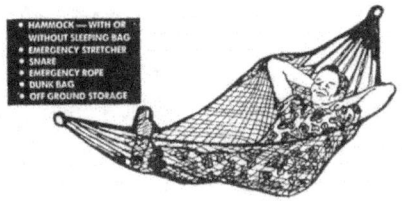

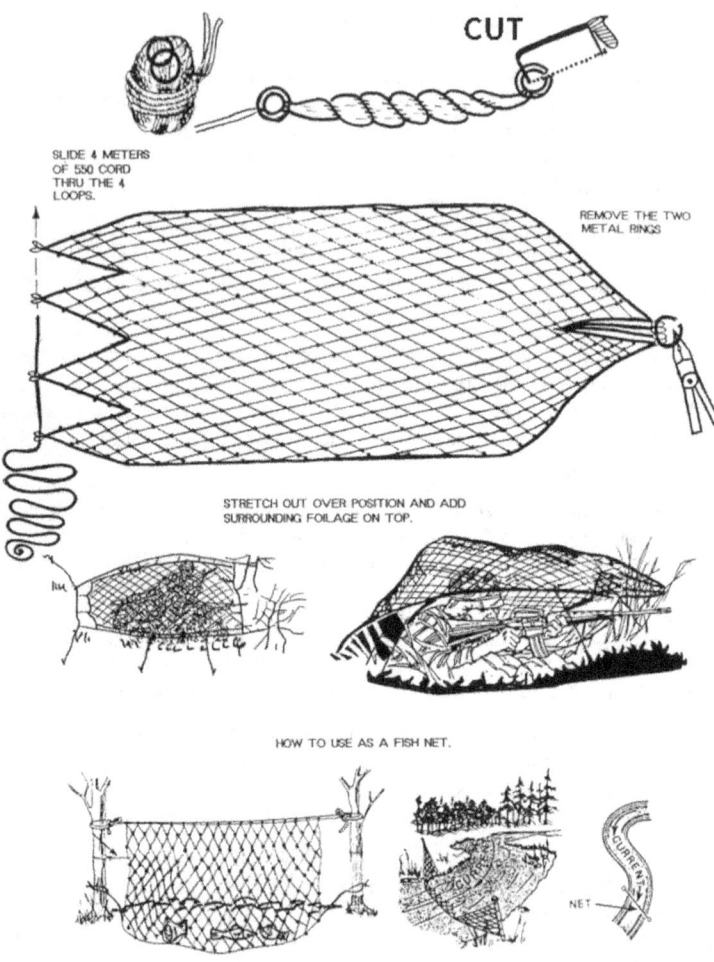

*The Complete Ranger Digest: Vol. VI*

# RAIN PONCHO TIPS

*Submitted By: SPC Robert G. Warmack*

Specialist Warmack writes...
Whenever I wore my poncho out in the rain, I would always have a problem in keeping my back and shoulders dry. At first I just thought there were some holes somewhere in my poncho, but when 1 checked it for holes, I couldn't find any. I soon realized that this wetness wasn't due to some rain leaking in, but due to my own body heat in producing sweat.

But I figured out how to solved this little problem. I cut a hole in the center of one of my towels, placed it over my head, and wore it underneath my rain poncho "Poncho Villa" style. Now I no longer have any problem with a wet, damp back or shoulders.(NOTE: Carry an extra towel or two as a backup.)

RANGER RICK'S COMMENTS: If your going to wear a rain poncho or "non-gortex" type of wet weather suit, you had better get used to being wet from your own body sweat. Especially if your gonna be on the move a lot. Because there's no way in hell your going to remain dry unless your sitting still or standing in one location.

The more you move, the more body heat you'll produce, the more body heat you produce, the more sweat you'll produce underneath that rain poncho. And if your in a cold weather environment, you can bet your sweet ass that your gonna freeze your ass off too.

SOLUTION: When wearing a wet weather suit, leave the pants and jacket partially open to allow some of the body heat to escape. When wearing a poncho, always wear it OVER the Ibe and ruck sack and NEVER underneath it.

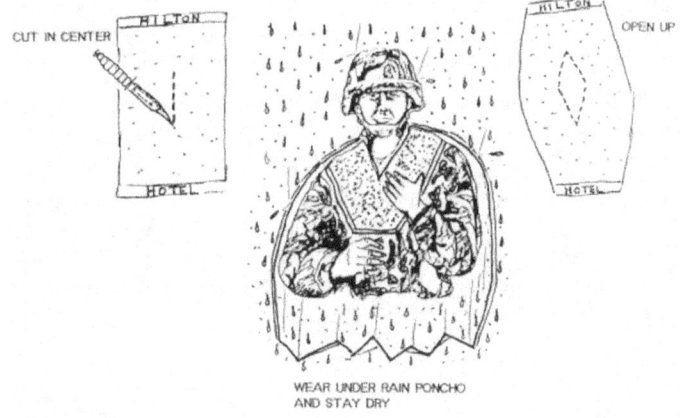

WEAR UNDER RAIN PONCHO AND STAY DRY

Rick F. Tscherne

# HOW TO DRY CLOTHES IN A GP TENT

Have you ever walked into a GP tent that had wet and damp clothes hanging from one side to the other? Wasn't it a pain in the ass to walk in and around these clothes? Not to mention smelling pretty nasty too, huh?

Well, maybe not all the units have or use these GP tents. But to those of you who do use 'em, here's a tip on how you can dry 'em out more quickly without 'em being in everyone's way.

Instead of hanging your clothes low to the ground from one side to the other and or around the support poles, hang 'em high up close to the tent ceiling.

Why? Well, because hot air rises, and due to this, the ceiling is always the warmest part of a tent. Therefore, if you want to dry out your clothes much more quicker, HANG'EM HIGH. Not only will they dry out faster there, but they'll also be out of everyone's way too. Makes sense, don't it?

# ACCIDENTAL WEAPON FIRE

Have you ever been on a raid or recon mission when some clown in your unit accidentally fired his weapon? What did your leaders do? Panic, or did they continue on with the mission like it never happened?

If your on a raid or recon mission and someone should discharge their weapon accidentally, your first option should be to abort the mission. Why? Well, for one thing, your element of surprise has been compromised and the enemy now knows your there or at least in the immediate area.

But if higher headquarters tells you that you gotta continue on with the mission regardless, then you have only one option left. Approuch and attack the objective from another direction at a different time.

But if your already in position when someone accidentally discharges his weapon, then you have no choice but to attack immediately. And depending on how well trained your men are and how fast they can move, the odds could still be in your favor in successfully overcoming the objective.

*The Complete Ranger Digest: Vol. VI*

## ON-LINE ASSAULT THRU THE DOOR

When I was a platoon sergeant assigned to "C" Company, 1/509th Airborne Battalion Combat Team in Vicenza, Italy (1981-1984). I learned a very useful tactic from my company commander, Cpt. Charles "Chuck" Busick. He taught me and other leaders how to get a rifle platoon or company quietly on line in position for a night attack. He called it his "Door Method." Here's how it worked...

Once you have pin pointed the objective, made your leaders recon, and determined the best "cover and concealed" route to attack the enemy position. You must then select a release point where you intend to set up your assault line. Which should be a location just slightly beyond the range of being seen, heard, and detected by the enemy.

Once you have briefed all the squads or platoons on the order of move, sectors of responsibility, etc. You must then select 4-6 good soldiers and insure they have either a flashlight, chem-light, or illuminating compass in their possession.

Leaving the main body of the unit behind until you have returned or called for them to come forward. Move forward with the 4-6 individuals and place two (2) of them at the release point (RP) standing, kneeling, or laying down about 2 meters apart. They must then be instructed to hold a red filter flashlight or chem-lite just slightly above ground level in the direction of where your unit will come from to get on line, (opposite of the enemy). This release point now becomes what is called the "DOOR" to forming the assault line and the two individuals now become the DOOR GUARDS.

Starting at the DOOR position and with the aid of a compass, walk 1 or 2 of the other soldiers the entire length of the "RIGHT SIDE" of the assault line and position him or them on the far right end. Then return back to the DOOR and walk the other 1-2 soldiers the entire length of the "LEFT SIDE" of the assault line and position them on the far left end. These positions now become the far left and right side SECURITY GUIDES for the assault line. They must also be instructed to hold a red filter flashlight or chem-light in the direction of the DOOR position and away from the enemy.

Returning back to the DOOR position, either call up or lead the remainder of the unit just short of the release point. Depending on their assigned positions or location within the assault line. Each squad and individual must then pass through the DOOR between the two DOOR GUARDS either take a 90 degree left or right turn in the direction of the glowing light and stop when they reach the left or right side SECURITY GUIDES. Then, if necessary, back-step all the way back to the DOOR until everyone is evenly spaced out.

Once everyone is in position and properly spaced and spread out, they must then wait until the platoon or company commander gives the final word to move forward and attack.

NOTE TO LEADERS: The "DOOR METHOD" can be successfully employed either during day or night time operations as long as it is used during pre-planned raid attacks.

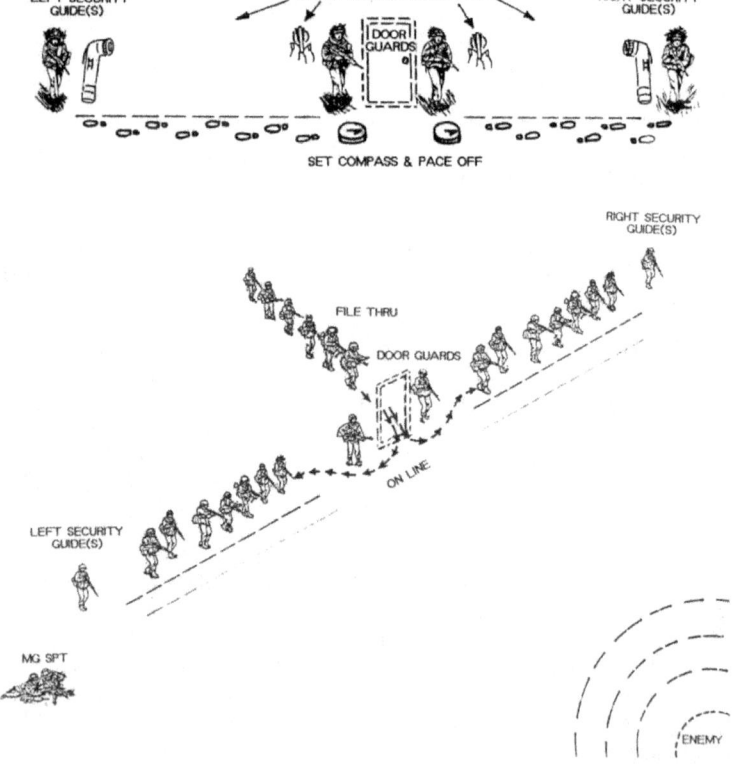

*The Complete Ranger Digest: Vol. VI*

# IMPROVISED MORTAR TUBE
## (Training Purposes Only)

I learned this next trick back in Ranger School, it wasn't exactly taught to us, but... And if your an 11C, you'll probably like this trick even more than the rest of us. But first let me tell you the story on how I '.stumbled onto it.

One day while on a Ranger patrol somewhere up in the mountains of Georgia, we came under a grenade & artillery "simulator" attack. The thing that was so strange about this attack, is that there wasn't anybody tossing'em at us. They were falling out of the sky and exploding all around us. We could hear a distanced "boom," but couldn't figure out how they were being launched or propelled.

Well, later on we had to go on a recon patrol to look for a suspected enemy mortar position, and low and behold we found out how they were doing it. The Ranger instructors had a thick metal pipe buried in the ground at an angle. They would pull the fuze on one artillery simulator, drop it down the pipe, pull the fuze on a second simulator and drop it down the pipe on top of the first one.

When the first simulator exploded, the force would propel the second unexploded grenade into the air like an actual mortar round. It didn't have the full range of a real one, but it sure the hell sent the simulator down range a long, long way. The simulator would explode either in the air or on the ground. In getting a better closeup look at this home-made mortar tube, here's what it looked like.

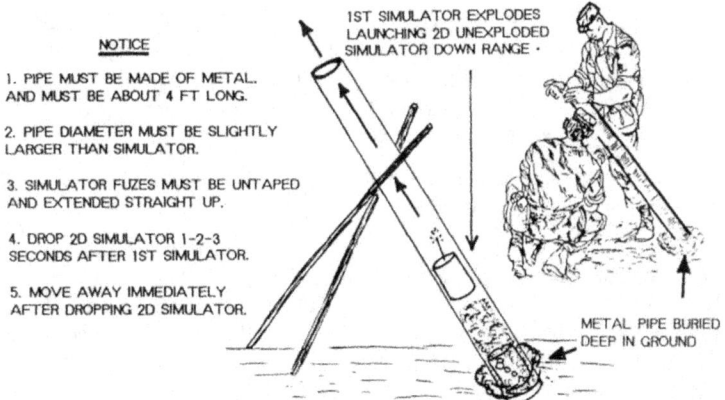

NOTICE
1. PIPE MUST BE MADE OF METAL. AND MUST BE ABOUT 4 FT LONG.
2. PIPE DIAMETER MUST BE SLIGHTLY LARGER THAN SIMULATOR.
3. SIMULATOR FUZES MUST BE UNTAPED AND EXTENDED STRAIGHT UP.
4. DROP 2D SIMULATOR 1-2-3 SECONDS AFTER 1ST SIMULATOR.
5. MOVE AWAY IMMEDIATELY AFTER DROPPING 2D SIMULATOR.

1ST SIMULATOR EXPLODES LAUNCHING 2D UNEXPLODED SIMULATOR DOWN RANGE.

METAL PIPE BURIED DEEP IN GROUND

WARNING: Using grenade or artillery simulators in this mode can be EXTREMELY DANGEROUS. Use extreme care, caution, and common sense when handling any type of explosives.

Rick F. Tscherne

# MORE MILITARY JOKES

A paratrooper from the 325th Airborne Battalion Combat Team was on guard duty at the Tuzla Air Field in Bosnia-Herzegovina when his platoon leader (a 2d LT. West Pointer) came by to inspect his guard post.

As the lieutenant came up to the soldier he was greeted with a snappy salute. The officer, not wanting others to think he was an inexperienced butter-bar LT., proceeded to question the soldier on his guard duties and responsibilities. The soldier was as sharp as a razor, he knew his general orders, rules of engagement, chain of command and much more than what the lieutenant expected.

Not wanting the soldier to think he was smarter than he thought he was, the LT. proceeded to quiz him a little more. "OK Airborne," said the officer with a big smirk. "What would you do if you saw a battleship coming at you from across this air field?" "I would fire a torpedo at it, sir," replied the soldier.

"And just where would you get this torpedo?" asked the LT. "From the submarine, sir," replied the soldier.

The lieutenant, now laughing and shaking his head in disbelief asks, "And just where in the hell would you find a submarine way out here in the middle of Bosnia?"

The soldier, smiling from ear to ear replies, "From the same f—ken place you got your G—damn battleship from, sir!"

---

A 1st Armor Division platoon sergeant was training his newly assigned platoon leader on how to conduct night time "defensive operations" in Bosnia. To get first hand experience, the Psg placed the newly assigned officer in a forward LP/OP position by himself. He then instructed the LT., "If you see anything move to your front, just fire 'em up."

"But sergeant..." asks the lieutenant. "It's pretty damn dark out here, how will I be able to see anything move?"

"Sir," asks the PSG, "can you see those trees to your front?" The officer looked and then replied "Yes, yes, I can see them, Sgt." "Count 'em, sir!" commanded the platoon sergeant. Puzzled, the officer did as he was told and carefully counted each of the trees. 'There's exactly 22 trees in front of my position, sergeant." replied the lieutenant.

'Very good, lieutenant," replied the platoon sergeant. "Now, keep counting them over and over and over again, and as soon as you count more than 22 of them - FIRE'EM ALL UP!"

*The Complete Ranger Digest: Vol. VI*

A couple of 2d lieutenants were sitting on a beach relaxing with their girl friends when they overheard one of them say, "Awww look at that poor dead seagull." The two officers quickly looked up into the sky and asked, "WHERE?"

## MALLET & SLEDGE HANDLE PROTECTORS

*Submitted By: Richard M. Dobson*

Those of you REMFs in a mech, armor, or combat support unit will probably appreciate this next tip more than us grunts.

When using pioneer tools (sledge hammer/ mallet, etc.), do the wooden handles usually become badly chipped, cracked or broken over a period of time? Well, this is usually do to some jerk "missing" the stake or ground rod when hitting it into the ground and instead hitting the wooden handle. Little-by-little or chunk-by-chunk it will begin to deteriorate until it splits or falls apart unexpectedly like on a major field deployment exercise.

To protect the wooden handles from ever splitting apart and needing replacement, here's what you can do to prevent it.

STEP 1: Get 2 X metal stakes either from your shelter half set or camouflage support kit and some 100 mph (duck) tape.
STEP 2: Take the two stakes and place them onto the wooden handle near the head of the tool (sledge hammer, mallet, etc).
STEP 3: Take the duck tape and tape them securely in place.

What this does is protect the wooden handle from splintering and splitting apart should someone miss hitting a wooden or metal stake or rod and striking the wooden handle instead. Pretty smart, huh?

Oh, don't worry about the tape coming off, it has to take an awful lot of beating and abusing before it'll come apart. And even if it does, wouldn't it be much easier to replace the tape than the wooden handle out in the field? Huh? Huh?

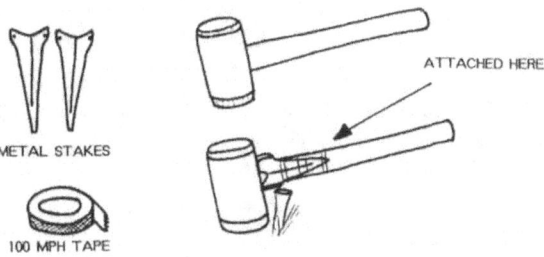

*Rick F. Tscherne*

# RAPPEL ROPE TIPS

### Submitted By: CAP C/MSG RYAN PRIMEKANO

Instead of carrying rappel rope the "Chuck Norris" way, across the chest and one shoulder, try carrying it across "both shoulders." This method will provide you a bit more comfort in carrying it and less hassles of it sliding down over your shoulder. When coiling the rope to carry it in this mode, leave a bit more slack so that you can comfortably wear it over, under, & around your shoulders.

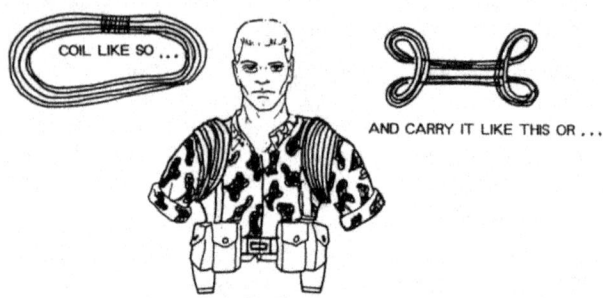

When carrying rappel rope in a ruck sack, place it between the top flap and the main compartment with the ends (the loops) dangling a little over the sides.

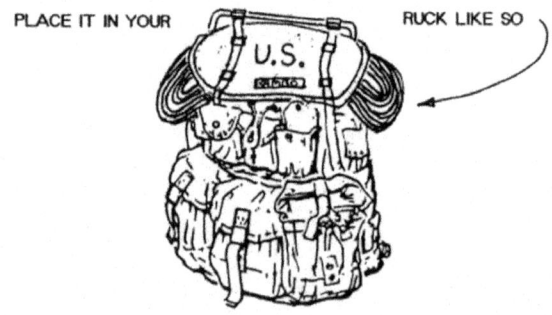

By using one or both of these methods, it will allow you to quickly unravel and use the rope in the event of an emergency.

*The Complete Ranger Digest: Vol. VI*

## REMEMBERING YOUR PACE COUNT

*Submitted By: Sgt. LeRoy Wolpert Jr.*

I use to have a difficult time in remembering my pace count, but not anymore. At least not since I wrote it down on a piece of white tape and attached it to the inside portion of my compass. Now if I forget, I just open up my compass.

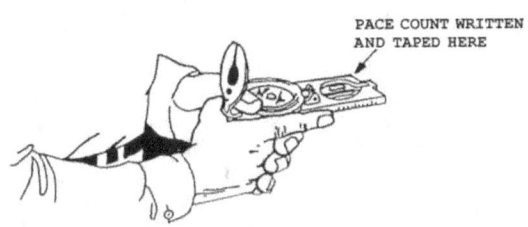

## 550 SLIP-KNOT PACE COUNTER

*Submitted By: Spc. Spurloch*

WHO - Leaders, compassman, paceman, etc.
WHAT - An 18-24 inch piece of 550 parachute cord.
WHERE - Attach it to the upper portion of the LBE harness.
HOW - Place one "slip-knot" in the cord per every 100 m.
WHY - To help you to remember the distanced that you traveled. When finished, pull cord hard from bottom and all the knots will come right out.

RANGER RICK'S COMMENTS: Use two different pace cords for long distances, one for IDOm increments and the other for kilometers. Remember to tie them separately on each side of the LBE harness and don't get 'em confused.

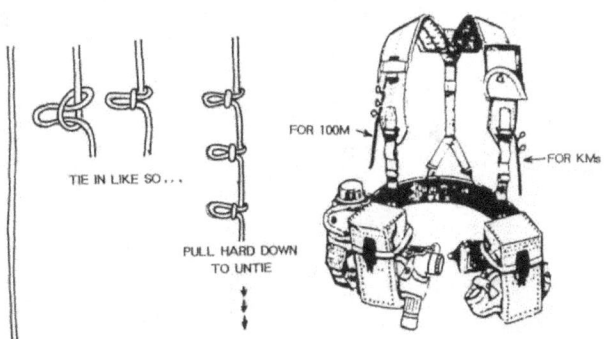

*Rick F. Tscherne*

# SUCK, BLOW, OR JERK

No, I'm not talking about SEX. But how to transfer fuel from a vehicle's fuel tank to a container or vice versa by the use of an ordinary garden hose. There's 3 X ways you can do this;

SUCK IT OUT - This method requires you to stick a hose inside a vehicle's fuel tank and suck on it until you get a "mouth full"of fuel before dipping the hose into an empty container.

BLOW IT OUT - To do this, you will need a hose and a plastic trash bag. Take the hose and place it inside the vehicle's fuel tank. Then take the trash bag and wrap it around the hose to securely close the fuel tank's opening in order to make it "air tight." (NOTE: If the tank cannot be securely closed and air tight with the plastic bag, it will not work.)

Then simply blow into the hose several good times folding or biting down on the hose in between breathes to prevent the air from escaping back out. Once you have built up enough air pressure inside the tank, just dip the hose over into an empty container and watch it flow.

NOTE: If the fuel doesn't come out, it's either because the opening of the fuel tank is NOT completely sealed or the fuel tank is very, very low. In either case, you will have to either reseal the opening and or blow more puffs of air inside the fuel tank to create more air pressure.

JERK IT OUT - This requires a "rapid" up and down hand motion with the thumb being placed over the opening of the hose when it is being raised and releasing it just when your "rapidly" bringing it back down into the tank.

When the thumb is placed over the hose in the upward motion, it traps or freezes the fuel in place in the hose. When you jerk the hose very rapidly downward while simultaneously releasing the thumb, it forces the fuel further up inside the hose until it reaches the end or the opening. Where by then when it finally reaches that point (after several rapid hand motions) all you have to do is dip the hose over into an empty container and watch it flow out on it's own.

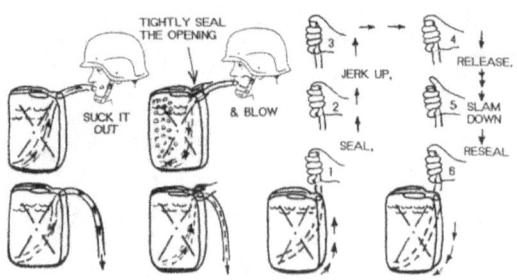

*The Complete Ranger Digest: Vol. VI*

# HOMEMADE VS-17 SIGNAL PANEL

A very useful military item that is used to signal flying aircraft and marking landing zones (LZ) and pick-up zones (PZ), is a two color panel (fluorescent red and orange) known as a VS17 (Visual Signal). Some units have them, some don't, and others just can't seem to get 'em, wellll, why don't you make your own?

They're not so difficult to make, you only need to visit a sewing supply center (Wal-Mart, Kmart, Sears, etc) and buy a meter or two of their brightest "orange & red" cloth. Then get yourself some 550 parachute cord or string, lay it along the edges of the cloth and sew them together back-to-back. Then add some 4-6 tie-down cords along the corners and sides and you now have a VS17 VISUAL SIGNAL PANEL look-alike.

If you really want to make it stand out like the real McCoy, buy a can of fluorescent orange and fluorescent red spray paint. Use this to paint one side red and the other orange. Works great, just as good or better than the real thing. (well almost...)

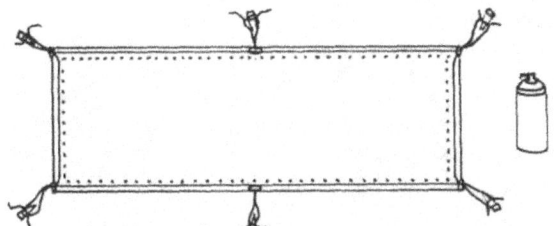

# ANTI-PESKY CRITTER POWDER

*Submitted By: PFC RYAN C. SHIPLEY*

"Hey Ranger Rick," he writes. Recently I was on an FTX with my platoon when our fearless leader ordered us to start digging in.... right smack in the middle of an infested ant colony. No matter where we turned to dig, those nasty critters were everywhere. When we asked permission to move to another location, he said not just no, but "Hell - F--- No!"

Well, to make a long story short. While I was sitting down putting on some foot powder, I accidentally spilled some on top of some ants. And boy, were those critters pissed. So then I placed some on the ground to see what they would do. Well guess what? Ants hate foot powder! So to keep them away from our sleeping area and gear, we just sprinkle a bunch of it all over the place. Including the inside and outside ot our sleeping bags. Well it didn't kill 'em, but at least it kept them away from us so that we could sleep at night.

## MAG LITE-TO-LBE

One day I was watching CNN on television when I saw a reporter interviewing a soldier in Bosnia, Yugoslavia. Attached to this soldier's upper lbe shoulder harness was a special pouch that had one of those mini-maglite flashlights.

Well, in visiting my local camping and military supply store, I found out that you can buy these special pouches for a mini-maglite flashlight. It even had an adjustable "velcro" belt loop that can be easily attached or detached to any leather belt or web gear harness loop.

This soldier had his mini-maglite hooked up on the lbe shoulder harness loop where most soldiers attach their military angle flashlight or first aid pouch. Not a bad idea, huh?

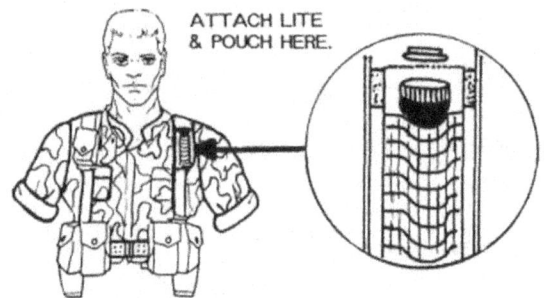

## TAPE IT TO THE FIELD

*Submitted By: Sgt. LeRoy Wolpert Jr.*

"Hey Ranger Rick," Sgt Wolpert writes. "How many times have you gone to the field and wish you had brought along some 100 mph masking tape? A bunch of times, right? Well, I've always carried and wrapped about 5-10 feet of this tape around several pieces of my equipment. Such as my flashlight, E-tool, ruckframe, knife sheath, etc. This way if I ever needed any tape in the field, I always had some at my finger tips. "Be Prepared" is my motto.

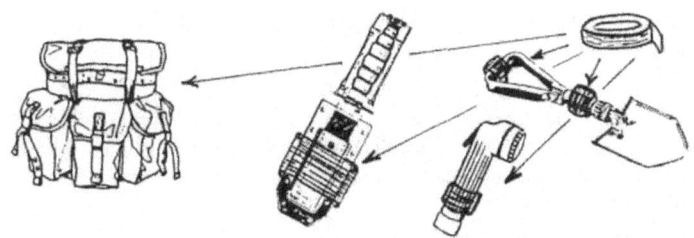

WHERE TO ATTACH EXTRA TAPE

## Excuses

Do you know what the lop 10 Excuses" that are given by soldiers who fail to carry out a certain task or mission? As David Letterman would say, "Hereeeeeeeeee we go...."

| | |
|---|---|
| Number 10 | "Hey, I just work here, I'm not paid to make f_ken decisions." |
| Number 9 | "I was so f_ken busy I plum forgot about it." |
| Number 8 | "I was f_ken waiting until Sergeant Dickhead got back." |
| Number 7 | "That's his f_ken job, not mine." |
| Number 6 | "I didn't think it would make a f_ken difference one way or the other." |
| Number 5 | "I was waiting on his f_ken ass to tell me (or us) what to do." |
| Number 4 | "No one f_ken told me to do it." |
| Number 3 | "That's not my f_ken department." |
| Number 2 | "I didn't know you were in such a f_ken hurry to get it done." |
| Number 1 | "That's the way we've always f_ken done it." |

These excuses may sound pretty funny to you, but if you think about it, they are the top 10 excuses' soldiers E-1 thru E-4 usually give for not accomplishing a certain task or mission.

If soldiers in a unit (team, squad or platoon) are constantly giving excuses, it's usually a warning sign that the individual or unit has an internal problem. And not necessarily with the enlisted soldiers, but with the leaders.

Soldiers make excuses either out of fear, lack of caring, or not fearing a leader (s) enough, When excuses are accepted by a leader, it sends a strong signal to soldiers that they can just about do whatever they want without fear of being punished for it.

Excuses occur because weak leaders allow them to occur. And if you're a weak leader that accepts excuses, you are doing nothing more than "surrendering control" to the soldier. And when excuses become part of an everyday workplace, it usually indicates that there are some other hidden problems too.

SOLUTION: The best way to eliminate excuses is by not accepting any and or by anticipating where the problems will occur and having a contingency plan for them. When soldiers realize their excuses are not going to be accepted, they will start to figure out on their own how to accomplish a certain task or mission. And when a work place is rid of excuses, you will start to see other problems disappearing too, and more unit cohesion, team work and better attitudes.

## MILITARY HANDBOOKS WORTH BUYING

Out of all the military handbooks that I've read and used myself during my career, I found these four books to be the most useful. If your serious about wanting to be the very best in your field, then I strongly suggest that you acquire these handbooks for your own library. Not only will they come in handy for training, but teaching and improving your unit's combat survival skills too. You don't have to go to school or wear the "TAB" to know everything that a qualified Ranger or SF soldier knows. Just study the books and put what you learn to good use.

### THE US ARMY RANGER HANDBOOK

This book is crammed with just about everything you need to know from basic tactics & drills to patrolling. Rappelling, hand & arm signals, how to make knots, issue an operation order, survival & evasion techniques, and much, much more. (there is a compact pocket sized edition too) An excellent book for the "wanta be" Rangers. http://www.amazon.com/Ranger-Handbook-Large-format-Official/dp/1780396597

## THE US ARMY SPECIAL FORCES HANDBOOK

A "No Nonsense" handbook designed for elite soldiers and leaders who want to learn everything that a Special Forces "Green Beret" knows. Covers topics and subjects in guerrilla warfare, tactics, demolition, communications, weapons, water, aerial, and psychological warfare operations and more http://www.amazon.com/U-S-Special-Forces-Handbook-Department/dp/1602391262

## US ARMY SPECIAL FORCES MEDICAL HANDBOOK

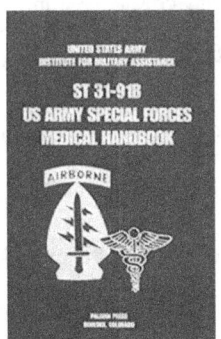

This compact, quick reference, medical handbook contains everything from diagnosing diseases to emergency surgery. To treating burn and blast victims to other medical emergency procedures in such areas as veterinary, obstetrics, pediatrics, orthopedics, and much, much more.
Update: This older edition has now been superseded by the newer edition created by

*Rick F. Tscherne*

SOCOM, see: https://www.amazon.com/Special-Operations-Forces-Medical-Handbook/dp/1492947520/

## SPECIAL AIR SERVICE (SAS) SURVIVAL MANUAL

Learn to survive anywhere in the world, whether it's on dry land or at sea and in any type of weather environment. This handbook is packed with thousands of detail drawings and illustrations that show everything from making a shelter to what types of plants & bugs you can eat. http://www.amazon.com/SAS-Survival-Handbook-Revised-Situation/dp/0061733199

## FIELD HYGIENE TIPS

Back in November'94 I received a letter from a Major in the Army Dental Corp. To be honest, I had to read his letter several times over to figure out if he was serious or just pulling my leg. I don't know if he wanted his letter printed or not, but here's part of it anyway.

An important habit that troops need to start doing in the field, is practice "washing their hands" before they eat and after they take a shit or piss. How simple it is to prevent the spread of germs, yet how we forget or refuse to do it.

Due to the lack of water in the field, I know it's not always possible for troops to wash their hands. But when it is available, unit leaders should have a few bars of soap and some water containers placed in front of the chow line so they can wash before they eat. This is extremely important for those who serve or handle food, as it's very easy for them to contaminate an entire unit as they pass through the chow line.

Once upon a time when I was assigned to the 7th Special Forces Group, I knew a bunch of "Rambo" SF troopers who refuse to wash themselves in the field. In fact, they considered diarrhea a "Badge of Honor" in showing how tough they were. When in reality, all they were doing was showing their "Stupidity" in not understanding that diarrhea is an illness that weakens the body. It's better to be a "wimp" in cleaning your hands before eating and after shitting than to be awarded the "SF Field Badge of Honor."

Medically speaking, I think it was the dumbest idea for the Army to add powdered drink mix to the MRE meal packet. When a drink mix is placed inside a plastic canteen, it can produce a potential harmful osmotic effect on soldiers if they fail to clean them correctly afterwards. A typical G.I. cleaning is usually nothing more than a quick rinse of cold water when they should be using hot soapy water followed by hot clear water.

RANGER RICK'S COMMENTS: The major wrote a lot more than this, but I picked out only what I could understand, as he used a bunch of $10 words that I couldn't even pronounce.

As for the washing of the hands, I have to admit, guys & gals. He's 100% right! It was reported that during the Persian Gulf War there were quite a number of illnesses traced to only one source, "improper personal hygiene." How could most of these illnesses been prevented? Simple, wash your hands before you eat and after you take a shit or piss.

To the Army Major in the Dental Corp who sent me these tips, I thank you for your input, sir.

## WATER PURIFICATION SYSTEMS & METHODS

Are you assigned to an elite Ranger, SF, LRSU, or RECON unit? If you are, then you may want to consider purchasing a few special survival items for yourself or unit. Why?

Well, as our real world missions start to expand and take us to far away places like Somalia, Haiti, Rwanda, etc, it's going to be very difficult at times to locate clean potable water. To insure you and or your unit are well prepared for those unexpected remote and exotic deployments, consider these items;

FIRST NEED DELUXE WATER FILTER: Economical, yet effect This microfiltration system removes Giardia, bacteria and other harmful organisms and chemicals. Comes with pump, canister, hose filter, suction hose and strainer Usually sells for around $50.

BASIC DESIGNS CERAMIC FILTER PUMP: Effectively removes from water Giardia, bacteria, and other harmful chemicals. Compact, lightweight, and fits easily inside a rucksack cargo pocket. Filters up to 500 gallon per cartridge. Sells for around $30.

ACCUFILTER 5TM CANTEEN INSERT: A self-contained compact filter system that fits inside the neck of a GI canteen Just fill the canteen with clear water, place the filter insert inside and drink. Removes easily for refills and filters up to 40 quarts of water. Sells for about $16.

ACCUFILTER STRAW: A self-contained water filter system (straw) that fits easily inside a cargo pocket. Filters up to 40 quarters of water. Sells for around $ 16. NOTE: When the Accufilter Straw or Canteen Insert becomes very difficult to draw water through, discard and replace it.

WATER PURIFICATION TABLETS: Commercial / military issued tablets are designed to kill all forms of harmful water bacteria and other organisms. Dosage per quart will vary, always read the instructions on the bottle before adding.

If you don't buy or have access to the above products, there are two other ways you can purify water to make it safe for drinking. But they should only be used under "Emergency or Survival Conditions Only."

CLOREX BLEACH - If the water is clear, add two (2) drops of household "liquid" clorex bleach per each quart/liter of water, four (4) drops if the water is cloudy. Stir or shake well and wait 30 minutes before consuming.

BOILING - If the water is clear, heat the water until it it bubbles and let it boil for at least ten (10) minutes before letting it cool and consuming.

## SETTING UP A 292 ANTENNA BY YOURSELF

*Submitted By: Cpl. Gregory D. Esttlake*

Q: How many soldiers does it take to set up a OE-292 Antenna? A: ONE! Provided he has his shit together & knows these steps

STEP 1: Locate the exact spot where you want to set up your antenna and then hammer the base plate into the ground. Grab the copper-tipped antenna sections, assemble and attach them to the piece that will be fastened to the very top portion of the antenna. Now place it off to the side until later on.

STEP 2: Each pole section is 3 feet in length, starting from the antenna base plate, assemble together 5 sections for a total length of 15 feet. Using this 15 foot pole section as a measuring device, place in the ground 3 stakes exactly the same distance apart (15 feet) in a "Y" shape pattern.

STEP 3: Take one (3 foot) pole section and place it on top of the 15 foot (5 X piece) section and attach your 1st guide line plate. Now assemble the remaining set of poles and attach the 2nd guide line plate to the appropriate section.

STEP 4: Now Lay the poles on the ground so that it forms a 90 degree angle from stake #1. (IMPORTANT: If you fail to place the pole at a 90 degree angle from stake #1, you will have a big problem later on when you try to stand it up)

STEP 5: Attach one set of guide lines (one light and one dark color) to the appropriate holes of both guide line plates. Run these guide lines to stake #1 and tie them off at the antenna base plate. Caution: DO NOT mix up the two different guide lines.

STEP 6: Now swing this entire pole section around until it forms a 90 degree angle from now stake #2. Attach the second pair of guide lines (again) to the appropriate holes of both guide line plates and run it to stake #2 and tie them off back at the base plate. (IMPORTANT: Again, if you fail to place the pole at a 90 degree angle from stake #2, you will have big problems later on when you try to stand it up.)

STEP 7: Move the pole until it is half way between stakes #1 and #2. Now attach the last set of guide lines and hook them to the remaining holes of the guide line plates. Then run them to stake #3 and "past" the anntenna base plate, DO NOT connect or tie them off. Take the antenna piece that you laid off to the side earlier and attach it to the top portion of the pole. Now connect the antenna cable and run it down along side the poles and tape it in place.

STEP 8: Pick up the last two "untied" guide lines from stake #3 and go approximately to the middle of the pole section. Simultaneously, lift up and start pulling on the set of guide lines and begin raising the pole. Now it's WALK, RAISE, TUG, & PULL all the way until the pole is in the upright position.

STEP 9: Once it's fully erected and in the upright position, tie the last set of guide lines to the base plate as you did the others. Then go around and make any necessary adjustments in the guide lines to insure the pole is standing straight and erect. Double and triple check each stake and guide line to insure their stable and firmly in place before hooking the antenna cable to your radio. And that's it!

NOTE: Be aware that the first few times you try this, you will no doubt have some difficulties. But the more you practice, the more easier and proficient you will become in setting it up without any help from your fellow co-workers.

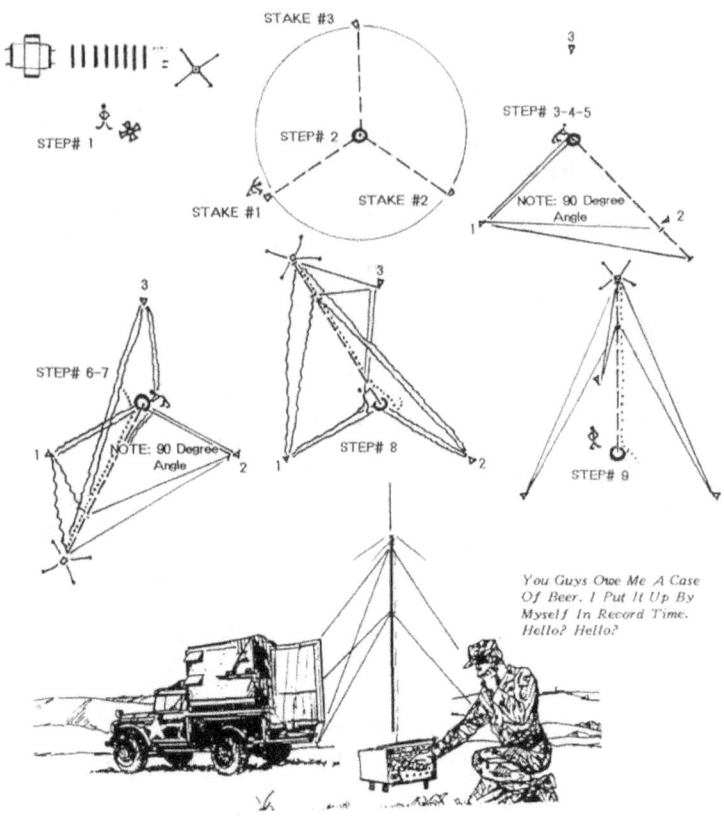

## LISTEN UP !

It's pretty funny how we overlook some of the simplest things in how to improve our military capabilities. How many times have you strained your ears at night in an LP/OP position trying to listen for enemy movement? Have you ever tried "cupping your ears?"

Now I'm not an ear specialist or a doctor, but I know it will definitely increase your hearing capability by at least 50% or more. You don't believe me? Try it!

The next time your in a night LP/OP fighting position and your trying to listen for some movements. Try cupping one or both of your hands up to your ears, trust me, it works!

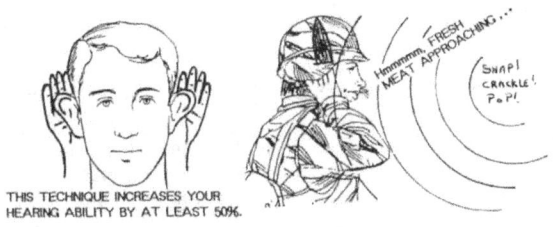

THIS TECHNIQUE INCREASES YOUR HEARING ABILITY BY AT LEAST 50%.

## GREAT BALLS OF FIRE

Need to make a "non-tactical" fire and you can't seem to get the wood to burn? Try pouring some good 'ol military insect repellent on it, the bottled type, NOT the spray type. This stuff gets it burning fast because it contains a large amount of alcohol.

Another little helpful fire starter to use is gunpowder from blank or ball ammo. Just pry open the crimp end of a blank round, or pull out the lead bullet from a ball round, and dump it in a nice "little pile" (don't sprinkle). Place some small pieces of dry wood on top of the gunpowder and ignite.

WARNING: When using gunpowder to light fires, be very, very cautious as it ignites and burns "extremely fast."

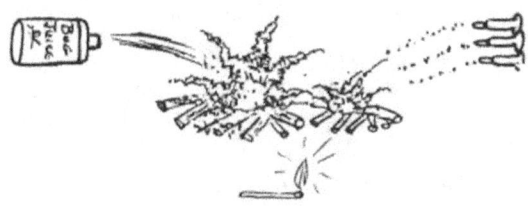

## BORING ROAD MARCHES

Did you ever wonder why we always seem to view road marches as boring, miserable, and painful? Well, besides being that they are, they don't necessarily have to be. Here's a few road march tips that seemed to work for me and maybe for you.

Instead of always road marching with combat boots, do it every so often with tennis shoes. Just because the Army puts out training standards doesn't mean you can't modify some of them. After all, if your allowed to take a "diagnostic" PT test, why can't you take "diagnostic" road marches too? Makes sense, don't it? Realistically, road marching with tennis shoes won't have any effect on the leg muscles or the load of the ruck, but it will reduce or eliminate a lot of blisters.

If your a leader, try doing more diagnostic road marches in tennis shoes than in combat boots. And if your chain of command won't allow it, try to convince them to allow the troops to carry them in their rucks. This way if you have any soldiers who can't keep up with the unit due to painful blisters, allow only those hurting to switch from boots to tennis shoes. Which is better, to finish as an entire unit or as a "partial" unit with a bunch of stragglers?

To motivate and entertain the troops while on a boring road march, have someone carry a portable stereo in their ruck. You'd be surprised at just how a little bit of music can effectively increase a unit's spirits and morale. And if your chain of command won't allow this neither, then carry your own cassette-radio and hide the ear plugs under your shirt.

Never break in new boots on a road march, they should always be gradually broken in back in the rear. And if your a leader, then it's YOUR responsibility to inspect your troops boots and insure they're NOT wearing new boots. Their blisters and pain will only slow you and your unit down.

When road marching, don't concentrate on the march itself, but on something pleasant like "SEX." Or talk with your buddies, crack jokes, make fun of someone, carry a water pistol, etc in order to help others to keep their minds occupied and off the march. By displaying humor and joking around, it will not only help lift your morale, but those all around you.

TRUE STORY: One time I was on a 12 miler with my unit (1/75th Rangers) when I noticed that one of my men had a small hole in the butt of his pants. I told him (jokingly), "Hey Ranger you better repair those damn pants at the next break or else"

Suddenly, he reaches behind to locate the hole and says, "I can't repair them, Sarge." When I asked, "Why not?" He then suddenly stuck his finger in the hole, tears his pants wide open displaying his entire ass and says, "Because I ain't got enough thread in my sewing kit to sew it up." Well, we completed that road march in record time laughing the entire way, because that damn Ranger wasn't wearing any underwear.

# CLEARING & DEFENDING HALLWAYS AT NIGHT

A technique that I learned in Belgium Commando School that you probably won't find in any of our military doctrine. Is how to clear and defend a hallway (or tunnel) at NIGHT, the Belgium Commando Way.

If you look though all the TMs, FMs, and other military manuals, you'll notice that they only teach the basic fundamentals in how to attack, clear, and defend a room or building during daylight hours. But what about at night? Hum.'

Well, some valuable lessons that I remember from the school, is how to improve your chances of NOT becoming a casualty when attacking, clearing or defending a hallway at NIGHT.

Lesson #1; When moving down a hallway at night, keep to the sides of the wall (obviously) and stay as low as possible to the floor. Either low crawl, high crawl, or run, dodge, & squat down behind something (if something is available).

REASON: Soldiers defending or attacking down a hallway at night will usually spray their weapons at waist level high. Therefore, your chances of becoming a casualty increases should you be standing, but decrease if you remain close to the floor.

Lesson #2: When firing your weapon down a hallway, fire it from a prone position close to the wall.

REASON: Soldiers defending or attacking down a hallway are taught and trained to always "hug" the sides of the wall. Therefore, by firing your weapon from a prone position, it will decrease your chances of becoming a casualty and increase your opponents chances of becoming one instead.

Lesson #3: When moving down a totally dark, blacken, hallway or room without the aid of a flashlight. Extend the elbow of the non-firing arm until it touches the side of the wall and at the same time raise the hand slightly forward and above the head.

REASON: This will allow you to safely feel your way around in the dark and at the same time protect your face from any low level beams, windows, or doorways and still be able to tire your weapon in self defense.

Once again, you probably won't find any of these techniques in any of our military or police SWAT team manuals. But if they work for the Belgium Commandos, why wouldn't they work for YOU?

Rick F. Tscherne

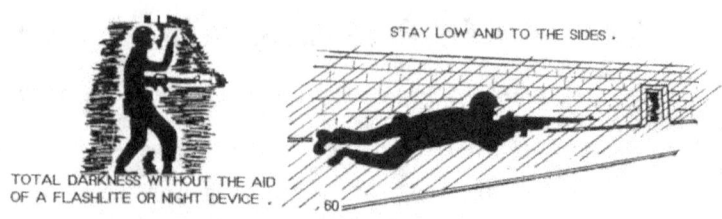

## M16 ASSAULT SLING CLIPS MADE FROM COAT HANGERS

Yep, I'm back again with another coat hanger tip. Except this time, how to make Assault Sling Clips for your M16 rifle. You will need the following three things:

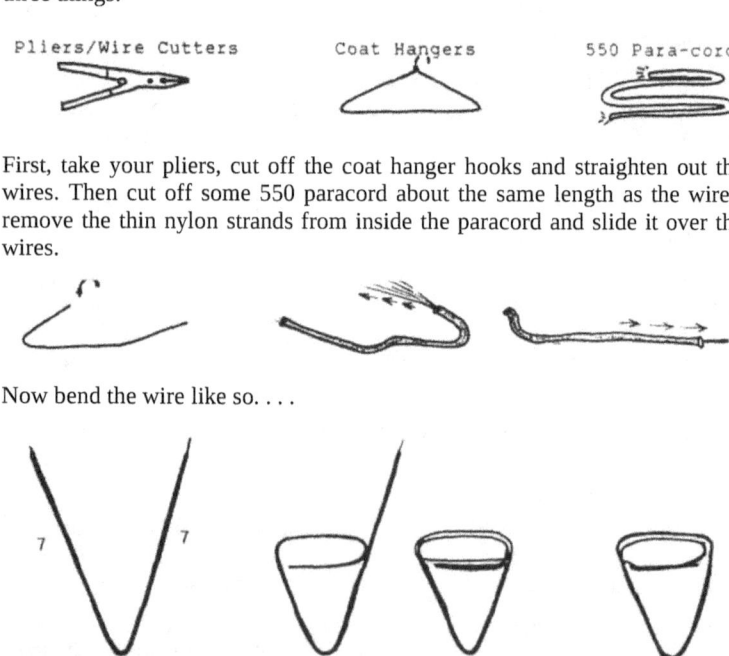

First, take your pliers, cut off the coat hanger hooks and straighten out the wires. Then cut off some 550 paracord about the same length as the wires, remove the thin nylon strands from inside the paracord and slide it over the wires.

Now bend the wire like so. . . .

Take a Bic or Zippo lighter and melt the ends of the 550 paracord to the clips, this will prevent them from unraveling. The purpose of covering the coat hanger wire with the paracord is to reduce the metal "glare" and to make the wire sound proof when it's attached to the weapon.

Now take these two Assault Sling Clips and attach them to your weapon like so (see drawing) and attach your sling. Not a bad idea, huh? Do you like this tip? Do ya? Huh? Huh!

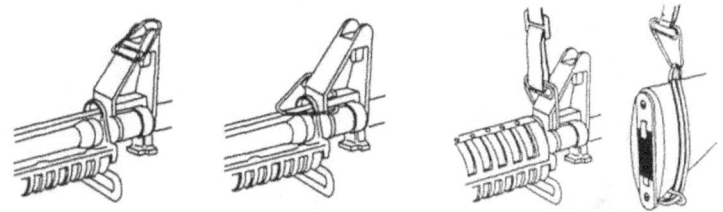

## RATTLE PROOF YOUR WEAPON MAGS

*Submitted By: David J. White*

An excellent way to rattle proof magazines in your ammo pouch, is to take an old bike inner tube and cut off a few pieces of rubber. Then slide the pieces over and around the bottom portion of the magazines. If one rubber band is not enough, then add a second one well below the magazine "slot lock" so that it won't interfere with the mag locking latch. (See below)

NOTICE: As per Mr. White's request, credit tor this tip goes to a "Fighting Firearm" magazine writer by the name of Mr. Ken Hackathorn, it was entirely his idea.

## FIREMAN & COMBAT READY

ALERT! ALERT! ALERT! MOVE! MOVE! MOVE! GO! GO! GO! GO! Have you ever been on a special alert status that required you to be fully dress and outside in a formation in less than 3 minutes from a dead sleep? Difficult, ain't it?

What did you do to save some time, sleep with your boots and clothes on? Well, it may be easier for you to do it this way, but it sure ain't too comfortable. Especially if it's a really hot day or evening.

Wanta know an easier way? Just do like the firemen do. Take off your boots, keep them half laced up and place them near your bunk. Take off your pants, place the pant legs over the boots and push them down flat to the floor.

Now if an alert is called, place your feet inside the boots, pull up on your pants and close the belt. Take the halt laced boots, pull up tight on the string, wrap the excess around the ankles, and tie it oft. Then whenever you have the time, such as in formati6n, start lacing the boots up the correct way. This method will allow you to sleep or rest a bit more comfortably and save you time in getting dressed.

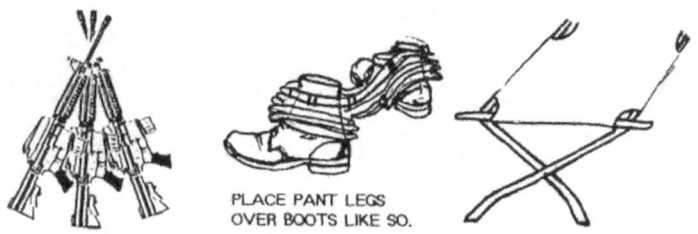

PLACE PANT LEGS OVER BOOTS LIKE SO.

## FOLDING GOLF CART CARRIER USES

*Submitted By: Sean P. Gilday*

Dear Ranger Rick,
Here's an idea that might help some "light infantry" grunts to transport some of their equipment more easily in the field.

If they get hold of a couple of used folding golf bag carriers and make a few slight modification to it. They can use them to transport such equipment as an M47 dragon, mortar tube, base plate, tripod, ammunition, etc. If it worked during WW I, WW II, and Korea, why wouldn't it work today.

Back in the old days, light infantry was considered "light" when you only carried a blanket, an extra pair of socks and a one day supply of food and water. Today, your not considered light infantry unless your carrying at least a 100 pounds of equipment on your back and a three day supply of rations.

## WATER-PROOFING WALLETS

If your looking for something really durable to water-proof your wallet, try a plastic MRE wrapper. You can either use it like it is, or cut it up and modify it like the way I use to carry mine.

The only thing you need is an MRE wrapper, some 100 mph tape, and a knife. Take out your wallet, measure it to the size you need, cut away the excess, and tape it up. Works better if you put a "lip" on the end of the tape before sealing it across the opening end of the wrapper. This way you'll be able to seal and unseal the wrapper a lot easier.

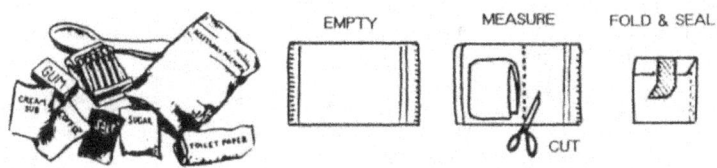

## ZIP-IT-UP SANDWICH BAGS

*Submitted By: Sgt. Lance Hefington*

Sgt Lance Hefington says: "Whenever I get ready to deploy to the field, I always buy a box of zip-lock sandwich bags for the guys in my squad. I want to make sure they keep their wallets as dry as possible, as you never can tell what the weather will be like on an FTX, nor what type of water obstacle you'll be crossing.

RANGER RICK'S COMMENT: I like the zip-lock bags called Presto's Re-closable Storage Bags. The 12 X 12 size is excellent for water-proofing BDU shirts and pants, not only will it keep your clothes clean and dry, but you can also store your dirty clothes inside them too.

## TACTICAL REAR VIEW MIRRORS

*Submitted By: Sgt. LeRoy Wolpert Jr.*

Have you ever had to drive a tactical vehicle on a busy highway and when you looked up in the rear view.... there wasn't any mirror there! If your like me, it's a hard and annoying habit to break, as tactical military vehicles don't come equipped with windshield mounted review mirrors, just side view mirrors.

But I solved my problem by purchasing one of those $1 cheapo, self-adhesive 2 inch round mirrors that truck drivers usually affix to their sideview mirrors. Except that I mounted mine to the windshield instead. Now whenever I look up, I can see everything directly behind me without turning my and taking my eyes off the road. Good idea, huh Ranger Rick?

RANGER RICK'S COMMENTS: It was a very good idea, Sgt Wolpert. I wonder why none of those Pentagon VIPS didn't think of this? Or maybe they did and they thought it would save us some tax dollars, huh?

## 550 PARACORD OTHER USES

*Submitted By: Sgt. Martin Dudel*

Sgt. Martin Dudel tells me, "One time I was driving late at night when the belt on my alternator broke off. I knew I wasn't going to find a garage open anywhere, nor did I want to take a chance on continuing to drive and letting my lights run down my battery.

Well, in the trunk of my car I just so happen to find some 550 paracord, so I decided to try to make an improvised belt. After making the first belt and seeing that it didn't work too well (due to lack of traction), I decided to try it again. Except this time, I tied a knot in the 550 cord every inch or so. Well guess what? Yep! Believe it or not, it actually did work and 1 got home safely that night too.

## WHEN SHOULD YOU BUY COMFORT ITEMS?

Soldiers have no excuse to complain about how cold it is in the field, they are issued the best damn military gear in the entire world - Bar None! And if it's still not enough to keep their little asses warm, then they should sacrifice a few lousy bucks and buy what they really need to stay warm.

Now most soldiers always seem to have to learn the hard when it comes to buying comfort items for the field. It's sorta like going grocery shopping for food on a full stomach, if your not hungry at the time, you don't usually buy much food. But if your hungry when you go shopping, you normally buy MORE FOOD than what you actually needed or intended to buy.

Well, it's pretty much the same way when it comes to buying comfort items for the field. When your out in the field freezing your ass off, you promise yourself that the next time your back in the rear, your going buy what you really need for the field. But as soon as you get back home and your nice and warm, you forget all about that promise. Right?

Well, the next time your getting ready to deploy, try sitting down and remembering all the times that you were miserable in the field. Then make of list of all the nice things that you could have used to stay more comfortable. Then GET OFF YOUR ASS and BUY THEM!

*The Complete Ranger Digest: Vol. VI*

## MAG-1 COMBAT GLASSES

*Submitted By: David J. White*

Mr. White says, "I have to wear glasses in order to see 20-20. But I absolutely refuse to use those optical inserts for the M-17 Protective Mask. So I ordered me a pair of those MAG-1 Combat Frames" from US Cavalry and use them instead."

Well, out of all the well known nationally advertised optical centers, the cheapest is no doubt Wal-Mart's Optical Center. It only costs $50 bucks for the lens and to have them installed into the combat frames, you can't beat that price. (I just thought you and your readers would like to know this little bit of info.)

MAG-1 frames: http://www.rangerjoes.com/Eyewear-Mag-I-Combat-Size-48-22-P516.aspx
MAG-1 modification with parachutist retention strap: http://www.combatreform.org/glasses.htm

Combat Glass Frames
Non-reflective, self-adjusting nylon frames accept prescription or sunglass lenses. Wear with Army M17 gas mask, tanker helmets and more. Adjustable rubber headband. Includes pattern. Weight 7 ozs. Made in U.S.A. Black.

## HOW TO SHARPEN A KNIFE

Now I don't claim to be a professional or expert knife sharpener. Nor do I waste my time sitting around sharpening my knives for hours like the "Rambo Cowboys" and dreaming of killing someone someday. If it's sharp enough to shave some hair off my arm or ass, then it's good enough for me.

Today you can find about a dozen different types of books and devices that can help you to bring that knife to a razor's edge. So I'm not going to waste your time nor mine, I'm only going to cover the basic stuff. So if you don't know anything about how to sharpen a knife, then read on. And for those of you who do know how, then you can go ahead skip this page...

The most important thing you need to remember when sharpening a blade, regardless of the type of stone (wet or dry) or the type of instrument your using. If you don't hold the knife correctly at the right angle, it won't sharpen.

The correct way to hold a knife is at about a 20 degree angle so that the edge of the blade rests evenly and flush against the stone or instrument, (see picture). While maintaining this 20 degree angle, slide the blade across the stone or instrument in one swift even stroke while maintaining pressure on the edge of the blade.

Repeat this motion several times and rotate the blade from one side to the other so that both sides get about the same number of strokes. After you have repeatedly stroked and rotated the blade several times, check it for sharpness. Should the blade remain pretty dull after numerous strokes, be advised;

**A.)** You may be holding the edge of the knife at the WRONG angle. Too much of an angle and it will cause the blade to rub "into" and not against the stone or instrument. Not enough of an angle and the blade will not properly touch or rub against the stone or device.

**B.)** You may be using the wrong type of stone or instrument designed for your knife. Knives made out of certain steel or metals require a special stone or instrument to sharpen them. Always follow the knife manufacturer's instructions on the type of stone or instrument to use on your knife.

**C.)** If the knife is NOT new and it was dull before you ever began and your sure your holding it correctly. Then the knife edge could have folded over producing what is known as a "hollow ground edge." If this appears to be the problem, then you should take it to a knife shop to have it sharpened with a high-speed grinder.

When sharpening a blade (knife, axe, or machete), remember...

1. All blades can be sharpened, it's just that some are more easier or harder than others. It depends on the type of steel or metal that the knife is made out of and the type of stone or device your using to sharpen it.

2. When sharpening a blade, be patient, take your time, it's a slow process that should never be rushed. Rush jobs usually produce carelessness that leads to blood, "YOURS!"

3. Never use a grinder to sharpen a blade unless it's being sharpened by an "expert" knife sharpener. A good knife can become permanently damaged if it's ground too much and overheats, which can cause the blade edge to become"brittle."

Take your time and practice, practice, practice until you become proficient in knowing how to sharpen a knife.

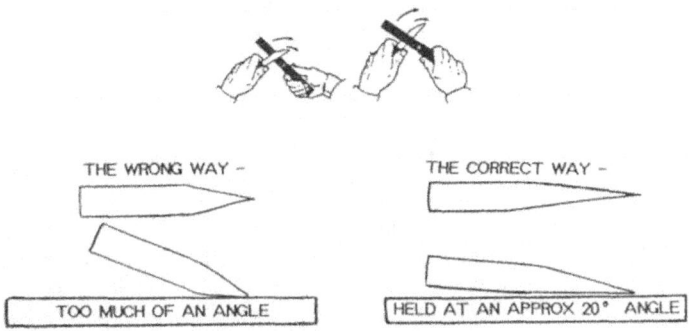

If your looking for a shortcut in sharpening your knives and you don't plan on using them for "hospital surgery." Then I'd recommend you buy yourself one of those kitchen knife sharpeners called the "CHEF'S CHOICE." It's a very handy little device that works pretty well on most knives. It has two slots, one for bringing the blade to a sharp edge and the other for honing it to a razor's edge. Works great!

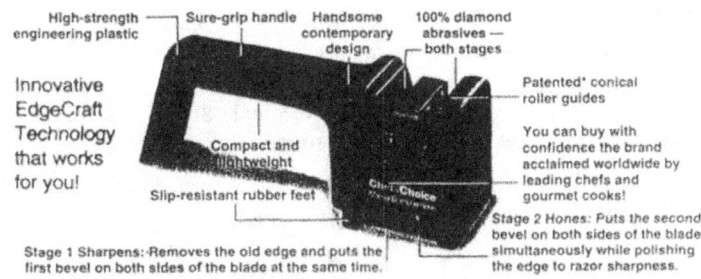

## MAP READING TIPS

*Submitted By: Sgt. LeRoy Wolpert Jr.*

Dear SFC 1/2 (Ret) Ranger Rick,

Congratulations on your handbooks. I'm in the Army National Guard and I don't normally get the opportunity to visit many army installations. But whenever I do, I always look for handbooks written by you, they're outstanding!

While attending PLDC, I had to help teach another soldier how to read a map, use a compass, and navigate. It didn't take me very long to realize that this soldier was kinda weak and incompetent when it came to reading and following a map.

Yet this soldier kept telling me that he never had any difficulties in following a road map when he was on the road with his truck. Which by the way, was what he actually did for a living, he was a truck driver.

Finally, an idea hit me. I told the soldier to plot his points on the map and to connect them by drawing a straight line from one to the other. After he plotted the points and drew the line, I then told him to make believe the line was a road and to follow it like he was in his truck.

Well, you might think that all this sounds pretty silly or stupid, but guess what? That soldier looked at his map, took out his compass and lead us right down the "road" to the first point. Coincidence? I think not, he repeated this procedure over and over again and got all of his points. Not bad for a rookie on his first land nav course.

Well, when it came time for him to take the final PLDC land navigation test, guess what? Yep, it took him a- little while to get his points, but he got 'em all within the required time limit. Coincidence? I don't think so.

My point is this, some soldiers, especially young inexperienced E-1 to E-4s get overwhelmed and confused the first time they get to use a military map and compass. To make it more easier for the slow learners to understand, you may want to try using non-military terms and teaching methods.

Don't be so fast to shove the "military ways" down a soldier's throat. Just because it's not taught at the school or written in the manuals doesn't mean you can't try something new. Just Keep-It-Simple-Stupid (KISS), and if it works, then it must not be all that stupid, right?

RANGER RICK'S COMMENTS: I love to hear training stories like tod job, Sgt. Wolpert!

# BUYING MILITARY SURPLUS (PDO SALES)

Oh, Boy! I think there's probably going to be a few military supply store owners pissed off at me when they hear that I let out one of their little secrets. (But who give a f---!)

Have you ever heard of a PDO or ERO sale? It stands for Property Disposal Office (PDO) or Equipment Reutilization Office (ERO). Their military offices responsible for disposing or selling of government property that's either excess, outdated, used or damaged. That is, according to their standards, NOT necessarily yours or mine. It's a great way to purchase some good stuff at "ROCK BOTTOM PRICES."

Items such as military chairs, desks, tables, shelves, closets, refrigerator, stoves, televisions, typewriters, computers, parachutes, ruck sacks, helmets, sleeping bags, tents, bikes, boats, survival gear, vehicles and much, much more. No I'm not bull shitting you, partner. It's true!

How do you find out about these sales? Easy, just contact any US military installation (Army, Navy, Marine, Air Force) and ask the military operator for the telephone number of the Property Disposal Office or Equipment Reutilization Office.

UPDATE: Govt. Surplus auctions online: http://www.govliquidation.com/index.html

Then ask the individual at that office when do they plan on holding they're next sale. Be advised that...

1. Some military installations may not have a PDO or ERO office, they may ship all their used gov't property to another nearby installation. Be that Navy, Army, etc.

2. PDO/ERO sales are not held every day, week, or month, they are usually held once every 2-3 months or twice a year. It depends on how much stuff they need to get rid of.

3. Those wishing to attend these sales may or may not need a military ID card. Some are open to the public and others are only open to private dealers and contractors.

4. Some purchasing restrictions may apply to certain items such as to cars, trucks, boats, and planes.

5. "Cash & Carry" is the way they like to do business. Though sometimes they'll take checks, but definitely NO CREDIT CARDS

When calling for information about their sales, keep in mind these five (5) questions:

   a. WHO can attend? (Military, civilians, etc)
   b. WHAT type of payment will be accepted?
   c. WHERE is the sale going to be located?
   d. WHEN is the sale, the date and time?
   e. HOW often do they hold these sales?

*Rick F. Tscherne*

# HOW TO FIND DIRECTIONS

Before departing on a military operation or patrol, unit leaders are suppose to issue their men what is called a "Mission Operation Order." It's a checklist that covers everything from what is supposed to be done, taken, and carried, to all sorts and .types of contingency plans. Including what you need to do should you become lost or separated from the unit, or most commonly referred to as the "escape & evasion" plans.

Leaders are also required to brief their men on the general direction the unit will be heading as well as the return route. So if in the event someone should become separated or lost, they will at least know which way to go. Which is usually to a rally or link up point to rejoin their unit or to head back to friendly lines on their own.

If you don't carry a compass, then I would strongly recommend that you buy one and start carrying it, as you never know when you'll need it. But if you refuse to carry one, then here's a few ways that you can still determine directions.

SHADOW-TIP METHOD
(A) Grab a stick, find a level area, and place it in the ground. (B) Find the shadow of the stick and place a rock at the very tip/end of the stick's "shadow." (C) Wait about 15-20 minutes until the shadow tip has moved a few inches. (D) Place again another rock at the end of the shadow tip. (E) Connect the two rocks together by drawing a straight line in the ground from one rock to the other and slightly beyond. (F) Place your heels on this line with the stick located to your rear.

You are now facing in the general direction of north, east is on your right, west on your left, and south is in your rear. If you have to travel in a northeast direction, turn your body slightly halfway right, pick out a landmark (hilltop, large tree, etc) and start walking towards it. Repeat the process as often as needed and don't try to guesstimate!

## WRIST WATCH METHOD

Depending on where you are in the world, those that are located north of the equator (Northern Hemisphere): (A) Grab a small stick, find some level ground, and place it into the ground. (B) Take off your wrist watch and place it on the ground near the stick until the stick's shadow is running along the "hour hand." (C) The general direction of south is now located between the hour hand and "12 O'clock" and north will be located in the opposite direction between the hour hand and 12.

If you are south of the equator (Southern Hemisphere): (A) Place the watch on the ground until the shadow of the stick is running along the 12 and the center portion of the watch. (B) The general direction of north will be located halfway between 12 and the hour hand with south being located in the opposite direction between the hour hand and 12.

NOTE: The term "general direction" refers to approximately, NOT precisely.

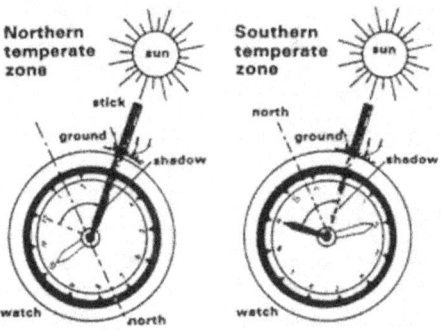

## NORTH STAR METHOD

Look up at the night sky and find the "big dipper." The last two stars in the "cup" point directly at the Polaris or commonly referred to as the "North Star." It's located about 5 times as far out as the distance between those two stars in the cup. If you face the north star, east will be on your right, west on your left, and south in your rear.

Rick F. Tscherne

## DID YOU KNOW...

While attending the Advance NCO Academy at Fort Benning (1980), I met a pair of NCOs who worked at Aberdeen Proving Grounds evaluating and testing new military equipment and weapons. One particular infantry weapon system that I was really interested in hearing about, was the old M202A1 Rocket Flame Launcher.

Because the M202A1 and M72 LAW both take the same size rocket rounds (66mm) but different pay loads, I've always wondered why the Pentagon never came up with an M202A1 that could fire H.E.A.T. rounds too. While the M72 LAW is a one shot-single barrel "throw-away" weapon, the M202A1 is a four shot-four barrel "reloadable" weapon system. I always thought that if the M202A1 could be converted to fire LAW rounds, it would make an excellent weapon for engaging "multiple" targets (tanks, buildings, bunkers, troops) simultaneously. Makes sense, don't it?

Well, after talking with these two NCOs, I was very surprised to find out that the Pentagon did convert one M202A1 launcher to fire 66mm (M72) H.E.A.T. rounds. And even though it worked extremely well, they claim the idea and weapon had to be "shit-canned." Why?

Well, due to the amount of money the Pentagon already spent or wasted in developing and testing a new anti-tank weapon system called the VIPER. They claim it would have been too costly to the Pentagon and Govt. to cease further testing and development to pursue a less expensive, more useful, and present weapon system. (Hmmm... it kinda makes you wonder sometimes if the brass at the Pentagon are really our smartest & brightest officers, don't it?)

### Keep Ranger school

I read the article on the Ranger training program and found it very interesting. It's not surprising that there are individuals around who think the best thing to do is close the school. I'll bet they never spent one minute in the swamps where these soldiers died.

Those who do not have a Ranger tab don't understand it's not the fact that you're in a swamp that's important. It's the sense of duty, accomplishment and camaraderie that being under extreme physical and mental pressure develops. This course cannot be duplicated; Ranger students want it to be tough because they don't want some rubber-stamp kindergarten experience that a lot of schools are.

Having a "tab" places you in an exclusive club in the military. The dues for this membership are paid in blood, sweat and tears of training areas like Eglin AFB, Fla. Those soldiers should not have died, but let's punish the guilty, retrain the remainder and drive on.

These four guys drove on when others would have quit long ago. They serve as

an example of the guts, determination and the backbone needed to earn the right to be called a Ranger. That's something to think about the next time you retreat to the comfort of a warm-up tent or building when you get a slight chill. The safety systems to prevent this type of incident are most likely in place; someone just neglected to enforce them. That's no excuse, but that's most likely the truth.

It's always easy to throw stones at something you don't understand. Ranger training is no exception. Those who went, know; those who haven't, don't. The proudest day of my life was when I graduated from Ranger school. I was 30 pounds lighter, beat up and exhausted, but I knew I had been challenged.

Let's not get politically correct or fashionable by lowering the standards, admitting women or closing the school. The Army and the nation need schools like this.

Diamonds are created by pressure and heat. Anything else is a lump of coal. Rangers lead the way.

Sgt. 1st Class Edgar W. Dahl
Operation Able Sentry, Macedonia

# DATE/MATE
# DANGEROUS WARNING SIGNS

I might be sounding a little like "Dear Abby" or "Ann Lenders," but this next topic is strictly for the female soldiers. Or soldiers (male or female) who know of a friend who is being abused by their military boyfriend or spouse.

Is your guy or husband:

1. Jealous of your time away from him such as when your with some of your co-workers, friends, or family members?
2. Trying to constantly control you by always asking where your going or where you've been? Does he insist on making personal decisions for you or telling you what to do?
3. Isolating you away from your family and friends?
4. Always blaming others for his problems or misfortunes?
5. Easily upset or angry when you ask him to do something?
6. Cruel or insensitive to animals and children?
7. Play strange games like pushing or pinning you to the bed so that you can't move, then forcing you to have sex when your not ready to do it? Does he force you to have sex even when your sick or ill?
8. Call you cruel, insensitive, humiliating names?
9. Have a Doctor Jekyll - Mister Hyde personality? One minute he's nice & loving, then next "angry & abusive?"
10. Hitting, breaking, or smashing things for no reason at all and then

apologizing later for doing it?
11. Ever threaten to bodily harm either you or someone else?
12. Ever physically abuse you during an argument such as slapping, choking, kicking, or shoving you?

Ladies, if your boyfriend or spouse has any of these symptoms, it's a warning sign that you could become a battered victim of abuse. And no matter what you say or do to him, ladies, he's NOT going to change. In fact, it will only get worse (not better) and escalate to a more dangerous level.

So if you've just started the relationship, break it off while you still can. And if your afraid or don't know how to get out of the relationship, then get some help or advice from either your chaplain and or call a women^ crisis hot-line number listed in your telephone book.

And to all of you guys who know of someone who abuse their date, mate, girlfriend, or wife, your a low-down scumbag if you don't do anything to help them. Don't wait until someone is seriously injured or killed before doing something, or by then it may be too late. And if your afraid to confront them head on about their problem, then at least make a photo copy of this page and send it to them "ANONYMOUSLY".

## 5.56 AMMO BANDOLEER USES

*Submitted By: Pfc Rich A. Fongeallaz Jr.*

PFC Fongeallaz says, "The next time you go to the rifle range, grab yourself a couple of those expendable 5.56mm Ball Ammo Bandoliers. Then buy a soap dish for each of the compartments and store some of your smaller field items inside, makes a great storage organizer. For example:

| | |
|---|---|
| MINI 1ST AID KIT | SEWING KIT SET |
| SPARE BATTERIES | SHAVING KIT/RAZORS |
| SURVIVAL KIT | POGGY SNACKS |
| CASSETTE TAPES | CIGS |

PLASTIC SOAP DISH     BANDOLEER POCKETS

## MRE EATING

*Submitted By: Sgt. Robert Robinson*

When they first came out with MREs, one of the complaints the troops had about the new meal (besides tasting nasty) was that the spoon was too short. There was no way f--ken way anyone could eat from the main meal packet without getting food all over their hand.

Then someone at the Pentagon finally said, "Daaah, I think we need make the MRE spoon a little bit longer." Great idea? I don't think so....

In order for the company (MRE manufacturer) to make the spoon longer, they need to add some more plastic to it, right? So if they add more plastic to it, that means they need to spend more (tax) money On it, right? Not to mention spending more money on a new plastic spoon wrapper too. Right?

Well, the Pentagon could have easily solved the problem and save us tax payers a little bit of money if they had done what Sgt. Robinson had suggested. He said, "If you take your knife and slit the wrapper the long way,, you can remove the food contents more easily without making a mess. Make sense?

RANGER RICK'S COMMENTS: Hummmm, I wonder why the brass with the PhD's at the Pentagon didn't think of this first. It definitely would have been easier and cheaper to put a "pull tab" the long way rather than the short way. Wouldn't you agree?

**CUT LONG WAY & EAT HEARTY**

 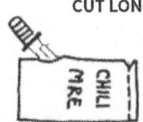 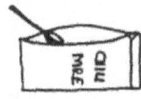

## GOT A HEAT CASUALTY?
## GIVE 'EM A BLAST OF SOME CO2

Heat exhaustion and or heat stroke can occur when an individual is exposed to too much sun or very hot weather temperatures. The symptoms are hot, dry skin (or excessive sweating), dizziness, feeling sick, rapid pulse, rapid breathing, and can be conscious or unconscious. If the body temperature is not lowered immediately, shock could set in.

The first thing that should be done is lay the casualty down under some shade, open up his clothes, and elevate the feet so their slightly higher than the head. Take some cool water and gradually pour it over the casualty's head and the rest of his body. If there's a nearby stream, pond, or creek, submerge the casualty in the water keeping his head above the surface until either the body temperature has been lowered or until he or she is feeling better.

If there is no cool water in the area, look tor a nearby military vehicle that has a $CO_2$ fire extinguisher. Though they're supposed to be used for putting out a vehicle fire, it can also be used to help lower a heat casualty's body temperature in an emergency situation. It contains a liquid gas called carbon dioxide, which when sprayed from the extinguisher it converts to a freezing non-toxic cold gas.

So the next time your on a road march and you have a sun or heat casualty with severe symptoms, try giving him a blast of some $CO_2$.

## M249 SAW DRUM USES

*Submitted By: ALAN FOSTER*

Hey Ranger Rick, here's a tip that some warrior might find useful;

FIRST, get yourself an empty M249 SAW Ammo Drum & Bandoleer. Remove the lid, run a piece of 550 paracord through the small holes and then tie them off at the ends to keep them in place.

SECOND, take some 100 mph tape and cover the entire bottom portion of the drum including the drain holes and ammo slot.

Now you can use this ammo drum as a little storage container for just about anything. To insure your items remain 100% dry and waterproof, place them inside a plastic zip-lock sandwich bag before placing them in the container.

RANGER RICK'S COMMENTS: Got another idea on what these empty SAW ammo drums can be used for? Well, get off your butt and send them into me. And if I accept your idea and print it, I'll send you a free copy of the next Ranger Digest Handbook.

## BLOOD AWARDS

What are "Blood Stripes, Wings, & Badges?" It's when someone is promoted or awarded a badge or medal for their skills and accomplishments and it's SLAMMED into your chest or collar through the shirt "without" the attached caps. Thus penetrating the skin and drawing blood, YOURS! And although this may be an old unit tradition among Airborne, Ranger, and Special Forces, it's UNAUTHORIZED.

Would you allow some medic to stick a hypodermic needle or IV into your arm after he accidentally dropped it on the floor? Or how about allowing some jerk to slam a couple of thumb tacks into your chest? Would you like that? Huh?

If your assigned to an elite unit that conduct these types of award ceremonies. Be advised it's entirely up to YOU (the awardee) and NOT the unit awarder in deciding how you would like your award presented. Don't feel obligated or pressured in accepting "blood awards" if your not comfortable with it. Remember, it's your body and your choice, NOT the awarder.

*Thanks for the Bloody Award, sir...*

Rick F. Tscherne

# BASIC WATER SURVIVAL TIPS

Not too long ago I was watching CNN and saw a story about a Marine who supposedly accidentally fell overboard from his amphibious assault ship. He claims he woke up one night, couldn't go to sleep, so he decided to get some fresh air out on the deck. When all of a sudden (he claims), the ship's door knocked him overboard into the sea. (Yea, sure, right!)

Now I've never been a sailor nor a marine, but I've been on a number of ships during training exercises, including an amphibious assault ship like the one he was on. And if you could see all the safety nets, fences, etc, that these ships have all over the damn place, you'd be puzzled as to how in the hell could some nerd or jarhead marine fall overboard. But the Navy & Marines and the "press corp" treated him like a hero when he was rescued a day later by some cargo ship.

Why then such a big fuss? Well I'm only guessing,, but I think it was probably to save the Navy & Marine Corp some embarrassment. Now don't get me wrong, I have no doubt his water survival skills did indeed save his ass. But to downplay his stupidity in falling overboard and then commend him on using his USMC water survival training skills is S-T-U-P-I-D.

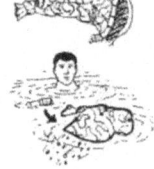

Be advised that his USMC Water Survival Training is no different than what most of us are taught in Ranger, Special Forces or Airborne School. It involves the following:

When jumping from a tall ship, bridge, or hovering helicopter. Keep your body straight, legs together, and head erect. If you are holding a weapon, either release it or raise it high above your head to prevent it from smacking you in the face upon water impact. If you do not have a weapon, pinch your nose and cover your mouth with your hands. Once in the water....

Remove your boots, take off your pants or shirt and tie a knot into each of your pant legs or sleeve's. Then force the air inside the legs/sleeves by pounding & splashing your hand in and out of the water at same time directing the air bubbles into & under the pants or shirt. Once they fill with air, then place the shirt/pants behind you with the two inflated legs/sleeves under each of your armpits. NOTE: In order to keep them inflated, you will need to periodically pound and splash air inside.

While waiting to be rescued, assume the HELP position, Heat-Escape-Lessening-Posture. This will help reduce the loss of up to 50% of your body heat and postpone the onset of hypothermia. Strange as it looks, it works.

# M80 DRAGON FIRING DEVICE

Just like the M80 Claymore Mine Firing Device, the M47 Dragon can also be rigged to fire the M80 TOW Blast Simulator. You will need:

1.) Wire, rubber coated, double stranded (4 ft)
2.) Electrical alligator clips (2)
3.) Battery container, must hold 4 "AA" size bat.
4.) Electrical switch, small durable type
5.) 100 mph "duck" tape.
6.) M80 TOW Blast Simulators

What's important when rigging an M47 Dragon to fire an M80 TOW Blast Simulator, is that you **MUST** rig it to fire in one of two ways.

a. Either have the "weighted metal sleeve" installed inside the M47. (Note: This is a thick metal pipe used to give an expendable Dragon the actual weight and feel of a live Dragon.) *Or*....

b. Drill out, cut, and remove only the cover portion of the round metal canister located in the rear of the Dragon. (NOTE: Do not remove the entire round canister, just the cover portion. Drill, hack, cut, and remove.)

The purpose of doing one or the other, is so that when you fire an M80 TOW Blast Simulator from inside the Dragon. It will not rupture and shatter through the fiber glass container and injure the firer.

The "weighted sleeve" and or the removal of the rear metal cover container will absorb the blast and not injure the firer. Failure to follow these instructions & drawings precisely will cause bodily injuries to both, the firer and those standing nearby.

**WARNING! DANGER! HANDLE WITH CARE!**

*Rick F. Tscherne*

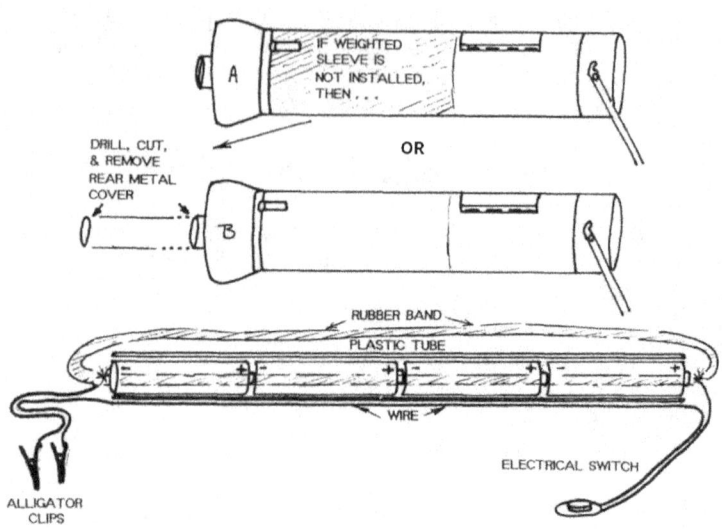

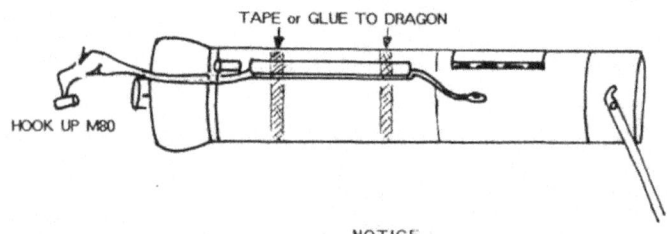

*Loose*
CANNON
ENTERPRISES

### NOTICE

Should M80 fail to fire
and all components are
operating correctly,
add additional batteries.

*The Complete Ranger Digest: Vol. VI*

# ROLLING & STORING 550 PARACORD

*Submitted By: CAP C-Sgt Joshua Berrier*

Dear Ranger Rick,
Recently I purchased a few of your Ranger Digest Handbooks and took them with me on a Civil Air Patrol Search and Rescue Exercise. In fact, I even shared many of your field tips and tricks with my fellow cadets.

Well, once I had a problem as to where I should store my 550 paracord so it wouldn't get all tangled up when I needed to use some. So here's what I did....

I took most (but not all) of the parachute cord and wrapped it carefully and evenly between my elbow and thumb. Then I removed it, took the last remaining lengths and wrapped it carefully around the main body of the cord and tied it off.

Then whenever I needed to use any paracord, instead of unraveling it, I just grabbed the "untied running end" and pulled off only what I needed. Works like a charm!

The best place to carry this paracord is around the bottom half of the ammo pouch snapped into the two grenade snaps (see drawing).

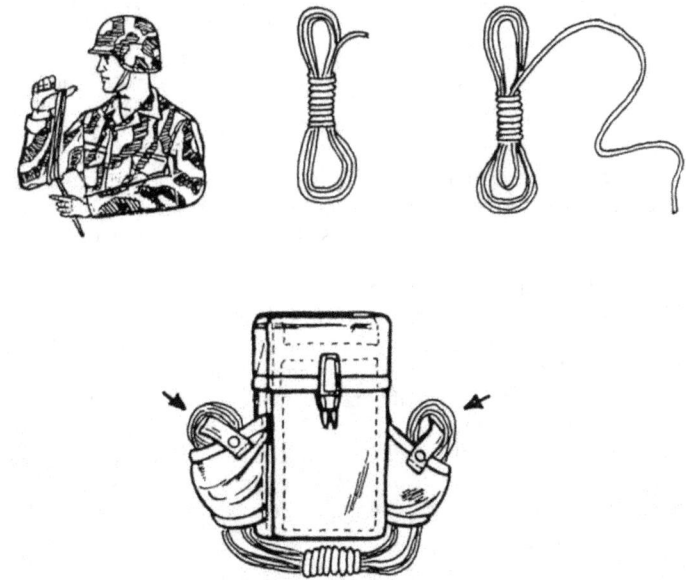

*Rick F. Tscherne*

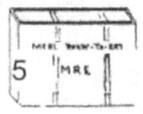 # MORE MRE BOX USES

*Submitted By: Sgt. Paul R. Headen*

The Department of Defense has announced that the US Armed Forces will closely follow all EPA regulations and become more involved in helping to keep our environment clean.

Soldiers who work or spend a lot of time in the motor pool know exactly what this means -"Clean Up After Every Oil or Fuel Spill." But what if your in the field and you got a leak or need to change some oil or other liquids. The answer? An MRE "Drip Pan."

Just open up an empty MRE cardboard box, place a large trash bag inside, secure it in place with some 100 mph tape, add some loose dirt, and "PRESTO"... an improvised MRE DRIP PAN. To dispose of, just send it back to the rear with the XO or one of the supply guys.

IMPORTANT: Never drain hot liquids directly into an MRE box or it could cause the plastic bag to melt and leak through.

# DEDICATED MILITARY TRUCK DRIVERS

Now I ain't never been a truck driver, nor have I ever really dedicated truck drivers who take pride in taking care of their vehicle. I've seen some drivers....

(1) Install a "pop-in/pop-out" am-fm radio cassette recorder underneath their dash board or seat. Some have even gone as far as adding portable speakers to the back of their truck so that their passengers could listen in on the music too.
(2) Installed or carried a portable CB so that they could talk with their "buddies" on the road. (Breaker-Breaker!)
(3) Installed fancy truck seat covers, cushions, floor mats, ash trays, soda can holders, etc inside their cab to give it a more "homey and comfy" look.
(4) Purchased a portable spotlight and or some warning lights to help load and transport cargo at night.
(5) Installed a few special lights in the rear of the truck so that their passengers could see at night "Now this I like, I don't know how many times I've jumped in the back of a truck at night to bitched about not being able to see a f----n thing."

Yep, when I see a truck driver go through all this trouble to improve his truck (which by the way is against military regulations). I say to myself, "Now there's one dedicated truck driver that I don't mind riding with."

*The Complete Ranger Digest: Vol. VI*

# WHILE YOU WERE SNOOZING & IGNORING MY ADVICE

In almost everyone of my *Ranger Digest Handbooks*, I dedicated couple of pages on where you should be investing your money or greater returns. And if you were snoozing or ignoring my investment advice, then you no doubt lost out on some financial gains. And I'm NOT talking about savings bonds, CDs, nor banks, but MUTUAL FUNDs.

You don't need a lot of money to make a lot of money, but you do need to invest wisely and be patient in letting it grow. And the way things are going today, our government will either have to take away some of our social security benefits. Or raise the retirement age well above 65 before anyone can retire and collect their benefits. (No BS!)

In either case, you stand a 50-50 chance in never, ever seeing your benefits materialize even though you paid for it. And you'll either have to work the rest of your life up to the day you die, or you can get off your ass now and start investing for your future. It's your responsibility to prepare yourself for a comfortable retirement, NOT the government's. And if you don't care about yourself, then think about your family.

To help you to see the light more clearly and get started in investing, here's a list of the "Top Mutual Funds For 1995." After reviewing these returns and you think it's a bunch of BS, then I strongly challenge and encourage you to call any of these toll free numbers and get the facts, info, and or an application. Remember, it's YOUR MONEY and YOUR FUTURE.

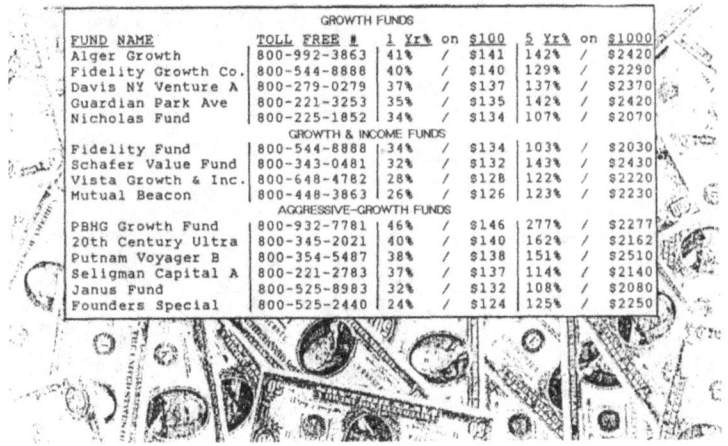

| FUND NAME | TOLL FREE # | 1 Yr% | on $100 | 5 Yr% | on $1000 |
|---|---|---|---|---|---|
| **GROWTH FUNDS** | | | | | |
| Alger Growth | 800-992-3863 | 41% | $141 | 142% | $2420 |
| Fidelity Growth Co. | 800-544-8888 | 40% | $140 | 129% | $2290 |
| Davis NY Venture A | 800-279-0279 | 37% | $137 | 137% | $2370 |
| Guardian Park Ave | 800-221-3253 | 35% | $135 | 142% | $2420 |
| Nicholas Fund | 800-225-1852 | 34% | $134 | 107% | $2070 |
| **GROWTH & INCOME FUNDS** | | | | | |
| Fidelity Fund | 800-544-8888 | 34% | $134 | 103% | $2030 |
| Schafer Value Fund | 800-343-0481 | 32% | $132 | 143% | $2430 |
| Vista Growth & Inc. | 800-648-4782 | 28% | $128 | 122% | $2220 |
| Mutual Beacon | 800-448-3863 | 26% | $126 | 123% | $2230 |
| **AGGRESSIVE-GROWTH FUNDS** | | | | | |
| PBHG Growth Fund | 800-932-7781 | 46% | $146 | 277% | $2277 |
| 20th Century Ultra | 800-345-2021 | 40% | $140 | 162% | $2162 |
| Putnam Voyager B | 800-354-5487 | 38% | $138 | 151% | $2510 |
| Seligman Capital A | 800-221-2783 | 37% | $137 | 114% | $2140 |
| Janus Fund | 800-525-8983 | 32% | $132 | 108% | $2080 |
| Founders Special | 800-525-2440 | 24% | $124 | 125% | $2250 |

Rick F. Tscherne

## US TROOP DEPLOYMENT
## BOSNIA INFO

### Bosnia gear

U.S. soldiers joining NATO's Implementation Force have clothing and gear designed for Bosnia's frigid winter.

- 1 M-40 protective mask
- 2 duffel bags or kit bags
- 1 barracks bag
- 2 waterproof bags
- 1 entrenching tool and carrier
- Parkas designed for cold, wet weather
- Trousers designed for cold, wet weather
- 1 poncho, wet weather
- 1 sleeping bag, extreme cold weather
- Body armor
- Insulated boots for cold and wet weather
- 2 shirts, cold weather, poly knit
- Underwear designed for extreme cold
- 2 wool sweaters
- 2 plastic canteens
- Vinyl overshoes
- Gloves with cold-weather inserts
- 1 pair leather work gloves
- 1 belt
- 1 wool blanket
- 1 pad mattress
- 1 nylon field pack with frame
- 1 case first aid dressing
- 2 cases ammo pouch

BOSNIA AND HERZEGOVINA
Newly accepted boundaries

### Checklist of important documents

- ☐ Powers of attorney
- ☐ ID cards (spouse and children)
- ☐ Passports (spouse and children)
- ☐ Stocks and bonds
- ☐ Social Security cards (spouse and children)
- ☐ Installment payment contracts
- ☐ Court orders for divorce, child support and child custody
- ☐ Phone numbers of family and friends
- ☐ Credit card accounts
- ☐ Vehicle titles and registrations
- ☐ TDY or PCS orders
- ☐ Organ donor instructions
- ☐ Burial and funeral instructions
- ☐ Emergency instructions
- ☐ Charge cards and records
- ☐ Dental and medical records
- ☐ Wills
- ☐ Marriage/adoption records
- ☐ Real estate documents (deeds)
- ☐ Leave/earning slips
- ☐ Citizenship papers
- ☐ Bank accounts
- ☐ Birth certificates
- ☐ Insurance policies
- ☐ Tax information
- ☐ Death certificates of family members
- ☐ Retirement plans
- ☐ IRA documents
- ☐ Mutual fund records

### Easing deployment stress

Many activities can help families of deployed servicemembers manage the stress caused by separation. The booklet Coping Right Now!, compiled by the Civilian Personnel Operation Center's Training and Career Management Branch, outlines a variety of strategies followed by military communities in Europe. Some of the strategies are:

- Form support groups.
- Provide access to professional services.
- Conduct predeployment interventions.
- Implement a telephone chain to receive and pass on information.
- Squelch rumors.
- Draw upon the resources of spouses who have successfully coped with separations in the past.
- Train spouses experienced in coping to facilitate support groups for less-experienced spouses.
- Help mothers to cope, thereby helping the children to adjust.
- Help children to adjust by:
  a. Allowing them to discuss their feelings with both parents before the deployment.
  b. Enforcing all of the same rules and routines during the separation as before.
  c. Writing separate letters to each child during separation.
- Refer children and families exhibiting severe emotional problems to mental health professionals.
- Incorporate a strong outreach program to combat social isolation.
- Assist families in working through the predictable and normal stages of grief: shock or denial; anger; guilt; depression and/or loneliness: tension, crying, irritability and insomnia; coming to grips, or despair and withdrawal.
- Provide morale telephone calls. Families should plan for these calls to make them less stressful.
- Families should come up with a realistic idea of how often they will write letters
- Plan for the family's reunion as the time for the military member's return draws near.

*The Complete Ranger Digest: Vol. VI*

# USEFUL TOYS FOR THE FIELD

**LISTENER 2000** — Powerful receiver-amplifier adjusts to hide behind either ear...picks up sounds as far as 100 feet away. Five volume levels let you turn a murmur into a shout or hear a pin hit the floor! With on/off switch. Uses one 1.5V battery (included). The Listener 2000 is not a medical device. It is not designed for people with impaired hearing. It is for anyone who wants to have better than normal hearing.

VERY USEFUL IN LP/OP POSITIONS

Set Only! $14.95 — Complete 5 piece set! Order #7JXX

Carol Wright Gifts, 340 Applecreek Rd, P.O. Box 8503, Lincoln, NE 68544-8503

## Soviet Spetznaz Machete

A tool that will out-perform most knives in the non-delicate work of chopping, hacking, and clearing vegetation...

can be used as a machete, knife, shovel, ice pick, saw, screwdriver, wrench, wire stripper, 90-degree compass, navigation sight, parachute cord cutter, ruler and much more. Survival supplies (fishing hooks, line and weights, matches) are included in a water-tight compartment in the polymer handle.

Each machete has a warranty until the year 2000. To order, send $69.95 plus $6.95 shipping and handling to: Sovietski Collection, Dept ASG, P.O. Box 81347, San Diego, CA 92138-1347; (800) 442-0002; international calls dial (619) 237-8000; fax (619) 237-8010. ●

The contents of the hollow handle include matches and fishing gear.

## UNIQUE TOTABLE TOOLS
*"Worth $50.00 When You Need Them!"*

The Classic SWISS ARMY KNIFE

Your Choice Only $7.95  Any 2 Tools only $14.95

Handy Pliers PLUS

Buy Now for Direct Savings from

**M.A.C By Mail**
903 S. Hohokam Dr, Tempe, AZ 85281

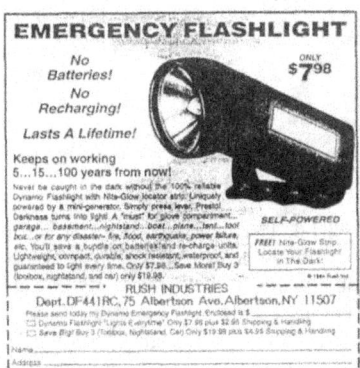

**EMERGENCY FLASHLIGHT**
No Batteries!
No Recharging!
Lasts A Lifetime!
Keeps on working 5...15...100 years from now!

ONLY $7.98

RUSH INDUSTRIES
Dept. DF441RC, 75 Albertson Ave, Albertson, NY 11507

Rick F. Tscherne

# RANGER RICK'S MULTI-PURPOSE ALL TERRAIN SURVIVAL KIT

QUESTION: Who do you think should carry a survival kit in the field?

(A) Infantrymen  (B)Clerks  (C)Mechanics  (D)Drivers  (E)Medics

ANSWER: All of the above. Regardless of your MOS or rank, if you go to the field on a regular basis, then you should always carry a small survival kit. As Forrest Gump once said, "Life's like a box of candy, you never know what ya gonna git." In this case, what will happen to you.

Now I know many of you are probably saying right now, "Aaaw your full of cow poop, Ranger Rick!" Right? Yep, I'll bet ya are. After all, you don't have any intentions of getting lost, wandering away, or separated from your unit, right? And besides, you don't need one because your not a "bad ass" Grunt, Ranger, or SF troopie. Right? *WRONG!*

QUESTION: Were you issued a ruck sack? Weapon? Sleeping bag? LBE? Cold weather gear? If so, then you probable go to the field at least a few times a year. Right? Well this is your basic military gear for surviving in a combat environment. But what do you have for surviving off the land itself? Nothing! Right?

This is why you need a basic survival kit, just in case you should ever become separated either from your unit or from your military equipment. And it can quite easily happen too.

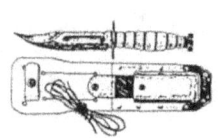 

Well, I'm a real tight wad when it comes to spending and buying things, just ask my wife, she'll tell ya. So I think you'll find my Multi-Purpose All Terrain Survival Kit pretty unusual and pretty darn cheap to make.

When putting together your own survival kit, remember to try to keep the amount of items needed down to the bare minimum. Just because you got the extra room in your ruck doesn't mean you should fill all those empty pockets full of shit. On the contrary, a survival kit should always be carried close to your body, such as on your belt, in an ammo pouch, or in one of your BDU cargo pockets. Because if you should ever need to jettison your ruck in an emergency situation (water obstacle, enemy hot on your ass, etc), by-by survival kit.

Here's a list of all the basic items that I carry in my personal survival kit, and I think you should too.

EMPTY MRE MEAL PACKET - A survival kit container shouldn't be any larger than the size of an MRE plastic packet. In tact, that's what the items can be stored in. As it will not only keep your items water proof, but the plastic packet will have numerous field uses too.

When preparing to use an MRE packet for your survival kit, remove the food contents by cutting the wrapper very carefully in the same manner as I described how to make a "MRE Water Bucket" in my Ranger Digest III. This way you can place your other survival items more easily inside the wrapper and still use it as a plastic water bucket too.

 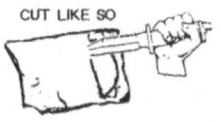 

ZIP LOCK SANDWICH BAGS - These are not only handy for water proofing your items, but when used in conjunction with the MRE wrapper they can be used as an improvised canteen or for carrying or storing food.

100 MPH TAPE (DUCK TAPE) - Most survival kits that I've seen in camping & military supply stores and catalogs are loaded with a bunch of band-aids. Now I'm sure they'll come in handy if you should ever have a bunch of "little cuts" on your little fingers. But what if you've got a big cut? The answer: Duck tape, or what the military calls 100 mph tape. This super strong multi-purpose tape can be used in many ways, including as a band aid, bandage, sling, or tourniquet.

A very convenient place to store strips of this tape is around the plastic MRE wrapper itself. It will not only make your survival kit container more durable, but waterproof too. Plus you'll still be able to remove it from the wrapper and reuse it on something else. Good idea, huh?

EMERGENCY THERMAL SPACE BLANKET - This item is a must if your gonna beat the odds in surviving in the outdoors, whether it be a hot or cold weather environment. There are two types, a Thermal Emergency Space Blanket and a Thermal Space Bag. Don't confuse these two items with any of those bulky multi-color fancy space blankets. They come in only one color and size, silver and pocket size.

Personally, I like to carry both of them in my survival kit. The bag for sleeping in and the blanket to be used as an improvised shelter, rain poncho or additional covering. But if you think only one is enough for you so be it.

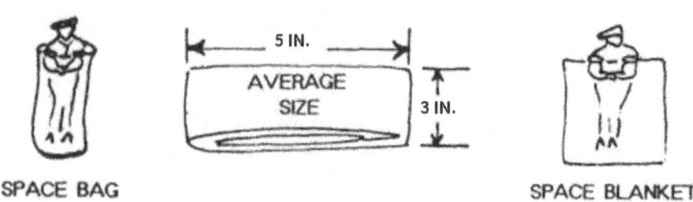

To convert a space blanket into an emergency rain poncho is very simply. Just open it up and place about 18-24 inches of 100 mph tape in the center of the blanket, both sides. Then take your knife or razor and cut a straight line right down the middle but NOT BEYOND the edge of the tape. Now try it on for size.

Once your sure that it fits comfortably over your head, remove it, take another strip of tape and close the hole so that you can use it again as an emergency shelter or blanket. (NOTE: If you so desire, you don't need to cut through the tape and blanket until it's actually needed in a survival situation.)

To use it as an emergency shelter, take eight (8) pieces of tape approximately 6 inches in length and attach them to all four corners and sides. Then puncture a small hole with a nail (not a knife) through the tape and blanket and attach some string. Now you can use it to erect a shelter as described in my *Ranger Digest I & V* handbooks.

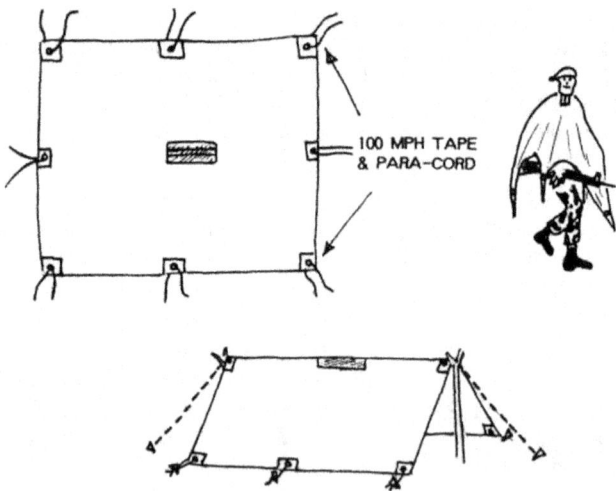

BIG GIG LIGHTER - Never carry in your survival kit wooden or paper matches, they won't last as long as a Bic lighter. You'll get several hundred lights from one Bic lighter vs only a few dozen from paper or wooden matches. Plus when the gas is turned off, it can then be used as a mini strobe light for signaling.

MINI PENLITE - The purpose a penlite in a survival kit is to either aid you in signaling or to help you to see something. You should NEVER use it unless you really need to. As a single or double cell battery penlite will only last up to about 1-2 hours of continuous use, at the most.

UPDATE: Nowadays you can select a small LED light that will give you much more light and battery life. Be sure to pick one with multiple-settings, especially a LOW one that can extend battery life on some for several dozen hours.

FISHING KIT - I've read, seen, and met many know-it-all survival buffs who claim that you only need to carry one or two types of fish hooks in a survival kit. Bull Shit! You should carry at least no less than several different size hooks, mini, small, and medium.

Everyone loves to picture themselves in a life or death survival situation sitting on some creek bank fishing for trout or salmon. Get real, wake up and smell the coffee, that only happens in the movies or in story books.

Though I'm sure you much prefer to catch trout, catfish, or salmon, you had better set your menu on something a bit more smaller and realistic. Like minnows and sunfish. Chances are, they'll be more plentiful and a lot easier to catch than a trout or catfish. That is, if you were smart enough to pack an assortment of fishing hooks in your kit. Do you get my drift there, Rambo?

WIRE, FISHING LINE, & PARACORD - All three of these items are very important if you intend to hunt or catch your meal. You can't fish without some fishing line, you can't trap or set a snare without wire, and you can't build a shelter or make a bow if you ain't got any cord. Right?

Well, the easiest way to carry and store all three of these items in your kit, is by wrapping them around some cardboard from an MRE box. Just tear off a square piece, make a few cuts along the edges and start wrapping. You shouldn't have any problems wrapping all three on one single piece of cardboard. How much you need will depend on how much you fell comfortable with. I suggest you wrap about 20-30 feet of wire and fishing line, and only about 15 feet of parachute cord.

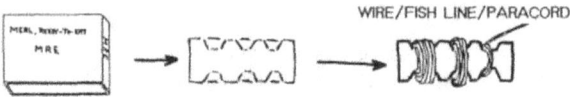

When choosing the wire, it should be thin, flexible, strong, and either black or green in color. Don't use aluminum or copper colored wire, as it will only shine and scare away animals if you use it to set up a trap or snare.

When choosing fishing line, don't just grab anything off the shelf, purchase fishing line that is at least 4-6 lb test. As this should be sufficient enough to hold a good size fish.

There's only two kinds of parachute cord, type 440 or 550, it's very, very strong and has many, many uses too. Not only can it be used for building shelters, making a bow, or as a first aid tourniquet. But if you open up the ends, you can use the 7 inner nylon strands for fishing, sewing, snaring, etc.

RAZOR BLADES - Even though you usually carry a knife, you should also carry a few razor blades too. What for? Well for such things as skinning or gutting your catch, minor first aid surgery, and other uses. The most durable and useful razor blades, are those that can be found in a hardware store and used for "scraping paint" off glass windows.

SIGNAL MIRROR - No doubt the Army or Air Force signal mirror is the ultimate choice, you don't necessarily need to spend that much money on one. You can get away with buying and using a simple, small, cheap cosmetic mirror, as long as it's, non-breakable or at least in a shock-proof container.

WATER PURIFIERS - There are three types of water purifiers that you can pack in your survival kit, Iodine Tablets, an Accufilter Drinking Straw, or an Accufilter Canteen Insert. My choice is the Accufilter Canteen Insert, why? UPDATE: ..or the Lifestraw purifier.

Well, to safely use iodine tablets, you need some sort of a container that will hold water. Plus, you will also have to wait 20-30 minutes for the iodine tablets to dissolve before you can "safely" drink the water.

The Accufilter Drinking Straw can be used immediately and directly from any water source (river, stream, pond) except salt water. It's just a straw with a filter that you suck on.

The Accufilter Canteen Insert can be used in both ways, as a canteen insert and also as a drinking straw too. (I'll bet you and the manufacturer didn't know this, did you?) Simply place the canteen insert in the water and suck, it works just as good or better than the straw. You'll not only get more of a mouthful of water, but it cost less too.

ACCUFILTER CANTEEN INSERT CAN BE USED IN ONE OF THREE WAYS

Both, the Accufilter straw & canteen insert filters up to 40 quarts or 10 gallons of water. I'd strongly suggest that you pack one or the other in your survival kit rather than the iodine tabs.

SPONGE - I know your probably thinking, "What in the f---en hell do you need a sponge for?" Right? Well, how else are you going to clean your butt hole in the woods or jungle when you gotta take a dump? It's re-useable too.

SOAK, THEN SQUEEZE TO DRINK

SIC! I was only joking and pulling your leg, dummy. But do you remember when that Air Force pilot (Captain O'Grady) got shot down over Bosnia, Yugoslavia in June 95? Well, one of the items he claims that helped him to survive his ordeal was to use the sponge to soak up rain water from the leaves and vegetation. Now think about it, that was pretty darn smart, wasn't it? It sure as hell beats trying to catch rain water in your mouth, huh? (It can also be used as a first aid dressing bandage too!)

MINI COMPASS - If you don't know how to use a compass, then it certainly doesn't do you any good to carry one in your survival kit, right? But for those of you who do know how to use one, purchase a cheap, compact, small compass about the size of a wrist watch. Remember, this is just an emergency back up, not your main one.

MINI SEWING KIT - Some of you are probably thinking that a sewing kit is a worthless piece of shit to carry in a survival kit, right? Not so fast, John Wayne. What if you got some badly torn clothing or a severe cut, how are you gonna repair yourself?

Rambo one time in a movie (First Blood) used his survival sewing kit to stitch himself up, so if it's good enough for Rambo, then it's good enough for you. You don't need an entire sewing kit, just a few different size sewing needles, safety pins, and some thread. (NOTE: The inner nylon strands from 440 or 550 paracord can double as sewing thread too.)

440/550 INNER THREADS

POCKET WIRE SAW - If you carry a multi-purpose knife such as a Marine K-bar, Air Crew Survival Knife, Bowie Knife, Short, Machete, etc, then you probably don't need a pocket wire saw.( As it's only needed for cutting down thick trees for making a hi-speed shelter or some wood for a fire.

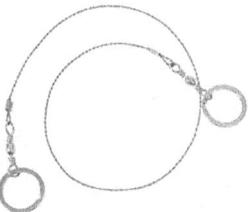

But if your only carrying a simple pocket knife such as a Swiss Knife, Leatherman Tool, etc, then you may want to consider carrying a pocket wire saw too. As the short saw on these tools or knives will eventually become very uncomfortable on the hands after prolong use. (Ouch! You know what I mean?)

SLINGSHOT or RUBBER BANDS - One of the easiest and most effective weapons you can make for hunting small game is a slingshot. The only thing you really need is either a slingshot sling or a few good strong thick rubber bands. Once you have this, then all you need to do is attach it to a strong wooden branch shaped like a "Y." If your using plain rubber bands, connect a few of them together and attach a piece of tape in the center to hold the projectile in place.

Well, that about does it. I'm sure you've seen many other survival kits with more or less of the same items. But if you pack your kit the same as mine, you will no doubt improve your chances of surviving in any outdoor environment. Whether you be in a jungle, desert, swamp, mountain, forest, or arctic environment. But most importantly, make sure you carry this survival kit attached to somewhere on your body.

COMPACT SURVIVAL KIT

Rick F. Tscherne

# A FEW MINUTES WITH RANGER RICK
## (A Ranger Rick Commentary)

Not too long ago these headlines appeared in the "Army Times" and the overseas "Stars & Stripes" newspapers. Maybe you've seen them, and maybe you haven't.

Faulkner gives it up as fellow cadets jeer

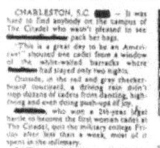

Unit pie-throwing gets nasty after sergeant creams major

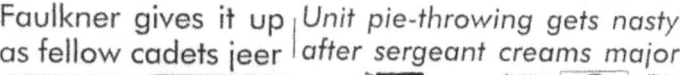
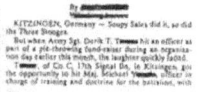

New commander, policies preceded Ranger deaths

After reading these articles, I decided to send a letter to the editors, but they never printed it. (Hmmmm, I wonder why?)

Dear Editor,
After reading the enclosed articles in your newspaper, I would like your military readers to know.....

ON SHANNON FAULKNER: Be advised that according to Webster's Dictionary, the difference between failing and quitting is...

   FAILING: Slight, insignificant defect in one's ability to pass.
   QUITTING: To give up, to abandon, to admit defeat.

Ms. Faulkner should have been practicing "push-aways" from the breakfast, lunch, & dinner tables to get in shape for the Citadel. There's no dishonor in failing, but there is in QUITTING.

As for the complaints about the male cadets celebrating her departure. Now lets be honest, do you seriously think that an all female academy would have celebrated any differently had the first and only male cadet quit the program? (Yea, right!)

AS FOR THE NCO PIE-THROWING INCIDENT: Your story was one sided and biased, your readers only heard the NCO'S version of the incident, and the officer's??? There's more than one side to this story, you know?

Being involved in a number of unit fund raisers myself, I can speak from personal experience in being hit, hitting, and witnessing others being clobbered with a pie. But some questions that quickly come to my mind are...

(A) Did this NCO throw the pie "beyond" arm's length?
(B) Did he smear the pie "within" arm's length?
(C) Or did he "forcefully" slam it at less than arm's length?

If this NCO forcefully slammed the pie into the face of the officer at less than arm's length, it's the same as punching someone except with an open palm instead of a closed fist. And if he did this, he took advantage of the situation and should be charged for "assaulting a military officer."

I'm no lawyer, nor am I trying to defend this officer. But...

1. Who'd benefit the most from an incident like this if someone reported it to a newspaper editor, the NCO or officer?
2. Who was the only one that gave his version of the story on what happened at the unit fund raiser, the NCO or officer?
3. Who's friends & co-workers were quick to open their months in pointing fingers and laying blame, the NCO or officer?
4. Who was pcs-ing and be in a better position in seeking revenge on someone they disliked without retaliation, the NCO or officer?
5. Who's picture appeared in the newspaper with a satisfied "Shit Eating Grin" on his face, the NCO or officer?

Well, after taking all these things into consideration, I would have to say it's clear that the NCO was the guilty culprit. And personally, if he ever did that to me, he wouldn't have any teeth in his mouth to smile for a camera.

REFERENCE TO THE DEATH OF THE FOUR RANGER STUDENTS; After reading about the Ranger students freezing and dying in the swamps of the Florida Ranger Camp (Eglin Air Force Base). I couldn't help but remember when I went through Ranger School.

Nor can I forget when I myself was an instructor (DI) and responsible for the health, welfare, and training of my soldiers. It's very hard for me to imagine and believe that any leader (NCO or officer) would be so stupid to risk getting their men seriously injured, ill, or killed in a training school environment or unit FTX.

Though I have to admit, I did one time serve under one such poor leader (Lt. Schertl) from the 82d Airborne (84-86) who would. And out of all his daring, macho, and dangerous escapades that he tried to do. I refused to allow him to lead our recon platoon across a well marked "Off Limits" Impact Area just to save some time in getting around some OPFOR.

While the entire platoon (NCOs & EMs) supported his idea and plan and NOT mine, I absolutely refused to allow it to happen. And though this one incident did cost me a favorable Enlisted Evaluation Report (EER) when I was reassigned, I stand firmly by that decision today. I rather go by GUT FEELING and COMMON SENSE that something is unsafe and wrong than risk getting one of my men killed or seriously injured.

Well, that's about it for now guys & gals, until next time...

**BOOK SEVEN**

# FOREWORD

Hup, two, three, four, back again with another book. I would have written this sooner had I not gone to Bosnia for about a year. I got offered a job that I just couldn't refuse, training the Bosnian Herzegovina Army.

Now before anyone decides to write and ask how much I got paid, what I did, where you can apply, etc. Wait until I finish writing my next book, *The Ranger Digest* "Special Edition." I got a whole bunch of stories and incidents to tell ya about and some of it ain't so pretty.

As you read this book you'll notice that I talk a lot about Bosnia, my fellow co-workers, and the techniques that I used to train the Bosnian soldiers. And though it wasn't always fun and games, I love teaching and training soldiers.

Now unlike most of my fellow co-workers, who by the way were mostly retired US Army "senior NCOs & Officers," I did NOT go to Bosnia for the money. Although I gotta admit, I did get paid very, very well, this wasn't the reason why I went there. It was the feeling of being wanted, duty, honor, and helping to train another country to defend itself against hostile aggression and atrocities. And this is the honest to God truth, no BS!

Would I have gone there for free? Sure, you betcha, but only for a limited time. Provided, of course, it was for a good cause and I was convinced that I was training the good guys.

Also, while I was in Bosnia I never knew how popular my books were until I met quite a few American IFOR/SFOR soldiers. Boy, what a great feeling it is to be recognized for doing something good for my fellow comrades in arms.

Not to mention, how envious and jealous some of my fellow co-workers, (mostly retired military officers) were of my accomplishments. Man, talk about a bunch of dishonest, money hungry, self serving, lack of integrity ex-military leaders. But hey, that's another story to tell ya about.

Anyway Ranger Digest readers, as long as you keep writing, then I'll keep on publishing these Ranger Digests. And as I keep telling ya, if you got the time, then why don't you drop me a line and tell me what you think of my books. Because your tips, tricks, and letters DO COUNT! Well, that's about it for now guys and gals, take care and hang in there.

—*Ranger Rick*

PS: Remember, if someone asks where you learned these tips and tricks from, you just tell them "from my buddy Ranger Rick".

AUTHOR'S DISCLAIMER: The author cannot be held liable for damages or injuries caused by the information contained in this book. Use at your own risk.

## SPECIAL THANKS

This page is dedicated to those who took the time to write and wanting to share their favorite tip, trick, and or ideas with the rest of us outdoor field soldiers. And if it wasn't for these caring soldiers and readers, well, there wouldn't be a *Ranger Digest VII* today. Thanks fellas!

### RANGER DIGEST VII CONTRIBUTORS

LTC MARTIN N. STANTON   1LT SHERAN L. BENERTH   1LT FRANK C. STEVENS

SSG SCOTT A. COOK   SGT DAVID J. WHITE   SGT JERRY D. MARTINEZ

SGT STEPHEN M. MICH   SGT CHARLES ROBINSON   SGT MICHAEL PILLANO

SGT JONATHAN HAVENS   CPL DANIEL D. LOGIE   SPC BYRON WALTER

SPC BRYAN MALTAIS   PFC SERGIO RAMIREZ   PFC BRIAN T. MARSHALL

PFC JEFF LOWE   DAVID YIM   THOMAS D. RUCKER

LES NISHI   CAP C. CALORUSSO   CLINT BOOMGARDEN

MIKE GILLES

AND AN EXTRA SPECIAL THANKS TO,
AIRBORNE SGT FRANK GILLILAND

Not only for his tips, but for volunteering his time and services to draw most of the illustrations in this *Ranger Digest VII handbook*.

### SPECIAL DEDICATION

To my two favorite fellow MPRI co-workers who supported and backed me up in Bosnia when I needed their help and assistance,

**Ranger (Msg) Steve Akana    SF (Sgm) Don Winwood**


Come 'on guys & gals, how about it? Can you give me a hand in keeping the *Ranger Digest* series alive? If ya can, for every field tip, trick, and or idea that I accept for the next Ranger Digest edition, I'll send you a free copy of the book and publish your name beside your tip. Fair enough? I hope so, I'll be anxiously looking forward to hearing from you soon.
(*NOTE – Submissions no longer accepted*)

*Rick F. Tscherne*

# RANGER DIGEST UPDATE

SHITTING POSITIONS - (Ref: Ranger Digest I)
SPC. BYRON WALTER says,"This is an old trick among us scouts and tankers. Instead of using the entire MRE cardboard box, just use the sleeve." If the cardboard edge hurts your butt, just place a few pieces of wood on top.

35MM EMPTY FILM CONTAINERS - (Ref: Ranger Digest II)
A few more uses an empty 35mm plastic container has....

For storing cut-up pieces of "Brillo Pads" so you can use them to clean up your canteen cup or mess kit when out in the field.

For storing fat chubby or skinny survival candles in case you need them for an emergency. (Don't forget to add some matches too.)

For keeping your life support system's light weight ear phones (not headphones) dry & from being banged up inside your ruck.

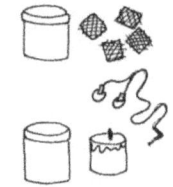

SPIT CONTAINER

Or, as one reader used his 35mm container for (name unknown). He wrote, "When you're in the field and or in a class room environment, some soldiers get pretty upset when they see you spittin on the ground or in a coke can. That's why I switched from a bulky coke can to a 35mm film container. I just pull it out, spit in it, put the lid back on and tuck it in my BDU pocket."

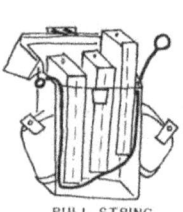

PULL STRING

MAG MODIFICATIONS - (Ref: Ranger Digest II)
PVT.BRIAN T. MARSHALL wrote, "Instead of adding a pull string to the magazine itself, why not just add a pull-string to the ammo pouch the same way batteries are placed inside radio's and cassettes? Then all you got to do is lift up on the pull string and out pops the magazines."

ASSAULT MAGS - (Ref: Ranger Digest II)
PFC JEFF LOWE wrote, "While trying to put three magazines side-by-side together like your book shows, the middle magazines move around a bit. So to rectify this problem, I added a couple of rubber bands to the top and bottom of the 2nd magazine to hold it firmly in place.

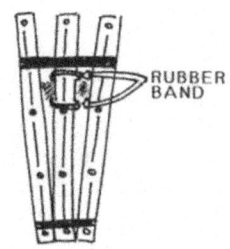

HOME MADE PACE COUNTER - (Ref: Ranger Digest III)
MR. MIKE GILLES wrote, "I thought the homemade pace counter was a pretty good idea. And after making a few for myself and some friends, I have a couple of improvements to suggest. Use "tapered" washers rather than flat washers, they are a lot easier to separate when they are together. If you use two different sizes, one for 100m and the other for kilometers, it'll make it easier to tell them apart. Or, just turn one set one way and the other set another way.

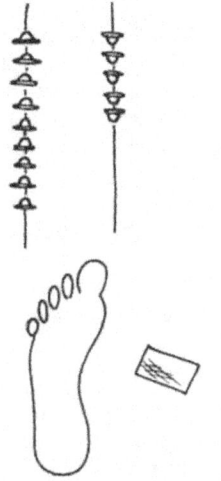

ANTI-BLISTER TIPS - (Ref: Ranger Digest IV)
THOMAS D. RUCKER wrote, "Being an avid hiker I've come upon an item that is ten times better than moleskin for treating blisters on your feet. The product is called Spenco Adhesive Knit. It's a woven fabric thinner and stickier than moleskin and won't bunch up or sweat off like moleskin because it stretches.

GROUND-TO-AIR EMERGENCY SIGNALS
– (Ref: Ranger Digest III)
SGT MICHAEL J. PILLANO tells me, "Hey Ranger Rick, your ground-to-air codes are outdated, here are the latest ones you need to publish."

MINI FLASHLIGHT TIPS - (Ref: Ranger Digest IV)
SPC BRYAN MALTAIS wrote, "At one time or another everyone has had to hold their mini mag-lite in their mouths so they could use their hands to read a map or something else. To hold the mag-lite more comfortably in between your teeth, add a few thick rubber bands around the bottom portion and then some 100 mph tape."

FIELD FIRST AID - (Ref: Ranger Digest IV)
PARK SERVICE RANGER DAVID YIM wrote,"A field expedient sling that you can use for a broken arm or wrist, is your shirt. Just elevate your arm, pull up on the shirt corner and pin it in place.

PULL UP & PIN

FIELD FIRST AID - (Ref: Ranger Digest IV)
CLINT BOOMGARDEN wrote, "If you've ever had a bug in your ear, you'll really appreciate this next trick that I learned back in PA school.

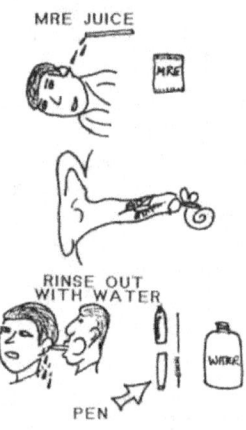
MRE JUICE

RINSE OUT WITH WATER

PEN

First, lay on your side keeping the ear with the bug in it facing up. If you have access to vegetable oil, good, if not, some oil from one of your MRE meals will do fine.

What you want to do is drown the bug and force him to loosen up his grip if he's holding onto something. And vegetable oil works a lot better than just plain water. When you feel the bug has stopped moving, it's time to sit up and flush him out.

Take apart a govt. issued pen, hand the lower portion to a reliable buddy. Instruct him to take a mouth full of water, place the pen up to your ear canal and gently squirt the mouthful of water inside of it. Repeat until bug and oil have both been flushed out.

## 550 PARACORD USES - (Ref: Ranger Digest VI)

1LT SHERAN L. BENERTH wrote; "Here in the Alaskan far north where it's extremely cold, it's difficult to build a shelter with 550 paracord while wearing gloves. Instead, we use "waxed dental floss." Simply wrap it around several times over, cut, and the wax dental floss will easily stick together without needing to tie a knot. Or, you can use a combination of both, first wrap on the 550 paracord and tie it off with some dental floss.

## TIPS & TRICKS - (Ref: Ranger Digest V)

While out in the field one day, there was a lot of rain and wind. So to keep a bit more dry, I decided to erect & sleep in a hammock.

As I stated in my previous books, I don't sleep very well in a hammock, especially when it's rocking back-&-forth. So what I did to stabilize it better and to make it more comfortable to sleep in, I made a frame and some legs for it. (Note: Make sure you have plenty of tie-down paracord and either a sharp axe or saw.)

## BLOOD AWARDS - (Ref: Ranger Digest VI)

According to the "Army Times" newspaper, there were about 50 X West Point cadets who were punished for presenting "blood branches" to each other. Now I can understand if a bunch of uneducated dumbshits or ignorant little kids did this, but a bunch of highly educated officers?

As far as I'm concerned, any soldier or leader who allows another individual to pin, stick, slam, or punch a pair of unsanitary pins into his, or someone else's body, you gotta be one dumb, stupid, #@&*?#$@. It takes a lot more balls to say "NO, not me" than to allow someone to abuse or harm your body.

*Special Message From Ranger Rick To The 50 X West Pointers*:
You're a bunch of stupid s.o.b.s who don't deserve your 2d Lieutenant bars. The next time you little boys want to prove how tough and bad you are, why don't you play with yourselves until it becomes hard and see how many "toothpicks" you can shove inside the hole. Hell, not even Sly, Chuck, or Arnold would have the guts to do this...

*Rick F. Tscherne*

## UPDATE: 2 X WAY COMMUNICATORS

In my first book, the *Ranger Digest I*, I talked about how useful and inexpensive a pair of 2 X way communicators can greatly improve a unit's capabilities. So, what I did before I went off to Bosnia, I ordered me a dozen pair of Maxon 2 X Way Hands-Free Communicators.

Because the Bosnians and me didn't speak the same language, I wore one of these Maxon hands-free communicators and gave another one to my translator. Then I divided the rest of the radios among the Bosnian Platoon's team and squad leaders.

At a safe distance away, and with the help of my translator, I was able to instruct these leaders step-by-step in how to correctly maneuver and engage an opposing enemy force. If a one of the leaders was screwing up, I didn't have to scream or holler at him. No sir, all I had to do was tell my translator what he was doing wrong over the 2 X way radio and he would translate and relay the information to that particular leader.

Not only did I enjoy using these Maxon Hands-Free radios, but so did the Bosnians too. In fact, many of them preferred to use my two-way communicators instead of their own. But I also taught'em never to depend on radios as their only source of communications, they need to have a back-up plan in case the radios go down and or become jammed.

Question: What is the most secure means of communications? Answer: Messenger, hand & arm signals, and telephone landline.

Hey guys & gals! Wanta know where you can save as much as 45% on these Maxon Hands-Free Two-Way Communicators? Contact:

EMPIRE MARKETING LTD
30 Cain Drive Plainview, N.Y. 11803
Fax: 516-753-1559 E-mail: EMP1492@worldnet.att.net

| Maxon 49SX | Maxon 49HX | Maxon 49FX |
|---|---|---|
| Single Channel | 5 x Channel | 5 xChannel |
| Headphone & Mike | Headphone & Mike | Earphone-Mike |
| * Box of 10 | * Box of 10 | * Box of 20 |
| $25.00 ea / $250 | $43.85 ea/$438.50 | $45.00 ea /$900. |
| (Reg Price $40 ea) | (Reg Price $75 ea) | (Reg Price $80 ea) |

NOTICE: As shown above, these Maxon 2 X Way Hands-Free Radios can only be ordered in quantities of 10 or more. Due to price changes, contact Empire Marketing LTD for latest price quote before ordering.

UPDATE: Maxon49 Handsfree radio do not seem to be available from the above supplier. Try eBay or Natchez Shooters Supplies: www.natchezss.com instead.

*The Complete Ranger Digest: Vol. VII*

## A MAP MARKER LIGHTER

### Submitted By PFC SERGIO RAMIREZ

Well, I'll be a son of a ...... If this don't beat all. PFC SERGIO RAMIREZ sent me a tip that I wish I had known when I was back in the Army.

He tells me that if you use a "yellow highlighter" on your map, it will glow in the dark (at night) when you shine a "blue light filter" onto it. So what's the big deal?

Well for starters, all military angle flashlights usually come with 4 different types of filters; clear, red, green, and blue. And like most leaders, I've never used a green nor a blue lens filter for anything, just the clear and red one.

But if I had known this little trick, I would have used it to mark my routes, positions, grid coordinates, etc, so I could locate 'em easier on my map at night without straining my eyes searching for them.

Thanks Pfc Ramirez, your tip will no doubt help many leaders during night time operations.

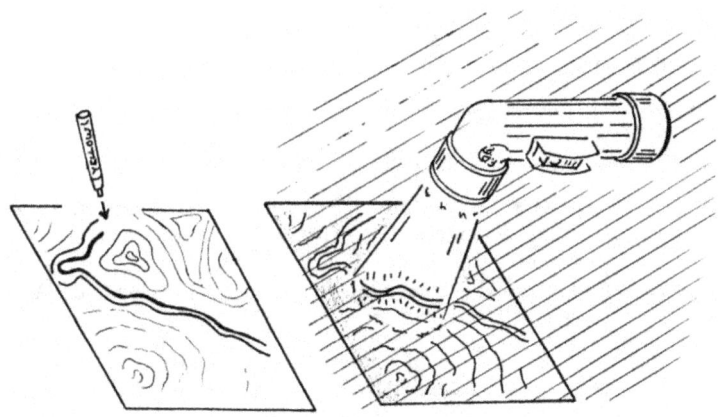

## THE M16 STORY

I'll bet there are very few soldiers who know the real story behind the M16 and how it was selected to be the standard infantry weapon for the U.S. Army. Well, believe it or not, the Army was not the first to use nor adopt this weapon.

Back in the early 1950's the Army began looking for a new type of weapon that would enable soldiers to hit their targets much more easily under combat related field conditions. Among other complaints, the Army argued that it took the average soldier too much time to carefully aim and fire their heavy M1 and M14 rifles.

So they decided it was time to come up with a new type of weapon system that would be lighter to carry and easier to hold, aim, & fire. But most importantly, it had to fire in the semi and automatic mode.

According to the Army's in-depth study, soldiers have a greater chance in hitting their target if they fire "several" well placed rounds instead of "one" carefully aimed round. In other wards, several quick "BANG, BANG, BANG" rounds fired in the direction of an enemy target has a greater chance in hitting it's mark than one carefully aimed "BANG" round. Thus the Army's "Salvo Program" was born and resulting in many types of experimental bullets, flechettes, and exotic ideas.

By the mid 1950's the Army had several prototype weapons to choose from, such as the AR-10 and AR-15 rifles. But by 1959 they lost interest in these weapons and instead pushed ahead with an entirely new different type of weapon system called the 6mm Special Purpose Individual Weapon (SPIW).

This might have been the end of the story had it not been for the good 'ol US Air Force, who they themselves were out looking for a lightweight weapon for their airfield guards. They took one look at the AR-15, liked it and ordered 8,000 rifles immediately for their guards in Viet Nam.

It wasn't long after the AR-15's arrival in Viet Nam that the Viet Namese Army also wanted them too. They liked the weight, the way it handled, and it was the perfect size weapon for their short-built soldiers too

Then in 1963 the U.S. Army finally decided to get their "heads out of their butts" and get rid of the 6mm SPIW program. And shortly there afterwards they also placed an order for about 85,000 AR-15s for their soldiers too. But it wasn't until some modification and improvements were made to the AR-15 first before it was finally accepted and renamed the M16A1.

So whether you infantry grunts like it or not, we gotta give credit and thanks to the good 'ol Air Force for our M16 rifles.

*The Complete Ranger Digest: Vol. VII*

# TO BE, OR NOT TO BE A

So ya wanna be a Ranger, huh? Well, not everybody can be one. And if you're not in a combat arms MOS, the less chance you have in becoming one. But I gotta admit, I did know a clerk, an MP, and even a Chaplain who went through Ranger school.

How in the hell did they get to go? The unit that they were assigned to had an authorized school slot for someone to attend. But, mind you, not many units have these school slots available to them. And usually only in Airborne, Ranger, and Special Forces units are they available.

But if you want to find out if your unit is authorized one or more of these school slots, visit your battalion S1/PAC and speak with the "School NCO." But don't be surprised if he laughs at you and tells you to get the hell out of his office.

If your unit is not authorized any of these school slots or allocations, then there's only two possible ways you can attend Ranger School. And that is to either re-enlist for it or to volunteer to be assigned to a Ranger unit.

If you want to re-enlist for it, you'll have to talk with your unit or post Re-Enlistment NCO to see if you qualify or not. If you can't re-enlist for the school itself, then maybe you can re-enlist for one of the Ranger Battalions. What's the difference?

Well, if you re-enlist for the school, you'll only be guaranteed Ranger school and not necessarily Airborne School nor a Ranger assignment. And if you should fail Ranger School for any reason, you will NOT be given a second chance to attend the course again. And more than likely you'll probably be sent back to your old unit and be the butt of jokes there. ("Hey Rambo, what happen? Couldn't take it? Ha, Ha, Ha.")

Plus it won't look very nice on your military record, as it will show you only attended the course and you didn't successfully complete it.

But if you re-enlist or volunteer for a Ranger assignment, you will not only be assigned to a Ranger unit, you'll also be sent to Airborne School too. The reason? Ranger battalions don't accept non-Airborne personnel, so you'll have to go to jump school first before being assigned to a Ranger unit.

And once you've earned your wings, then you'll be assigned to a Ranger unit, (l/75th - Fort Stewart/HAAF, 2/75th - Fort Lewis, 3/75th - Fort Benning).

After this, you'll then have to prove to your unit chain of command that you're mentally and physically ready to become a Ranger, which won't be easy.

How will you be able to prove it to them? Well, not a single day will go by that they won't be watching you. You'll be observed during PT, road marches, field exercises, and so on. And if you don't appear to be mentally and physically ready to be or to attend Ranger school, then they won't send you. So it's up to you to prepare yourself mentally and physically

Quite often I receive a lot of mail from soldiers asking me how they can become a Ranger, attend the school, etc. And I try my best to answer their questions even if I've been out of the military and retired for more than 5 years. And the most common questions that I've been asked...

HOW CAN I BECOME A RANGER or ATTEND RANGER SCHOOL?

As discussed previously, either by re-enlisting for it or by volunteering to be assigned to a Ranger unit.

WHAT MOS DO YOU NEED TO HAVE TO BE A RANGER?

Though there is no particular MOS prerequisite for attending Ranger School, there is if you want to be assigned to a Ranger unit. Which are;

11B, 11C, 31C, 31U, 31Z, 35E, 54B, 63B, 71D, 71L, 71M, 73D, 75B, 75H, 88M, 91B, 92A, 92G, 92Y, 96B, 96D, 97B.

NOTE: Soldiers in MOSs 11B, 11C, 11Z and 13F in the ranks of E-5/Sgt and above must possess a skill qualification identifier "V" (Airborne Ranger). All volunteers must be either Airborne qualified or be willing to attend Airborne School and a Ranger indoctrination and orientation program before being assigned to a Ranger unit.

HOW CAN I PREPARE MYSELF FOR RANGER SCHOOL or A RANGER ASSIGNMENT?

PHYSIOLOGICALLY - You must be able to endure pain, mentally and physically. And you must be able to go without sleep for as long as 24-36 hours straight before getting a "few hours" of rest. If you can't take pain, then there won't be any gain.

MENTALLY - You have to accept the fact that it's not going to be easy, but if you put your mind to it - YOU CAN DO IT! Motivation, determination, and teamwork play a big part in becoming and being a U.S. Army Ranger.

EDUCATIONALLY - You don't need to be a genius nor have any educational degrees, just common sense. The hardest part for most enlisted and NCOs is being able to prepare, write, and issue a small unit operation order. (Which can be overcome with my "*Do-It-Yourself Warning & Operation Order Handbook.*")

PHYSICALLY - You must be able to run and travel further and faster than the average soldier. If your weak in this area, then you'll have to build yourself up by jogging, running and road marching with a 70+ lb. rucksack on your back at least several times a week. You should be able to run or jog 5 miles in 40 minutes or less and road march 12 miles (with a ruck on) in less than 3 hours. You don't need to max out the APFT, but you should be able to score at least 250 pts or better.

HOW CAN I ATTEND A FOREIGN MILITARY SCHOOL?

Well, unless you're assigned overseas, you won't be able to attend a foreign military school nor a training program. I know of only two overseas units that have a foreign military training and exchange program, the 10th Special Forces Group in Germany and the 1st Bn/508th Airborne in Italy. Provided these programs and exchanges have not been cut back due to down sizing and lack of available funds, the unit chain of command determines who gets to go. And only the best soldiers get selected to go based on their daily job performance, physical endurance, motivation, and other deciding factors.

WHAT IS THE BEST UNIT TO 8E ASSIGNED TO IN THE US ARMY?

This is an unfair question to ask because I have never been assigned to all the units in the United States Army. Plus I'm partial to a selected few units that I've managed to be assigned to more than once during my military career. Such as the 1st Bn/508th Airborne located in Vicenza, Italy, which use to be the 1st Bn/509th ABCT, 4th Bn/325th ABCT and 3rd Bn/325th ABCT. This is the only elite American Airborne unit that trains and deploys on real world missions throughout Europe, the Middle East and Africa.

I've also enjoyed being assigned to the 2d Infantry Division in the Republic of Korea (ROK) and patrolling the famous DMZ. Though I prefer not to name the worst unit that I've been assigned to, primarily because there are no bad units in the United States Army, just poor (NCOs & Officers) leaders.

Well, I hope I've enlightened some of you on what it takes to be a Ranger. And if you want to know more about how to become a Ranger or how you can be assigned to a Ranger unit, contact the Ranger Regimental PSNCO at DSN 835-3790/5673 or commercial 706-545-3790/5673. Or send a copy of your DA Form 2A, DA From 2-1, and a completed DA Form 4187 to:

Commander, PERSCOM<br>
Attn: TAPC-EPMD-EPK-1<br>
(Ranger Team) Alexandria, VA 22331

*Rick F. Tscherne*

# MODIFYING & IMPROVING A MILITARY LENSATIC COMPASS

Hey! Hey! Hey! Here's an idea in how you can modify your military lensatic compass so you can use it at night easier.

While training the Bosnians, one of my many classes that I was responsible for teaching was how to use a military lensatic compass and map. Not to mention, I was also tasked to set up a couple of land navigation courses too.

Well, to make a long story short, there weren't a whole hellova lot of Bosnian leaders (NCOs or Officers) who knew how to correctly use a compass and read a military map. Not because they didn't see a need for them, but because they were scarce and hard to acquire during the war. In fact, the average Bosnian battalion had only about five compasses in their unit, which meant they were only issued to the unit commanders.(And even most of them didn't know how to use 'em.)

The leaders that I taught were very anxious and eager to learn how to use a military map and compass, which I was very happy to teach1em. But some of them had difficulties in navigating and following their compasses at night.

To assist them in navigating at night better, especially during (straight) long distance "dead reconning." I took some luminous tape and cut out a few long narrow "half arrow" patterns and glued them on the left and right side of the sighting wire. (See drawing.)

Well, believe it or not, this little trick appeared to have worked. Because after placing these illuminating half arrows on the compasses, none of them had any more difficulties in following their compasses at night.

Now listen up all you Gomer Pyles. I said long narrow half arrows, NOT thick short wide arrows. Read my lips, (or I should say my writing), use ONLY long narrow 1/2 arrows. Trust me, I know, I experimented, I tested, and I found these to work better than any other pattern. The thicker the arrows, the more distracting they became, the thinner and narrower they were, the easier they were to follow at night. Trust me, I know.

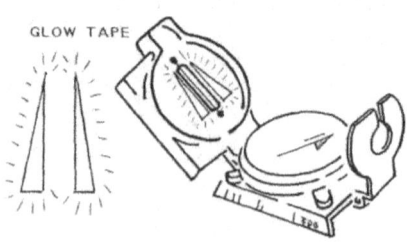

Because I didn't have enough military lensatic compasses to go around for everyone, I was forced to purchase a dozen or so inexpensive civilian compasses made by "Brunton." The only problem with these compasses, is that they're designed only for "day use" and not for night time use. Because they have absolutely no illumination whatsoever, not even to see the magnetic "North" seeking arrow.

So, being a "McGyver" kind of a guy that I am, I had to figure out a way as to how to make these compasses useful for night time training. And then it hit me, WHAM, I had an idea.

I measured the bottom of the compass dial and cut out a piece of luminous tape about the same size and shape and attached it in place with some sticky clear acetate.

Now I tried an assortment of different size shapes and patterns to see which one worked best, but it really didn't make a difference. What is important, is that when you attach the illumines tape to the bottom of the compass with some sticky acetate, you should be able to turn the dial. If it sticks, then you either have to trim it or keep turning the dial until it no longer sticks to the sticky acetate.

Regardless which type of compass you own, as long as the bottom of the compass is a clear "see-thru" plastic, you can add luminous tape to it and use it at night.

ATTENTION LEADERS! Encourage your soldiers to purchase and carry their own compass to the field, even if it's just a "cheapo" type. Then show 'em how to modify it so they can use it at night, it could save someone from getting lost, captured or killed. Think about it.

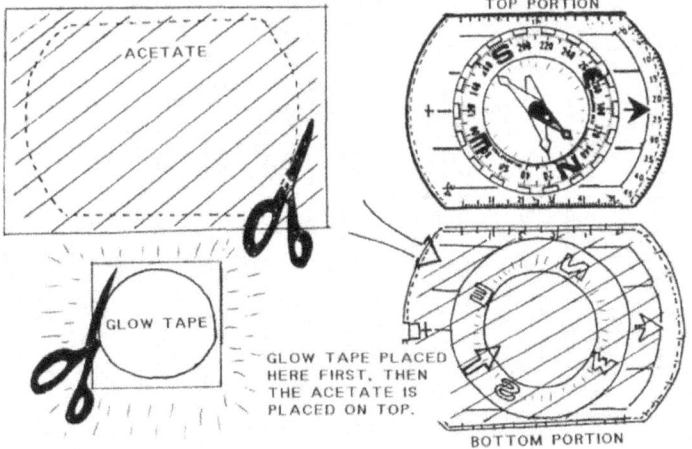

123

*Rick F. Tscherne*

# FIELD EXPEDIENT BOOT CRAMP-ONS

*Submitted By Spc. Byron Walter*

Spc. Walter writes, "A simple way to keep from falling down and busting your ass in the motor pool on those cold, icy, winter days, is to place some short screws on the bottom of your boots."

Well, after reading how he did his boots, I tried it myself and modified mine just a little bit differently;

First, find yourself about 16 X oval-head shaped short screws and insure the "screws-in portion" (minus the oval-head) is no longer than the thickness of the boot soles themselves.

Then screw 'em into the bottom of your boots in a square or oval shape pattern making sure you screw them in all the way down without leaving any space between the head of the screws and the sole of the boots.

ATTENTION: If you think the screws are a bit too long, then screw in only one and try it on for size. If you don't feel a small bulge or a sharp point protruding through, you're good to go. If you do feel a slight bulge or tiny protruding point, either replace the screws or place a padded foot cushion inside the boot.

WARNING: Never wear these crampon boots inside your quarters or barracks or your 1SG or wife will "kick off on your ass." When winter is over, simply remove the screws with either a screwdriver, pliers, or both.

NOTE: These cramp-on screws work best on "thick sole" boots.

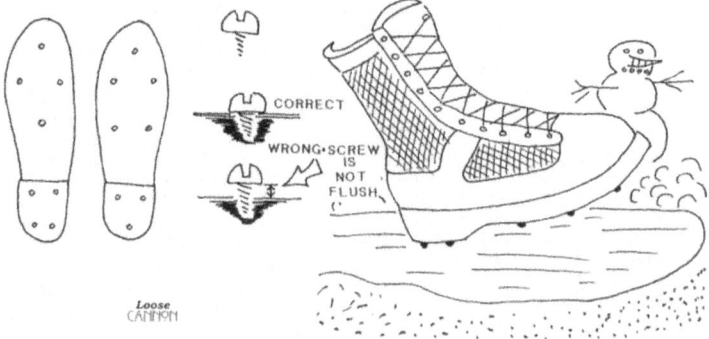

# HAVE YA EVER LOST A KNIFE?

When I was a young and naive infantry grunt, like everyone else, I lost my fair share of knives in the field too.

I remember my first boot knife, it was a beautiful double edge British Commando Knife. I was so proud of that knife, that whenever I went to the field, I carried it securely either on the outside or the inside portion of my boot.

And I owned it for about oooooohhh, a few months before I lost it somewhere in Grafenwohr (Germany). I still had the sheath, only the knife was missing.

Now I thought the snap-on buttons on a sheath are suppose to keep the knife securely in place. But somehow mine popped open and the knife came out without me noticing it.

That was the first and only "boot knife" that I've ever owned and lost in the field. After I lost this knife, I bought me one of those US Air Force "Survival Knives." And even though it was extremely strong and durable, it was NOT lost-proof neither.

Yep, you guessed it, I lost this knife too. I had it connected to the upper portion of my LBE the same way aviators carry theirs, upside down. But sometime during the night while out on maneuvers the sheath's snap-on button also popped open and gravity took it's toll. Where did I lose this one? Where else, in f——— Grafenwohr again.

Well, after losing these two knives, I finally wised up and started wrapping a rubber band around both, the knife and sheath. This way if the snap button came undone, the rubber band would hold it in place and I wouldn't lose the knife.

Which by the way, thanks to this little trick, I never lost another knife again in the field. Now I broke a few, but I never lost anymore.

The moral of this story? Learn from Ranger Rick's mistakes and don't be a dumbass and lean the hard way. No matter what type of knife you carry to the field, secure the snap-on button with a rubber band.

Rick F. Tscherne

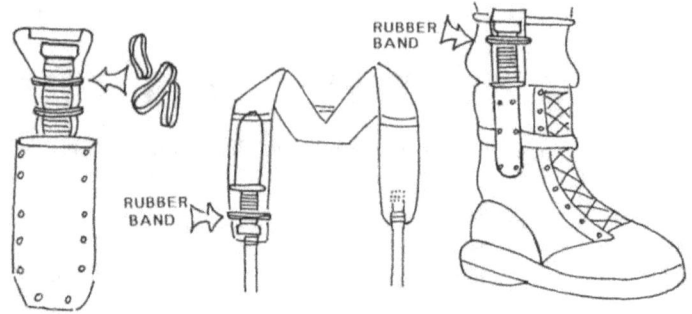

## MAG LIGHT TIPS & TRICKS

*By SSG. SCOTT A. COOK & SGT. JONATHAN C. HAVENS*

Boy, talk about getting max use out of a mag-lite, here's a few tips that I'm pretty sure will amaze you. (It did me...)

SSG. SCOTT A. COOK writes, "the mini mag-lite pocket clip and velcro carrying case are both worthless. As many times as you use the mag-lite in the field, either you wear out the case and or you bend & lose the pocket clip. Not to mention, the rubber lens cover too."

The solution? Simple, just take some 100 mph and an extra "belt clip" and tape it to the mini mag-lite. Now you can wear or attach it to your LBE without worrying about losing it. To reduce the chances of losing the rubber lens cover, just wrap some tape around the edge of the flashlight to make sure it's snug & tight.

Now this next tip is one of the best tips that I've ever publish. It's the "cream of the crop" when it comes to how to get maximum use out of your mag-lite.

One day I was studying how to attach my mini mag-lite flashlight to the side of my ol' kevlar helmet. Then by coincidence, a few days later, I received a letter in the mail from SGT. JONATHAN C. HAVENS showing me how he mounted his mini mag-lite to his kevlar.

Now there are several ways in which you can attach it to your kevlar. One way is to go to your local auto parts store and purchase either a metal adjustable hose clamp or a rubber insulated (non-adjustable) hose clamp. Both of which are used to clamp on or attach hoses to a car engine.

Before you can mount it to your kevlar, you'll have to drill a hole into the clamp as close as possible to the screw. But be careful, because if the hole is drilled too far away from the screw/ you'll have a hard time mounting it to the kevlar.

If you purchase a rubber insulated (non-adjustable) hose clamp, you will have to make sure you get the right size for your mini mag-lite. If it's too big - the mag-lite will fall out. If it's too small - you won't be able to fit the mag-lite inside.

Or find yourself a rubber tire "inner tube" and cut out a piece of rubber to the size & shape that you need for your mag-lite (see drawing). Then attach it either to the left or right side of the kevlar helmet by one of the two "chin strap screws" with a couple of mini washers.

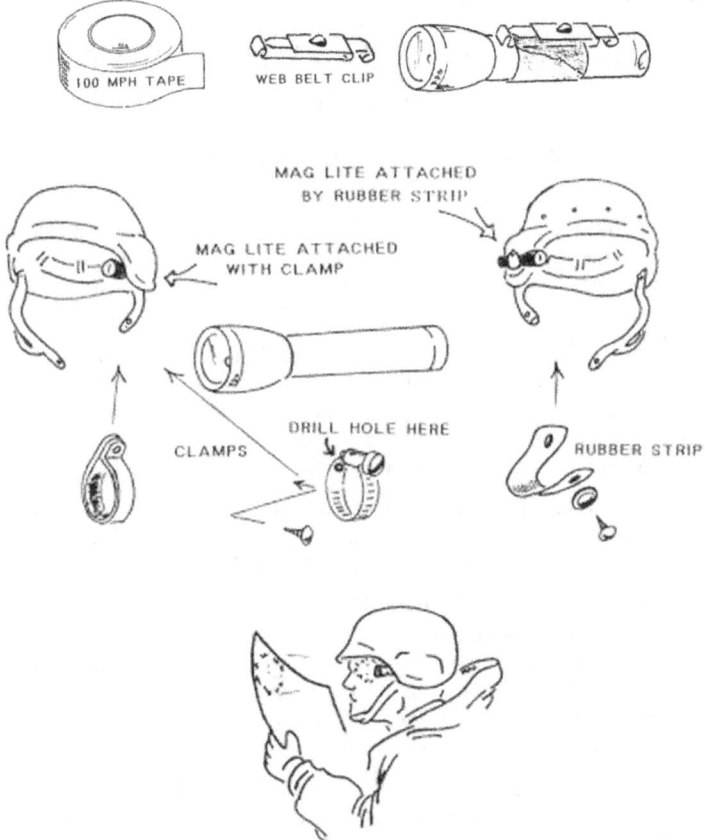

NOTE: When attaching a mag-lite to either the inside or outside portion of your kevlar, use either the Mini or Solitaire Mag-Lite.

Now all ya gotta do is put it on and adjust the light to where you want it. What is fantastic about this, is that you can use it just like a coal miner's headlamp. Any which way you turn your head, you got the light exactly where you need it. If you want to read a map (or playboy book) at night, just turn it on and tilt the mag-light slightly downward.

Don't waste your money buying one of those velcro mag-lite headband holders, this is the way to go. In fact, try wearing it with your night vision goggles on. But make sure you got your red filter lens cover on before switching on your night vision goggles or you'll blind yourself.

UPDATE: Dedicated helmet-mounted lights as now also available. (see the Princeton Tech MLS)

## HOW TO MAKE A FIELD EXPEDIENT STOVE & LANTERN

I know what you're thinking, "why in the hell would I need to know how to make a field expedient stove or lantern?"

Well for starters, flashlights, batteries, light-sticks, heat tabs, trioxane fuels, etc, may not always be available. Therefore, should your unit run out of these comfort items or forget to bring enough with them, you'11 know how to improvise.

Or if you're a smart field soldier like me, you'll always carry some kerosene inside of a container just in case ya need it. I always carried mine in a small (mini) "Johnny Walker" bottle and wrapped some rubber bands and 100 mph tape around it to keep it from breaking. I also kept a small rag inside of it so it could be used as a lantern or small stove.

Now I know a bunch of you Rambo Cowboys are probably saying, "Get the f--- outta here. Ranger Rick!" But seriously guys, if you use it sparingly, this little bottle of kerosene will last quite awhile. No BS!

When selecting a small whiskey bottle, make sure it's made of glass and NOT plastic. Don't worry about breaking it, they're pretty durable provided you wrap some rubber bands and 100 mph tape around it. And when adding a wick, make sure it's a piece of clothe or rope made of "cotton" so the fuel can be absorbed and fed to the flame. Don't use non-cotton synthetic material or it will NOT absorb nor burn the kerosene.

Also, you'll need two "metal" bottle caps. One for closing it up (so it won't leak out) and a second cap modified with a "slit" cut into it so the wick can pass through it.

When you're not using it, keep the wick always down inside the bottle and close it with the good cap. But when you need to use it, remove the good cap, replace it with the cap that has the slit, pull up on the wick, slide it through it and light it.

WARNING-DANGER: Never attempt to light nor burn the wick without using the metal cap with the slit in it, or the mouth portion of the glass bottle will become hot and cause the bottle to shatter and spill kerosene & fire all over the place. Use strictly only kerosene and nothing else, or KA-BOOM!

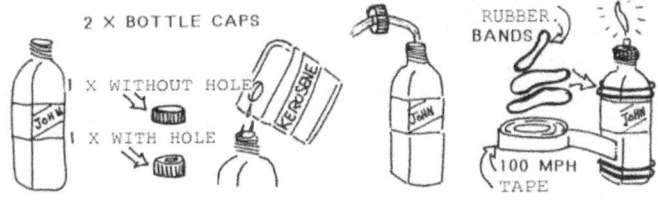

## OVERHEAD PROJECTOR TIPS

While in Bosnia, most of the instructors who were once officers always depended on overhead projectors and viewgraph slides to give their classes. If they didn't have these, most of them were lost and or had difficulty in giving their classes from memory and experience.

Now when I was back in the Army I never used an overhead projector to train my soldiers. It's not that I didn't want to, it's just that no one taught me how to use one effectively.

Well, after observing a couple of my fellow instructors (Ltc "Square Frank" and Cpt "Know-It-All Paul") and seeing how they presented their classes. Who by the way were a couple of professional REMFs (Rear Enchelon Mother!#%&*?) and staff weenies in the Army. As much as I hate to admit it, I learned from them how to effectively use an overhead projector.

Now most unit commanders usually brief their troops on an upcoming mission with maps, terrain models, drawings, etc. But there's a much easier and less time consuming way this can be accomplished, and that's with an overhead projector. Which is not hard to acquire, all you gotta do is sign for one from your unit or base training aids office.

Once you've got the projector, simply load the blank viewgraph (vgt) slides into a xerox machine and photocopy the map or pictures that you want to use for your briefing. Then place them on an overhead projector and show it up on a portable screen or bedsheet so that everyone can see it.

Another use? OK, try this... Take the map that you want to use and run it through a xerox machine onto a vgt slide. Tape some large pieces of paper or a bedsheet onto the wall and turn on the projector. Put on the slide, focus & adjust the picture so it's centered, take a magic marker and begin tracing the picture onto the paper or bedsheet.

*Rick F. Tscherne*

Once you have finished tracing it, turn off the projector and see how it came out. The traced map or picture will look like a professional artist did it. Really! Even a "bozo" without any artistic skills can do this. Now you can use this sheet of paper or bedsheet to brief your troops "without a projector."

NOTE: Once you begin tracing, don't touch, move, or turn off the projector until you are completely finished or it will be almost impossible to adjust & realign the picture again.

## SPECIAL OPS MINI PACK

*Submitted By: Sgt. Frank Gilliland*

Items needed: 2 x shoulder straps from a LC2 ruck, 1 x butt pack, 1 x canteen w/cover, and 24 inches of 550 paracord.

Attach the shoulder straps to the top and lower portion of the butt pack (as shown in the drawing) by the rings, velcro and or with some 550 paracord.

Grab the canteen cover & clips and attach it to the handle of the butt pack which is locate on the flap. Take about 12 inches of 550 paracord, tie a big ol' knot in the middle of it and run these lose ends through the bottom of the canteen cover inside-out.

Then take these two lose ends and attach them to the top rings of the butt pack to prevent the canteen from flipping around & over.

RANGER RICK's COMMENTS: Another item you can attach to the butt pack handle is your E-tool. But make sure you run some 550 paracord through the bottom of the E-tool case to prevent it (also) from flipping around & over.

If you got two extra canteens, and or one extra canteen and an E-tool, try attaching them along the sides of the butt pack. If you are only going to carry one of these items, then attach it to the handle of the pack handle so it won't be uncomfortably off balance.

A few more ideas that Sgt. Gilliland had....

If you attach a small loop onto the lanyard of the military issued compass, you'll be able to open and close the compass pouch a bit more easily in a cold weather environment while wearing gloves.

A good place to keep some 550 paracord, is around the metal frame of your rucksack. Just wrap it carefully around it and it will always be there whenever you need it. Plus it will also absorb and dampen any clanging sounds should your weapon or other gear come in contact with it.

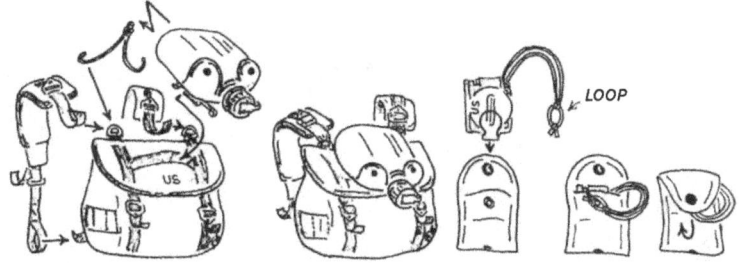

## MULTI-PURPOSE DOUBLE-HOOK SLING USES

The multi-purpose double hook strap has an assortment of uses. Personally, I like these straps because they're wide, o.d. green in color, and they have two snap-on hooks. I've used my straps for...

...as a sling for my weapon

...for carrying my butt pack across my shoulder

...for attaching & carrying ammo cans

..for attaching & carrying extra equipment to the outside portion of my rucksack.

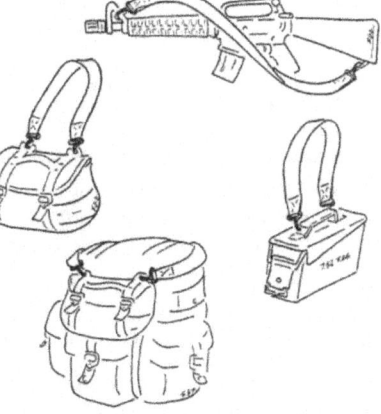

If you got a couple of these multi-purpose straps handy and some 550 paracord, you can use'em to rig up a handy-dandy stretcher support system. *How?*

First, take some 550 paracord and tie it to all four handles of the stretcher. Then, run it through the hooks of the multi-purpose straps. Then, take the straps and either slide it over your shoulders or across your rucksack as shown in the drawing below.

NOTE: The straps should be adjusted to a comfortable length so that the handles of the stretcher are no higher than waist level high or no lower than arms length down. Oh you'll still need to use your hands to steady the stretcher. But at least you'll be able to move faster and travel further without having to switch hands and or personnel so often.

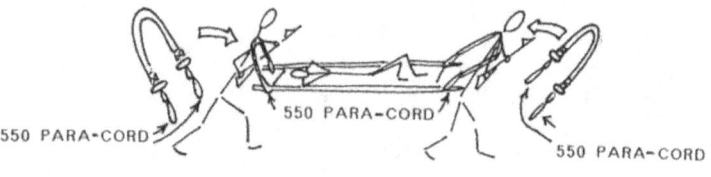

## CROSS YA HEART LBE SUSPENDERS

Well, I don't know how far this next tip will go with your chain of command, but if you're in a Special Ops or high speed unit, you might get away with it. And if not, well, maybe you can use it on your next hunting, hiking or fishing trip.

While in Bosnia the company that I worked for (MPRI) issued us some survival gear. Some of it was high-speed and some of it was just plain ol' military issued GI stuff. Though we were issued a web belt and two canteens, we were not issued any suspenders for it. Have you ever tried carrying two canteens full of water on a pistol belt without any LBE suspenders? It's not very comfortable, at least I don't think so.

Now we were issued a load bearing vest (LEV) or more commonly referred to as a "survival vest." But I didn't like it because it didn't feel comfortable while wearing it with the pistol belt and canteens.

Anyway, I manage to rig-up my own set of suspenders with a pair of double hook slings called the multi-purpose straps. They sell'em in the AAFES Military Clothing Sales Store and most of the off-post military supply stores and mail order catalogs. Though this strap was originally designed for carrying the military issued two quart canteen, someone at the Pentagon got smart and figured it could be used for other things too. (Duuuh.)

Well, while playing around with my two straps one day, I figured out how to rig'em to my pistol belt. And though I tried several different ways, it felt much more snug and comfortable to wear when I crossed the straps across my chest and back. (Hmmmmm, not bad.)

So then I added some weight to it and attached my two canteens, my two-way radio, my butt pack, knife, compass, and my strobe light too. And guess what? It felt much more comfortable to wear than the LEV and the military issued LBE. *Why?*

Well, the military issued LBE suspenders always seem to feel too loose or too tight, you know what I mean? And the load bearing vest (LEV) on the other hand, feels kinda bulky. But not my improvised cross-ya-heart suspenders, it kinda hugs you and feels much more comfortable. And I'm not just talking out the other end of my face neither, ya know what I mean?

Now the only drawbacks to using my improvised cross-ya-heart suspenders is....

a) It takes a little bit longer to put on (and take off) simply because you're locked and wrapped into it.

b) Should you need to ditch it in a hurry to save yo ass from drowning during a water crossing operation, you're @#!&%*!

c) Your chain of command will probably not let you wear it simply because you'll be Out of uniform with the rest of the unit.

But I challenge you, No, I dare you, No, I double and triple dare you to try it just once to see how it feels. And if you don't think it feels much more comfortable than the standard issued military LBE, you can send me a nasty letter and bitch me out. OK? So how about it, try it and let me know how it feels.

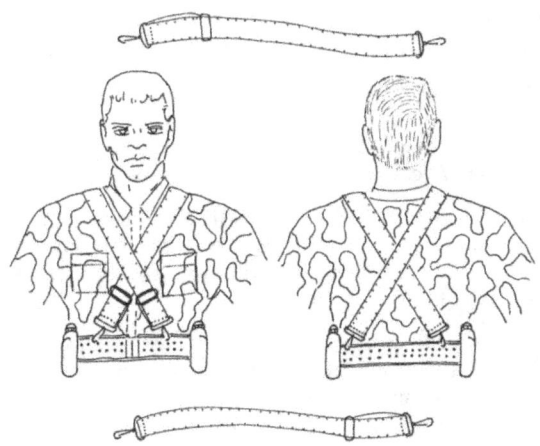

Rick F. Tscherne

##  LET 'EM SLEEP AND REST

As a squad leader and platoon sergeant, my No. 1 standing order for my men was:"When not engaged with an enemy force or directly involved in the preparation of a mission, get as much sleep and rest as you possibly can."

Why? Well, a lot of missions don't always go as planned. And if your men are not fully rested, they could be up for prolong hours and even days before the "Z Monster" captures them.

By allowing your men to squeeze in and get as much sleep and rest as possible, they will be better prepared and rested for those "unexpected" long missions. You know, the ones where your platoon leader or CO says will be a "piece of cake."

Staying awake to engage an enemy force won't be a problem, but staying awake & alert during the lull in fighting will. I don't care if you're an elite infantry grunt, Ranger, SFer, or a bad ass member of Delta Force. If you don't get enough rest and sleep, the Z-Monster will eventually creep up on you.

Here's a few tips on when you can let your men get max sleep and rest....

STAGING AREAS: When sent to a holding area to prepare for an upcoming mission or deployment, spread the duties and responsibilities around to everyone so you can accomplish them faster. Then (almost) everyone will have the opportunity to get the same amount of sleep and rest.

BATTLE POSITIONS: When not engaged or expecting any enemy confrontations, allow your men to sleep and rest wherever they're at. If your men are well trained and disciplined and you have good leaders, you should be able to allow 2 out of every 3 soldiers (66%) to sleep and rest.

Note: In the old days we use to allow 1/2 our men to sleep while the other half stayed awake & alert. But most leaders today prefer to allow 2/3 of their men to sleep and the other 1/3 to remain awake & alert. Or 1/3 sleep, 1/3 on alert, and the other 1/3 can either eat or clean their weapons.

LONG VEHICLE/AIRCRAFT MOVEMENTS: Unless it's a tactical movement, encourage your men to stretch out and sleep anywhere they can inside the vehicles/aircraft. Not only will this help them to pass the time away, but get plenty of rest too.

Now most senior leaders in my unit didn't approve of my No. 1 standing order, but the troops certainly did. And I'm proud to say (and brag), that I've always had the best infantry squad & platoon in the company. Now in the battalion, if we weren't the best, then we were no less than 2d best. **Whoaah!**

*Sent in by a Ranger Digest Reader...

Rick F. Tscherne

# OTHER NEAT TIPS & TRICKS

If your field watch has a "shiny" glass covered lens, luminous markings, and or a built-in light switch. You may want to take a permanent red magic marker (or alcohol pen) and mark over the lens to reduce the chances of the enemy spotting the light or glare, (day or night).

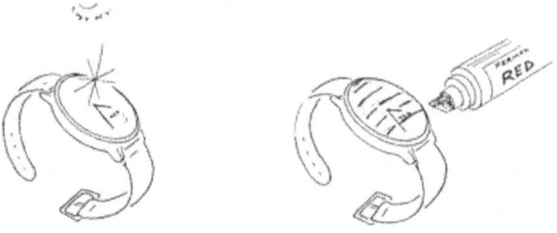

One cold winter day while refueling my apartment "Toyostove" kerosene heater, I discovered a nifty little device that can be used for filling up canteens in the field. Yea I know some of you are probably thinking, "BFD" (Big F-—n Deal), Right?

Well, instead of pouring water straight from a 5 gallon container into a canteen and carelessly spilling and wasting it on the ground. All ya gotta do with this device is stick one end in the 5 gallon container and the other end in the canteen and squeeze the water out. NOTE: You only need to squeeze it a few times to get the water flowing continuously.

Unit supply sgts and company XOs who usually make resupply runs for food, water, etc, should consider purchasing some of these devices for their unit.

The device is called a Manual Fuel Siphon - Part No.20450028 and can be ordered from: Toyotomi USA, Inc, PO Box 176, Brookfield, CT.06804-0176.

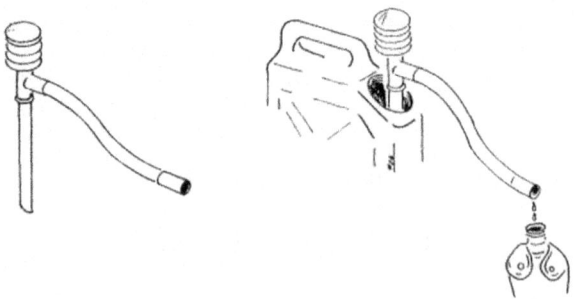

UPDATE: A similar manual pump is available from Northern Tool $39.99 http://www.northerntool.com/shop/tools/product_137382_137382

A nifty storage container that can be used to keep your extra cassette and radio "AA" batteries dry while out in the field, are paint ball containers. They're narrow long plastic tubes that can hold a total of 3 X AA batteries.

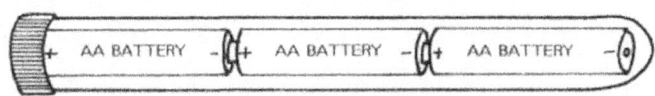

A similar container: http://www.countycomm.com/tubevault.html

Situation: As a member of an elite Ranger unit deployed behind enemy lines on a reconnaissance mission, you become separated from your unit (or vice versa). After searching for their whereabouts for several hours, you have no other choice but to E & E (escape & evade) back to friendly lines on your own.

Just then, you spot a low flying US Army helicopter approaching in your direction. You frantically search all your pockets for something shiny so you can signal it for help, but you can't find anything and it flies away.

Of course if you were a lot smarter, you should have purchased or made a signal mirror.

As an alternative signaling device, you could have used the inside portion of an MRE aluminum cracker wrapper. Though it's • not as nearly as shiny as a real mirror, but as long as there's some sunlight, you can use it as a field expedient 'short range" signaling device. Just cut open the wrapper, flatten or stretch out the wrinkles and shine it in the desired direction.

LES NISHI says, "when wearing gloves in a cold weather environment, it's sometimes difficult to grab the little zipper that's attached to a jacket, sleeping bag, etc. The solution? Take about 10-12 inches of 550 paracord, remove the inner strands, burn & melt the ends just a little bit to prevent them from unraveling and double lace it back through the metal loop of the zipper. Then, tie a knot at each of the ends so it won't come undone and you'll have something to grab onto."

SGT JERRY D. MARTINEZ says, "as a field soldier, you need to learn how to improvise and be resourceful when it comes to living out in the boonies. And the most important thing you need to look out after is your personal health. My dad taught me this next trick when he was a Boy Scoutmaster in Troop 95, Roswell, New Mexico."

We went on a camping trip one time and ran into a dilemma with our cooking utensils, pots, and pans. So my dad taught the explorers how they could clean them with some dirt and water.

First, you get a small hand full of dirt and mix it with a little water. Then you smear it onto the utensils, pots, & pans and rub it in until all the grease and grime has come off. Then thoroughly rinse them with some clean water.

NOTE: If you got any soap, it's better to wash and rinse the utensils, pots, and pans immediately after you have removed the stubborn grease, grime, and dirt.

The lesson here is, "dirt is not always your enemy."

"DOC" DAVID J. WHITE says, here's a tip on what else you can use a SAW Ammo Pouch for."

You know those plastic first aid kits that are suppose to be kept in every US military vehicle? Well, it'll take a little bit of stretching, but it'll fit inside of a Saw Ammo Pouch so you can then use it as a fire team or squad first aid kit. Or, if you happen to find an "empty" plastic first aid kit laying around somewhere, you can use it to store just about anything you want. (Radios, cassettes, CDs, electronic games, etc.)

Another useful item for medics, is the 5.56mm ammo bandoliers. They can be used for storing first aid items such as field dressings, drive on rags/slings, gauze, band-aids, and an assortment of other items.

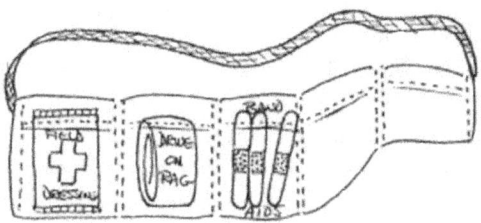

1LT FRANK C. STEVENS says, "in one of your *Ranger Digests* you mentioned why small unit leaders need to designate a "piss tree" when out in the field. Well, I'd like to expand on this subject a little bit more."

During cold weather training when everyone is up to their butts in snow it's pretty easy to get dehydrated. In fact, some people believe you can only get dehydrated in hot weather, but they're WRONG!

Whether you're in a hot or cold weather environment, the "dew point must be at or below the ambient air temperature." Basically speaking, this means the air will suck moisture just as quickly away from your body in cold weather as in hot weather.

How can you determine if you or someone in your unit is getting dehydrated? An obvious clue is the color of their urine. While a light yellow is normal, a dark yellow means you are becoming dehydrated. Now if you pee a bronze color or darker, you are very dehydrated. (Note: Daily vitamin takers usually pee a bright yellow, this is normal for them so don't worry about it.)

Though most young leaders will probably not go out of their way to check the color of their soldiers urine, they should preach, teach, and remind them what the warning signs are and how to prevent dehydration.

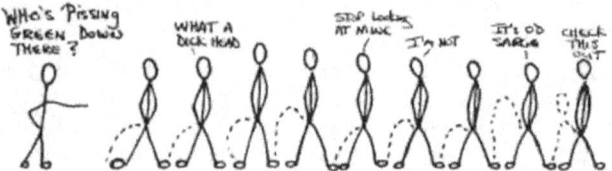

SGT "DOC" DAVID J. WHITE says, "Here's an MRE spoon trick that I haven't seen anyone send into Ranger Digest yet."

Last year my unit went to Panama for jungle training (JOTC). And while we were there, we did some cross training with some Venezuelan officers.

Well, while out in the jungle one of them came to me with a fractured finger. I wanted to send him back to Fort Sherman to have it taken care of, but he said he couldn't go back until he finished the training or his CO would punish him for wimping out.

Because I didn't have any finger splints in my aid bag, I used two plastic MRE spoons instead. I wrapped it with some gaze and tape to hold it in place until he got back to the rear. And yep, it worked!

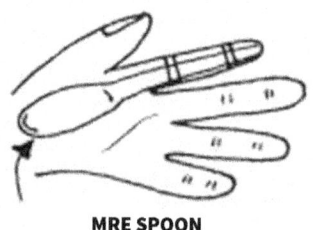

**MRE SPOON**

CPL DANIEL D. LOGIE wrote to say,"I have a good idea for drying clothes in the field."

Last fall (1997) my unit was sent to Kuwait where we lived in GP tents for about four months. And every time we hung up a clothes line inside our tent, someone was always banging into it. So then I took the line outside and lashed it in between our tent rope and no one ever banged into it again. (Not even the 2d lieutenants in our unit.)

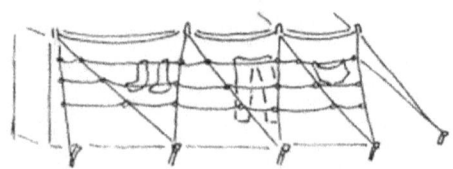

## CORDWHEEL or ROPEWHEEL

*Submitted By: CAP CHARLES CALORUSSO*

This next tip was sent to me by a member of the New York Civil Air Patrol (CAP). Though I have never seen nor used this tip (or product) before, it looks like it would work. And according to CAP Charles Calorusso, it does work.

He writes, "As a member of the Civil Air Patrol, we often train for search and rescue missions. And one thing that gets pretty annoying, is trying to keep the repell ropes from tangling up when you need to use them for a mission. The solution? An outdoor extension cord reel called a CORDWHEEL or KORD-O-WYND, it's made by Doskocil."

It only cost about $5 (+/-) and it's designed to hold about 150 feet of electrical cord. However, the length and diameter of military issued rappel rope (7/16 inches / 120 feet) fits perfectly onto the reel.

To roll up the rope, single or double strand, you just attach it to the reel and grab one handle and turn the other. Once you've rolled it up, you just set it down on the reel's feet.

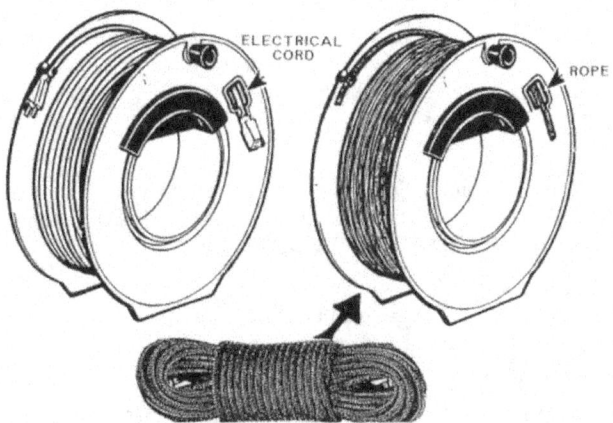

RANGER RICK'S COMMENTS: It looks like a winner to me, guys & gals. It makes rolling and storing rappel rope a lot faster, easier, and fun. (Well, maybe not fun....) And most important, it's pretty inexpensive. Hey, why hasn't the Army bought any of these yet? The CORDWHEEL is made by: Doskocil Manufacturing Co. Inc, P.O. Box 1246, Arlington, Texas 76004-1246

Rick F. Tscherne

# A LETTER THAT NEVER GOT PUBLISHED IN THE EUROPEAN STARS & STRIPES NEWSPAPER

Out of twelve (12) letters that I've mailed to the European Stars & Stripes Newspaper's "Letters to the Editor," only two have never been published. This is the second letter...... (Hmmmm, I wonder why?)

When I heard the United States was going to send American troops to Bosnia,! wished I was back on active duty again and going there myself. I retired from the US Army on 01-01-93.

In wanting to do my part to help my fellow comrades in arms, I wrote to about a dozen military community commanders throughout Europe suggesting they start what I call OPERATION MAGAZINE DROP or "MAG DROP."

What is it? It's the placing of empty boxes at designated locations throughout a military community such as in front of Post Exchanges, Commissaries, Libraries, etc. And asking people to either buy or donate their used magazines for our deployed troops. Then at the end of each day, collect, wrap, and ship them off to Bosnia via MPS. (Note: MPS is only available in the military european theatre.)

Well, I only received a few replies back, which didn't surprise me, (community commanders don't like to hear ideas or suggestions from retirees). Some said they either started something similar, or they would take my suggestion into consideration. (Yea, right. Translation: @*%# you, retiree!)

Then one day I read in the European *Stars & Stripes Newspaper* that people could send letters and packages to our troops in Bosnia by writing on the letter or package "TO ANY SERVICE MEMBER."

So then I began to write letters to the troops and send them an occasional box of books, magazines, and other reading material. But after boxing, wrapping, and filling out the required custom forms, I was only able to mail a few before being stopped by a female APO postal clerk supervisor.

After showing her my military retiree ID card, she then said to me; "As a retiree you are not allowed to ship letters and packages weighing more than 16 ounces."

After explaining to her that the packages contained books and magazines for our troops in Bosnia, she still insisted that I could not mail anything over 16 ounces from her APO.

I tried to convince her to allow me to ship these boxes and told her, "Listen, these packages are for our deployed troops in Bosnia...." She said, "It doesn't matter, you are a retiree and you are not allowed to ship anything over the weight of 16 ounces." She then instructed the mail clerks behind the counters not to accept any of my packages.

I could tell by the facial expressions on those postal clerks that they didn't have any objections to me forwarding these packages. After all, it was for a good cause, for our troops in Bosnia. But obviously, this !*%?&# enjoyed following postal regulations right down to the letter,"Rules-are-Rule."

Not wanting to give up so easy, my wife and I asked some of our friends here in the USASETAF community (active duty military & civilians) if they could help us ship these packages. Which they all eagerly volunteered to do.

I suspected this !*%?&# had probably warned her APO staff about me, so I had to reduce my letter writing and packages to avoid losing my retiree mailing privileges. And even today when mail arrives with my name on it, I suspect the APO goes out of their way to weigh it. Just so they can have the pleasure of putting a note in my box stating my mail was "returned to sender" because it exceeded 16 ozs.

Hey, as long as I've got friends stationed here in Vicenza (Italy), I'll always be able to ship and receive packages regardless of what the APO regulations state. "Friends help friends," right? *You betcha!*

When the European *Stars & Stripes newspaper* said "ANYONE" could send letters and packages to our deployed troops in Bosnia, I thought that meant retirees too. After all, it's for a good cause - For Our Deployed Troops In Bosnia. Unfortunately, APO 09630 prefers to follow their rules regardless if it's for a good cause.

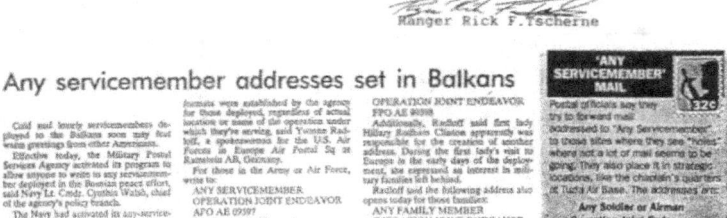

Rick F. Tscherne

BUT THIS ONE MADE IT...

## So you want a job? No lie?

RICK F. TSCHERNE

Send in your column, plus photo of yourself, to: **Be Our Guest**, SUNDAY magazine, The Stars and Stripes, APO AE 09211 or Postfach 11 14 37, D-64278 Darmstadt, Germany. If mailing disk, save text in ASCII and include printout. Columns are edited for clarity and length.

When I first decided to write this article, I said to myself, "Boy, there's probably gonna be a few people mad at me when they read this." No, I take that back, "There's probably going to be a whole lot of people mad at me."

But as my sergeant major once told me, "You know, Rick. Some people like to walk around puddles; some like to leap over them. But you, you like to jump in them with both your feet and get everybody wet." Well, I guess that statement kind of sums up what kind of guy I am.

Because I'm self-employed as a free-lance writer, and also the author of several military handbooks, I really haven't tried that hard to get a job through my local Civilian Personnel Office (CPO). It's not because I didn't want to work, it's just that there's not many jobs calling for an experienced (retired) Airborne-Ranger, Drill Instructor, Infantry Platoon Sergeant. You know what I mean?

But one day I just so happened to hear about a interesting job that would soon be opening up. And just my luck, it even matched my ol' military skills and qualifications, a rifle "Range Officer." So I left my cup of coffee (still hot) at Burger King and had the SF 171, AE Form 690-300, etc. Wouldn't it make a whole lot more sense to just rearrange these two questions and add a third line stating:

NOTE TO THE APPLICANT: If you answered yes to questions 1 and/or 2, you need not complete the remainder of these forms, as you are not eligible to apply for any job position with a GS rating.

This questionnaire form (worded as it is now) makes as much sense as a convicted felon filling out an application to become a police officer and the last question on the form asks, "Have you ever been convicted of a crime?"

• Individuals with have either "lied" on their job application or falsely exaggerated their skills and qualifications are supposed to be either denied a job and/or fired from their present one. CPO (and SJA) are supposed to verify an applicant's legal residence, skills qualification, diplomas, certificates, etc.

But according to my CPO, U.S. citizens with an Italian residency permit are not allowed to apply to hold a job position that has a GS rating. When an individual applies for a job, their names are supposedly submitted through the Italian community or authorities to verify their legal residency. The bottom line is this, if they get caught lying, they're not supposed to be hired. And if they are presently holding a job that has a GS rating, they are suppose to be fired and not given the opportunity to withdraw their residency.

In so many words, this may be what the regulations state, but CPO hasn't always enforced it. If an individual was hired "before" his residency was verified, CPO has been using the "quiet approach" in notifying the individuals to pull their residency if they wish to hold onto their present job.

This may be a nice gesture in trying to help someone who has intentionally lied or falsified information on their job application. But it actually hurts the other applicants who were honest in filling out theirs and were not selected for the job nor given that same opportunity to withdraw their residency.

Now come on, wouldn't it be much more fairer to just tell all the job applicants that if they are currently holding an Italian residency permit, they must be prepared to withdraw it just prior to being hired for a GS position? This way no one violates any U.S., Italian or CPO rules or regulations. Makes sense, doesn't it?

CPO may do an initial residency check on individuals who apply for a job for the first time, but they fail to do any follow up checks. What stops individuals from applying or reapplying for an Italian residency once they've landed a job on the base? Nothing!

• Individuals who were discharged, barred from re-enlistment, or "forced out" of the military due to financial, family, alcohol- or drug-related problems, should be closely scrutinized by CPO, and in most cases denied employment. For if these individuals could not perform their duties to military standards, then they should not be given a second chance or career working for the military. It's a privilege to work for the U.S. government; if you abuse it, then you should lose it or be denied future employment opportunities.

CPO needs to either change or update their current policies, or START ENFORCING THEM.

Thank God I still got my job as a free-lance writer. If this article should get printed, I can just about kiss my chances of ever finding a job through CPO. But hey, that's the breaks, it comes with the territory in being a writer.

*Rick F. Tscherne lives in Torri Di Quartesolo, Italy.*

**OPENING LAUGH**

## CONTENTS

**BACK IN UKRAINE** ........... 4-7
*Ex-Soviet officer learns to adjust.*

**IN TUNE** ........................... 9
*Sly's an actor.*

**THIS WEEK'S FILMS** ......... 10-11
*Chuck Vinch on "Judge Dredd."*
*Military circuit listings.*

**TELEVISION LISTINGS** ....... 12-15
*Local, cable and satellite schedules.*

**HOME ENTERTAINMENT** ..... 13
*"Death and the Maiden" on video.*

**BOOKS** ..................... 36 ...... 17
*A look at filmmaking.*

**SIDELIGHTS** ................... 20
*Getting late for civil rights.*

**VOICES** ........................ 21
*Joe Bob on Bob Dole.*

**WORDS & WIT** ............... 22
*Tony Kornheiser visits Dad.*

**ON PARADE** ................. 23
*Melanie Griffith's love affair.*

**LAST LAUGH** ................ 24
*Dave Barry manages his stress.*

SUNDAY is a weekly supplement of The Stars and Stripes

Editor:
Cover story design:
Production:

# HEAD GEAR TIPS

How many of you soldiers know that a BDU Hat, a Beret, and a Kevlar Camouflage Cover ain't waterproof? Did you know that? That's right. In fact, they suck and soak up water like a "SPONGE."(Well, almost....)

Now I thought they were suppose to provide some form of protection against the weather, didn't you? Oh well, I guess Uncle Sam had to cut corners somewhere to save a few lousy bucks. So I guess if you want 'em waterproof, you'll have to do it yourself like I did.

Among several water repellent products that are available on the market, I found that a can of spray called Kiwi "Camp Dry" works pretty good. But like any repellent, it has to be reapplied from time to time. A 12 oz. can costs about $5 bucks, which is a small price to pay to keep your head dry. To apply....

(a) First, wash and clean your head gear thoroughly of all foreign dirt and oily substances, and DON'T just shake it out.
(b) Second, hold the can about 7-10 inches away and spray the entire surface of the head gear very thoroughly. Not just once, but several times to build up a good coat.
(c) Third, then sit back and wait for it to dry, which will take about 4-6 hours.

NOTE: As you begin to spray you'll notice the liquid will start to build up and become very glossy and shiny. Don't worry about it, when it dries the gloss shine will disappear.

Now before you put it on, you had better test it out to see if it's completely waterproof. Take your head gear and run it under a spray facet or sprinkler to see if it repels water. If the water does not run off smoothly, then you either did not (a) wash it thoroughly (b) spray it properly. If you had, the "pores" in the material would be closed thus repelling the water. If it doesn't, start over again.

Does the "duck beak" sun visor of your BDU hat always look f———d up and out of shape? Then when you're not wearing it, try rolling it up and placing a rubber band around it so it maintains it's oval "sharp looking" shape.

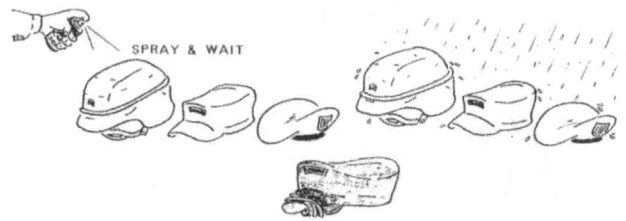

*Rick F. Tscherne*

# BOOT TIPS

There are so many different types of military boots on the market today that it's almost impossible for a soldier to decide which one is the best to buy for the field. Just because it's the most expensive, it doesn't necessarily mean it's the best or better made boot.

I use to maintain several pairs of boots for different occasions and military functions.

My INSPECTION BOOTS were only worn for morning inspections and or for ceremonies, parades, and other special events. I always kept them highly spit shined and cleaned at all times.

My GARRISON BOOTS were put on immediately after the early morning inspections as long as we remained in the unit area. Though they were NOT kept as highly spit shined as my inspection boots, they were always highly brushed shined.

My FIELD BOOTS were only worn in the field and I kept two different types. My Jungle Boots were my light pair which I usually wore during the summer time and or in hot weather environments. And my Cortex or Matterhorn Boots were my heavy duty pair which I wore in the winter time and or in cold weather environments.

My MOTOR POOL BOOTS were a pair of old beat-up boots that showed a lot of "wear & tear." Because a motor pool has a lot of oil, grease, fuel, and a bunch of other stuff that can really f---up your boots. I only wore these when I was going to be around some vehicles or tracks.

When it came to real world missions and deployments, depending on where we were going, I only took the necessary boots. If we weren't told where we were going, then I took all my field boots with me.

Out of all my boots, I treated and took care of my field boots a bit differently than I did my other boots. Such as...

a. Replacing the boot laces with some o.d. green 550 parachute cord. It's not only a lot stronger, but you can use the seven nylon inner strands as emergency survival string.
b. Replacing the "cheapo" inner sole cushions (that most new boots come with) with a more comfortable & durable cushion.
c. Treating them with a water repellent shoe polish and or spraying them with some KIWI'S "Camp Dry" heavy duty water repellent spray. (Note: Must be applied and reapplied often.)

# A FEW RANDOM NUGGETS OF INFO

By LTC MARTIN N. STANTON

Dear Ranger Rick,
I enjoy reading your books because they contain a lot of useful information for the field. Well, here's a couple of "nuggets" that maybe you can use in your next Ranger Digest...

Most operation officers and NCOs use an acetate map for their situation board, so they can draw on it with an alcohol pen to show where the friendly and enemy units are located. But every time you need to make some fine changes to it with some alcohol, it usually drips, smears, and you have to redo it all over again. But thanks to MSG SCOTT HARD-CASTLE, he showed me an easier way. Just take a dry eraser pen and erase over the symbol, brush it off and then make your necessary corrections and changes.

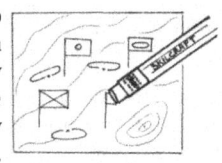

If you are in a type of unit that is always being deployed somewhere in the world. A good source of information and knowledge can be acquired from most tourist guide handbooks. I was amazed at finding a book called "*The Lonely Planet Guide to East Africa*" which showed more accurate maps of the smaller towns in Somalia than what we were issued. These useful handbooks are a good source of info for Bn. and Bde level S-2's who want to get as much INTEL on their new AO prior to deploying.

When packing goodies for a family member who is deployed on a military operation, don't pack the box with styrofoam or other useless packaging to cushion the load. They are not only worthless to a soldier who is on a deployment, but it may be hard for him or her to get rid of it too. The solution? Cushion the box with small plastic food additives such as those found at the fast food restaurant (mustard, ketchup, relish, etc) or rolls of toilet paper, magazines, & newspapers. Something useful, edible and or trade-able.

*Rick F. Tscherne*

# TRAINING WITH TOY SOLDIERS

As a little kid I always enjoyed playing with my toy soldiers and setting 'em up in different battle field situations. Then when my parents allowed me to buy a BB gun, it became even more enjoyable. As I would set 'em up in my backyard and shoot 'em down one-by-one. (Boy, that was a lot of fun...)

Well, before I went off to Bosnia, I bought me a whole bag full of toy soldiers so I could use 'em as training aids for teaching and explaining battle drills & formations. In order for these toy soldiers to stand up better, I attached and glued them to some bottle caps and pieces of wood.

I used them in explaining how, when, and where they should use a particular formation or battle drill. In fact, before they actually went outside to practice these drills and formations, they first had to demonstrate and prove to me in the class room how and when they would use them.

I gave all the students a toy soldier, selected the element leaders, and then instructed them to organize themselves into fire teams, squads, and platoons.

After this, I'd create battle field scenarios and point to one of the student leaders and ask (for example), "Ok, what would you do if you got hit from this direction...." In which either the leader or a student that I pointed to had to show me with the toy soldiers what he (or they) would do.

These toy soldiers were not only an excellent training aid, but due to a language difference, it made it easier for both of us to understand one another. All we had to do was use the toy soldiers to explain what we were trying to get across.

When it came to teaching actions on enemy contact, crossing danger areas, react to flares/artillery, etc, I built a couple of multi-tactical situation kits. One was a roll-out and the other was a fold-up. The roll-out was made from an old white bedsheet and the fold-out was made from a large cardboard box.

Both were colored and painted with magic markers and or spray paint to look like different types of terrains that they would more than likely encounter on a mission or tactical operation. Such as green for woods, black for roads, blue for water, etc.

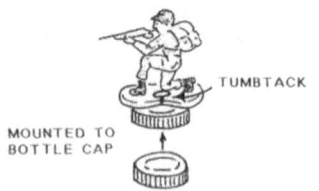
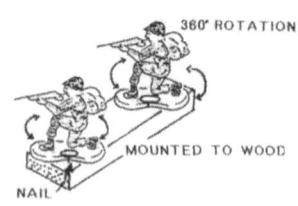

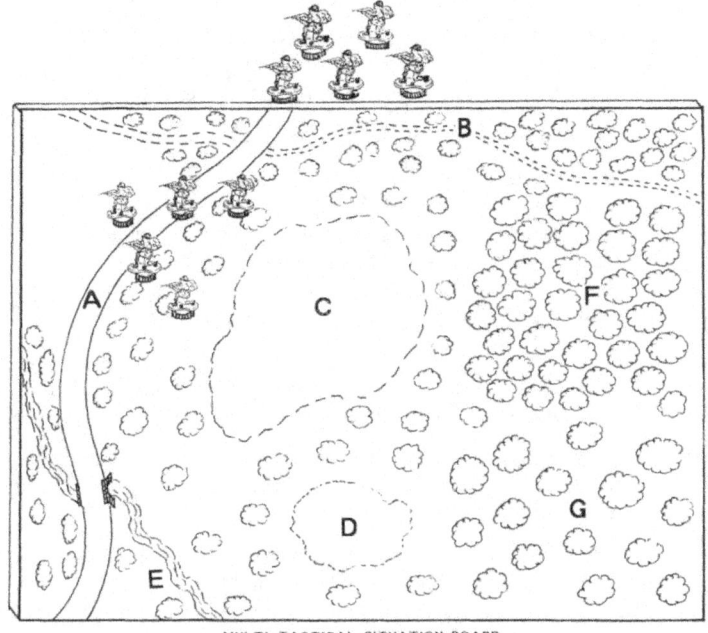

MULTI-TACTICAL SITUATION BOARD

An example of the different types of tactical situations that they would probably encounter on a mission or on a patrol;

| area | type of terrain | what it can be used for |
|------|-----------------|-------------------------|
| A | Road | Crossing Linear D.A./Staggered Form. |
| B | Trail | Crossing Linear D.A./File Formation |
| C | Large open Area | Moving Around or Crossing Large D.A. |

| | | |
|---|---|---|
| D | Small open Area | Moving Around or Thru Small D.A. |
| E | Stream/Creek | Crossing Water Obstacles/Linear D.A. |
| F | Thick Woods | Types Of Formations/Enemy Encounters |
| G | Light Woods | Types Of Formations/Enemy Encounters |

What you will also need to complete this scenario kit....

1. Red colored toy soldiers for enemy encounters/snipers.
2. Red colored buttons/chips for encountering enemy mines.
3. Red colored plane/helicopter/tank for enemy encounters.
4. Black colored wooden blocks for built-up areas.
5. White cotton balls for incoming artillery/mortar rounds.
6. White filtered flashlight for reacting to enemy flares.
7. Green toy soldiers for the "Good Guys."

NOTE: This is just an example of how I made my kit, you can either add or delete some of these items from your tactical situation kit.

When using a bed sheet as a multi-tactical situation kit, you can create rolling hills, mountains, and valleys simply my placing something underneath it, such as grass, pine tree needles, etc.

If you want to use the multi-tactical situation kit as a portable terrain model kit for the field. Just spray paint the bed sheet light green and don't add any roads, trails, streams, creeks, etc, onto it. Instead, use different colored "knitting yarn" to create the terrain features. Such as red yarn for roads, brown yarn for trails, blue yarn for water, etc. This way you can tailor your terrain model kit to your actual mission.

Or, if you don't want to use a bed sheet as a multi-tactical situation kit, you can always use your "camouflaged" poncho liner. But don't forget to carry enough colored yarn for the terrain features.

## UNBREAKABLE HOMEMADE MAP PROTRACTORS

When we first started teaching the Bosnian Army in how to read a military map, we didn't have any US Army issued GTA map protractors. So naturally, we had to improvise. Luckily I brought along a couple and I was able to make some paper copies at least for classroom teaching purposes. Then later on I was able to acquire some acetate and make them a bit durable and re-useable.

Well, even though these acetated paper copies worked pretty well, we were in need of something a bit more durable for outdoor use. So one night I sneaked on down to our training room and stole (or "borrowed") a box of Xerox blank overview slides and I tried to reproduce a copy of a GTA Map Protractor onto it. And guess what? Yep, it worked! So I used up that entire box of Xerox VGTS and made about 200 repro GTA map protractors.

IMPORTANT: Before mass producing and running off a bunch of these GTA map protractors, place a US Army issued map protractor up to the first xerox copy to see if it's to size. If the repro is slightly smaller or larger than the standard issue GTA map protractor, you may have to enlarge it or reduce it anywhere from 1%-3%.

Well, even though they were a bit thinner and less durable than the real McCoy, once cutting 'em out of the plastic and down to size. I then took some acetate and placed it on both sides of the repro GTA map protractors to make 'em a bit more durable and flexible. Then I used a straight edge razor and cut out the rectangle 1/100,000, 1/50,000, and 1/25,000 meter scales so they could be used effectively on a map.

These repro GTA map protractors worked so well I prefer to use one of these rather than the US Army issued GTA map protractor. Why? Because it's more flexible and durable and can take a lot more wear & tear than the real thing. If ya don't believe me, go ahead and make one and see for yourself.

Don't forget to modify it like the way I taught ya in one of my previous *Ranger Digests*. What? You ain't got any of my other books? Well then, you better go buy 'em if you wanta learn how to modify these map protractor some more. Because I almost never repeat the same tip or trick twice in any of my books. Sorry about that, buddy...

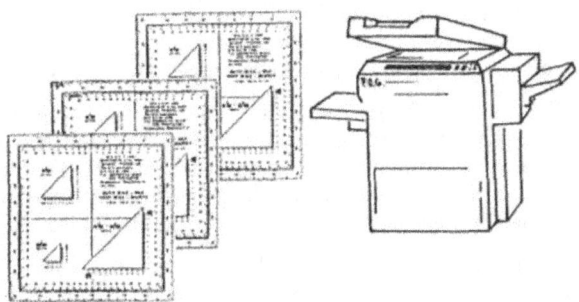

## TRADING, BARTERING, & GIVE-A-WAYS

Now I've lived, trained, and visited a lot of countries in the world, and Bosnia wasn't any different when it came to trading and bartering for things.

And every time I went to a foreign country, I always took with me a few popular trading items. What for? Well, not only for trading, but for services needed, good will gestures, and for bribing (mostly military, custom, and police) officials.

Now even though you might think there's nothing out there that you want from another country, you'd be surprised in what you can get with a few popular American made products.

A few of the things that I took with me to Bosnia to trade... (Note: Items listed below are in sequence of popularity.)

#1 Zippo Lighters - In the PX/BX they sell for about $8-$25, but in most countries, including Bosnia, they sell for three, four, and five times (or more) than that amount.

I rate this a #1 trading item because they're small, compact, inexpensive, and a very, very much in demand in almost every foreign country. And the more fancy looking it is, the bigger and better things you can trade something for it.

NOTE: The shiny silver smooth Zippo lighters with or without the fancy engraving ($12-$15) are the best ones to buy.

#2 Mini Mag Lites - Though there are several types of Mag-Lites, the most popular one for trading is the MINI-MAG LITE (AA battery). It comes in two colors, black and camouflage, the later being the most popular and in demand.

#3 The Leatherman Tool & Buck Knives - Now we're getting into a higher price range, but well worth purchasing if your looking for something really special. Don't bother wasting your money on a super-duper knife such as the SOG Paratool or Power Plier, the cheapo Pocket Survival Tool will do fine.

GIVE-A-WAYS - A poplar and inexpensive item that you can give away to both, adults & children, are small American flags. They cost only about 30 cents a piece and you can use them to make friends. In fact, whenever I gave these away in Bosnia, I usually got something back in return. Such as a free drink, food, war souvenirs, and meeting some pretty Bosnian ladies.

What did I trade the other items for? Well, to name just a few things; bayonets, helmets, night vision goggles, inert anti-personnel mines, and a few other things. Though I was offered some weapons, I did not trade anything for them. No BS!

# BOOT CRITTERS

*Submitted By: Sgt. Charles Robinson*

Dear Ranger Rick,

"Back in my old jungle days," Sgt. Charles Robinson writes. "Before we went to bed at night we use to roll up our socks and place them inside our boots to keep the little night critters out. And it worked pretty well too, but it wasn't always 100% anti-critter proof."

Then one day while out in the field with my unit, I saw a PFC Jessica Cuckler take a pair of socks and place them over the tops of her boots. She said she learned this trick back in A.I.T., it not only kept out the critters, but it helped air and dry out the socks too.

RANGER RICK'S COMMENTS: Another technique in keeping those pesky flying and crawling little critters out of your boots at night, is to spray some insect repellent inside of them. Or, tie the tops of your boots closed or place one boot inside/over the top of the other.

## CANDLE MATCHES

One day I was playing around with a candle and I accidentally dripped some wax on some paper matches. And when I lit one of them, I noticed the match burned a lot longer than normal. So naturally I began experimenting.

I went to McDonald's and got me a handful of plastic drinking straws and then I bought me a box of wooden matches. If you find and buy the type of wooden matches that light when you strike 'em against a rock or a piece of metal, good. If not, the wooden matches that come in a little box will do just fine.

Next, I got me a small pair of scissors (a razor blade will work too), an empty plastic 35mm film container, and a candle. I took about 10-15 wooden matches and cut"em down to the same size and length as the 35mm container. I then grabbed the straws and cut them down to the same size and length as the matches and 35mm container.

Most of the plastic drinking straws have a colored line that runs down the entire length of the straw. After you have cut these straws down to size, slit 'em along this line.

Now take your candle, light it, and allow some wax to build up on a flat surface. Then take the head of the wooden matches and place them into wax and hold 'em in place until they can stand straight up on their own. Then grab the short straws and slide 'em over each of the wooden matches.

Then take your candle and melt some wax inside each of these straws. When all the straws are full of wax, stop. Then very carefully remove the straws & matches from the flat surface and rub off any excess wax from around the plastic straws. Then smooth off the ends with your fingers and place 'em inside the 35mm film container for safe keeping.

OK boys & girls, now when you need to use these, all ya gotta do is "peel off" the plastic straws (where you have slit it along the colored line) and light 'em up.

What can you use these waxed matches for? As survival matches, you bonehead! They'll burn 5 X times longer than regular matches. Try it and see for yourself.

NOTE: Before lighting a candle-match, make sure you remove the plastic straw. If you forget, the wooden matches, wax, and plastic straw will burn up very rapidly. Also, if you're using the wooden matches that come in a box. Don't forget to remove the "sandpaper" from the side of the box and place it inside the 35mm container. Or else you won't have anything to ignite the matches with.

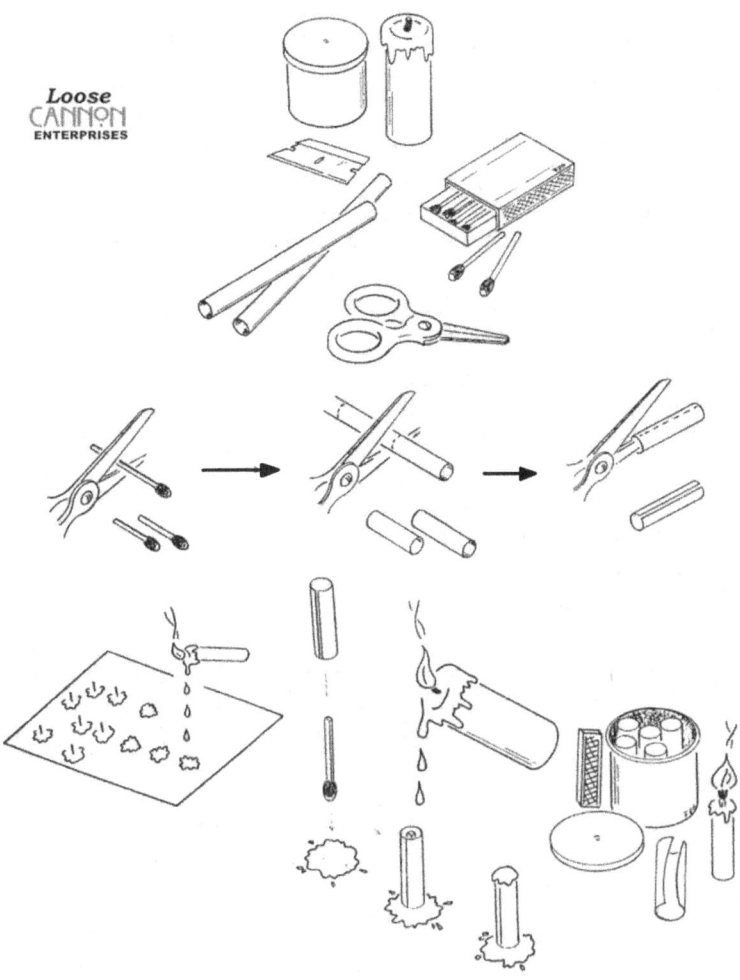

## MRE HEAT PACKET TIPS & TRICKS

When I was back on active duty in the Army, from 1972-1993. We went from heating our C-Rations by burning the box that it came in, to heating the MREs in a canteen cup of boiling hot water with a heat tab.

Well times have changed, now you guys & gals got a fireless & smokeless bag of heat for warming up your MRE meals. You just add a little bite of water inside this plastic bag, slide it beside your meal, wait about 10 minutes and PRESTO - an instant hot meal. Cool! Or I should say HOT!

Well, even though this was after my time, I was able to acquire some of these MRE heat packets and do a little bit of experimenting. And here's what I came up with that I think you'll find useful and or at least interesting.

COLD WEATHER HAND & FEET WARMER: Yep, you should have thought of this way before me, as it was obvious that it could be used for this. But, instead of following the instructions on the packet in how to use it, try this instead...

(A) Remove the o.d. green and gray heat packet from the plastic bag/pouch and cut it up into 3-5 narrow strips.

(B) Place the strips in the bag one-at-a-time, add a little bit of water and squeeze it so the water spreads around and gets absorbed, then close up the bag. In a few minutes you'll have instant heat.

Now place it either in your pocket, in your hands and or near your feet. If you want two of these for the price of one, just cut the bag in half and place an o.d. green and gray heating strip inside each bag half.

Now, the only draw back to this, is that these heating strips don't last very long. So as the heat dies down, you'll need to replace it with another narrow gray strip of heat & water.

It's better to replace the strips of heat one-at-a-time than to use the whole damn thing at once. Why? Well not only will you get longer use out of it (heat wise), but it won't get so @!#%& hot that you won't be able to hold onto it. Thus wasting valuable heat, ya know what I mean? You see my point?

To use it as a foot warmer, you'll have to remove your boots and place your feet on or near the heat packet itself. But watch out, if you're not careful you'll burn your little toes.

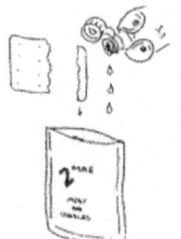 

If you're like me and hate putting on cold boots in the winter time. Instead of trying to warm 'em up with a cigarette lighter, match, or a candle, try using one or two MRE heat packets. When your boots are defrosted, don't throw away the packets of heat, just place it inside your BDU pockets for extra body heat until it cools off.

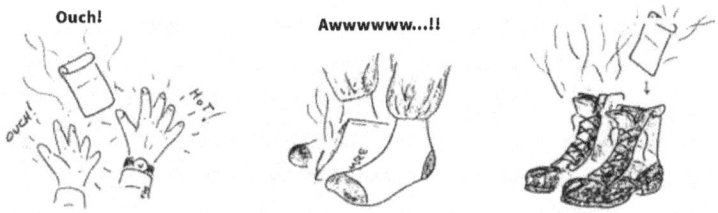

HOT CHOCOLATE/COFFEE HEATER: Now I know you wouldn't have thought of this, and before you criticize it, don't knock it until you've tried it. Or at least until you're desperate for some coffee and you can't make a fire...

If you're in a tactical situation and you can't make any fires, such as in a bunker, LP/OP position, etc, and you want to have a cup of coffee or cocoa. Well, here's one way you can warm some up without a fire. BUT, the hotness or warmness of the coffee and or cocoa will depend on how many MRE heat packets you use or have available.

ITEMS NEEDED: Several MRE heat packets and 1 X Zip-Lock Bag.

A) Take the zip-lock sandwich bag and fill it about 25% full of water (and no more) and then zip-lock it closed/shut.

B) Take an MRE heat packet and follow the instructions on it.

C) After following the instructions, place it underneath the zip-lock bag of water and wait for it to heat up.

IMPORTANT - Depending on how much water was placed inside the bag and how cold it is outside, it may take 2-3 MRE heat packets to warm it up. Though it won't become real hot, a nice warm cup of coffee or cocoa is better than a cold one. Try it!

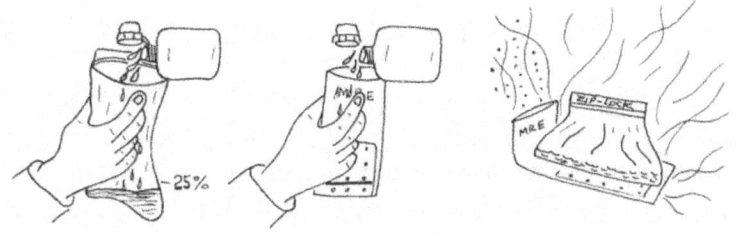

*Rick F. Tscherne*

# HOMEMADE WRIST CEOI CODER

When I first saw one of these in a military supply catalog, I said to myself, "What the #@^%?" But after I thought about it for awhile, I guess it's not such a bad idea after all.

Now when I was a platoon sergeant I always carried an acetated piece of luminous tape attached to some 100 mph tape. And on it I would write down the call-signs, codewords, frequencies, etc, so I could read it in the dark without using my flashlight. And it worked pretty good too.

Of course, there were some leaders in my unit who tried to copy what I did, but they screwed theirs up by not placing a second layer of acetate on the luminous tape. Why the double layer? Just in case the alcohol pen permanently stained the first layer, then all I had to do was remove and replaced it with another piece. If you try to write on the luminous tape without any acetate, it'11 permanently stain and mark it.

SCREW THIS! → **THE WRIST CODER**™
A Smart Tool For Tactical Leaders
Originated for use by the Israeli Defense Forces to eliminate the "Fog of Battle" by keeping code names, call signs and other data ready for instant reference. Worn on the arm above the wrist, the clear plastic window allows easy viewing of data. Features a second, internal pocket for storage of other materials. Snap-closing flap. Attaches with two hook-and-loop straps. Measures 5 ¼" x 5 ½" (13.3 x 14cm). [2 oz/57gm]
**MAKE YOUR OWN!**

To make a homemade CEOI wrist coder, here's what ya do;

Take some 100 mph tape and measure the width of it, then cut out a piece of luminous tape about the same exact size. Take this piece of luminous tape and "acetate" both sides. Yep, both sides, the sticky side too. Not once, but TWICE.

Take about 12 inches of 100 mph tape and lay it down on a flat surface with the "sticky portion" facing up. Take the acetated piece of luminous tape and place it in the center of this piece of tape with the luminous portion also facing up.

Now take this piece of tape (with the attached luminous tape) and measure it to your wrist. But don't attach the two ends together yet, you're just measuring it for size. Make sure this tape is about 3-5 inches longer than what you need, this excess amount will allow you to fold and overlap the two ends together. Then cut off the excess tape and lay it back down on a flat surface with (again) the sticky portion facing up.

Once you have finished measuring and cutting off the excess tape, take two other pieces of tape and place them on top of the "sticky portion" except for one of the ends. Leave this one end (your choice which one) exposing 3-5 inches of sticky tape.

Be careful when placing and attaching these two tapes one-on-top-of-the-other or they'll bunch & ball up together. When placing and attaching the tape near the "acetated luminous portion", just allow the tape to barely overlap and touch it. "THE WRIST CODER": A Smart Tool For Tactical Leaders

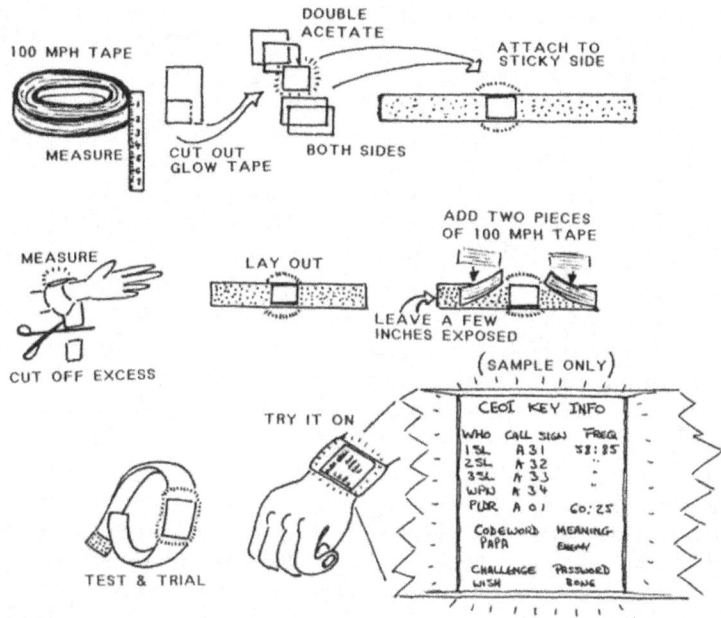

You'll want to be able to remove this luminous portion easily so you can dispose of it should you become captured. Next, take the 3-5 inch sticky portion end of the tape and fold it in about 1/2 of an inch, this will leave you about 2-4 inches of sticky tape remaining.

Pick up the tape, wrap it around your waist and then attach the sticky end to the non-sticky end of the tape. The 1/2 inch folded-in piece of tape is a "pull tab" quick release so you can remove and attach it to your wrist freely. If it should wear out, no need to make a new one, just attach another piece of tape on top of it.

Now all ya gotta do is grab your alcohol pen (and not a grease pencil) and write down your radio call signs, codewords, and frequencies on the luminous portion so you'll be able to read it at night and also during daylight hours.

Rick F. Tscherne

# RANGER RICK'S FAVORITE TOILET GRAFFITI

THEY PAINT THE WALLS TO STOP MY PEN, THE SHITHOUSE POET, HAS STRUCK AGAIN.

TO EAT, OR NOT TO EAT SHIT. 100 TRILLION FLYS CAN'T BE ALL THAT WRONG.

I CAME HERE TO SHIT & THINK, BUT ALL I DO IS FART & STINK.

IT'S BETTER TO LAUGH WITH THE SINNERS, THAN TO CRY WITH THE SAINTS.

HEY BUSTER, STAND CLOSER! THAT F——N THING BETWEEN YOUR LEGS AIN'T NO DAMN WINCHESTER.

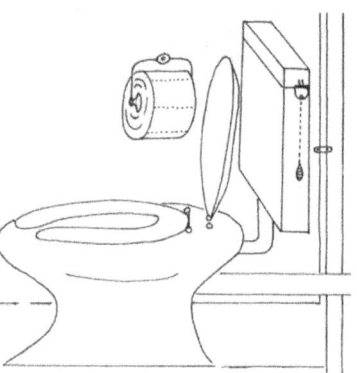

THE MOON WAS HIGH AND SO WAS I, AND THERE SHE WAS A WAITING FOR ME. SHE HAD A BODY AS FINE AS COULD BE, FOR SHE WAS HOPING THAT I'D BE SWEET. AND SWEET I WAS AND SWEET I'LL BE, AS LONG AS SHE IS SWEET WITH ME. I TOUCHED HER GENTLY AND WHISPERED TO HER SOFTLY, HOPING TO RELAX HER AND NOT TO SCARE HER. SHE SPREAD HER LEGS AND BRACED HERSELF, AND I BEGAN TO POSITION MYSELF. AS I TOUCHED HER BETWEEN THE LEGS, SHE BEGAN TO SCREAM AND POUT. FOR I KNEW IT WAS HERS JUST AS MUCH AS IT WAS MINE. THAT THIS WAS MY VERY FIRST TIME THAT I'VE EVER MILKED A COW.

IF YOU LIKE SEX, THEN YOU'RE IN THE RIGHT OUTFIT, WE GET F——D EVERYDAY.

THE ARMY'S SURGEON GENERAL HAS DETERMINED THAT HEMO-ROIDS ARE ONLY CONTAGIOUS IF TWO ASSHOLES COME IN CONTACT WITH EACH OTHER. (KEEP AWAY FROM THE 1SG)

THE SOLDIER WHO CLEANED THESE TOILETS & URINALS HAD A "DICK JOB."

WARNING! DUE TO A LACK OF GOV'T FUNDS, THE ARMY ORDERED THINNER TOILET PAPER. FOLD TWICE OR MORE BEFORE USING.

I'LL SOON BE THROUGH WITH ALL THIS SHIT, AND THAT WILL BE THE END OF IT. I'LL GO HOME AND THERE I'LL STAY. REENLIST? NO F——N WAY!

*The Complete Ranger Digest: Vol. VII*

# HAVE YA EVER HEARD OF GRAF?

## (YOU WILL IF YA EVER PCS TO EUROPE)

## A FEW MORE LAND NAV LAYOUTS

A few times I was tasked to set up a land navigation course for the Bosnians. But due to IFOR/SFOR restrictions, the area that we were given permission to lay it out in was a lot smaller than what we had wanted it to be. Unfortunately, there wasn't very much we could do about it.

Operating under these strict rules and restrictions wasn't my only problem. Because of who I was (Ranger Rick) and what I've accomplished, writing and publishing these *Ranger Digest* handbooks. There were some MPRI senior trainers, (retired Ltc."Square Frank", Col."Alzheimer Joe", & Col."Mike Slush") and a few SFers who were always anxiously looking for me to f--- up doing something. Which I never (or rarely) ever did.

Why were they so anxious to see me screw up? Oh, I guess like all envious and jealous leaders, so they could tell their fellow officers and SF buddies that good 'ol Ranger Rick f———ed up. or so they could say,"Hey Ranger Rick, that's not the f———ing way we did it back at Bragg or in SF."

But thanks to these fellas, because I knew they were always watching me like a hungry hawk out searching for food. I would always triple check all my work to deny them the pleasure of seeing me make any mistakes. And not only did the Bosnians learn a lot of training tips fit tricks from me, but so did these senior trainers and SF dudes. Which I know they'll never admit to anyone that they have because their too proud and think they're better than everyone else.

Remember this Ranger Digest Readers; It doesn't matter if you're an Officer, NCO, Enlisted, or if you wear a Special Forces, Ranger, Airborne, or "no tab" at all. It's how you use your brain and impress the troops that makes you stand out among the rest.

And when you do, there will always be some leaders who will try to put you down or make you look bad so they'll look better. F——— these guys, don't worry about 'em, the troops you train and lead know who are the good and bad leaders and instructors. Believe me, they know.

OK, now lets get down to business.

In my previous Ranger Digests I showed you a few ways in how you can set up a land navigation course for your unit. Well, here's a few more layouts that you might want to try that worked pretty well for me in Bosnia. But before you decide which one will work best for you and your unit, determine....

A. How much time will be available to set it up. B. What kind of terrain is in the general area of operation. C. How many key terrain features can be used as landmarks. D. How many hazardous obstacles are in or around the area. E. How many roads, trails, creeks, and or rivers are in the area and can be used as recognizable boundary limitations.

These are just a few and not all of the things that you should take into consideration when deciding which layout to set up for your unit. Whichever one you choose, always keep in mind SAFETY. Never send a soldier out by himself, always pair 'em up in teams of two or more individuals.

Now this first land nav layout I call the HALF MOON. Because there is only one starting point, and every time a point is located, the individuals or teams must return back to the same location before proceeding to another point. If the points are placed out at equal distances, the points can double as boundaries. This layout must be walked, measured, and verified for accuracy no less than three (3) times.

How the points are laid out:

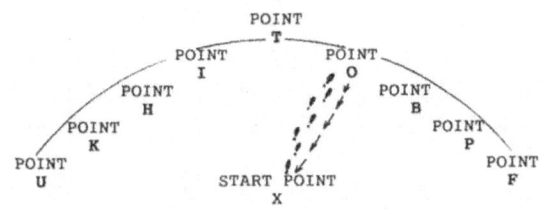

| LANE GRADERS ANSWER & SCORE SHEET | | | | | | |
|---|---|---|---|---|---|---|
| LANE | COMPASS AZ | METERS | L - | POINT | - R | GRID COORD |
| #1 | 56° | 1000 | - | U | K | 12345678 |
| #2 | 64° | 1000 | U | K | H | 12345678 |
| #3 | 70° | 1000 | K | H | I | 12345678 |
| #4 | 76° | 1000 | H | I | T | 12345678 |
| #5 | 80° | 1000 | I | T | O | 12345678 |
| #6 | 84° | 1000 | T | O | B | 12345678 |
| #7 | 89° | 1000 | O | B | P | 12345678 |
| #8 | 92° | 1000 | B | P | F | 12345678 |
| #9 | 96° | 1000 | P | F | - | 12345678 |

↑ Left of   ↑ Correct Point   ↑ Right of

This next one I call the PARALLEL or RECTANGLE LAYOUT. And provided these points are placed out at equal distances, again, they can also double as boundaries. And just like the other layout, it must be walked, measured, and verified for accuracy no less than three (3) times.

How the points are laid out:

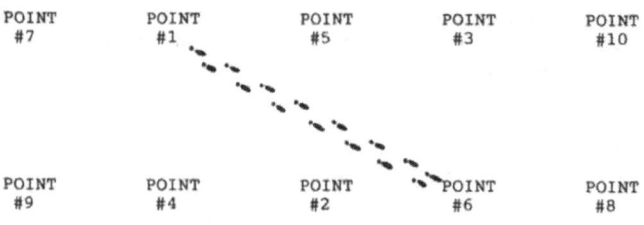

| LANE GRADERS ANSWER & SCORE SHEET | | | | | |
|---|---|---|---|---|---|
| LANE/START PT | COMPASS AZ | L | -POINT- | R | GRID COORD |
| # 1 | 191° | 8 | 6 | 2 | 12345678 |
| # 2 | 27° | 1 | 5 | 3 | 12345678 |
| # 3 | 200° | 2 | 4 | 9 | 12345678 |
| # 4 | 20° | 5 | 3 | 10 | 12345678 |
| # 5 | 205° | 6 | 2 | 4 | 12345678 |
| # 6 | 16° | 7 | 1 | 5 | 12345678 |
| # 7 | 208° | - | 8 | 6 | 12345678 |
| # 8 | 23° | - | 7 | 1 | 12345678 |
| # 9 | 198° | 3 | 10 | - | 12345678 |
| #10 | 25° | 4 | 9 | - | 12345678 |

Left of ↑ ↑ Right of
Correct Point

Now this last layout is one that is very popular with Special Forces and Ranger units. Though it's much easier to set up than all the others, the novice beginner and inexperienced will find it a bit more difficult to negotiate. Especially if they have never (or rarely) used a map and compass before.

The teams are given all coordinates to all the points and then instructed to plot and locate as many of them as possible within a certain time period. It doesn't matter what sequence they find them in, just as long as they record what's written down on the point for their assigned lane.

Unlike the others, this layout DOES NOT get walked nor measured for accuracy. But it does have to be triple checked to insure the grid coordinates are correctly recorded and the points are placed in the right locations.

*The Complete Ranger Digest: Vol. VII*

How the points are layed out;

```
START              POINT #1              POINT #6
POINT              LANE 1-A              LANE 1-FF
  X                LANE 2-F              LANE 2-GG
                   LANE 3-C              LANE 3-HH
                   LANE 4-D              LANE 4-II
                   LANE 5-B              LANE 5-JJ
                   LANE 6-E              LANE 6-KK

POINT #4                         POINT #2
LANE 1-S                         LANE 1-G
LANE 2-T                         LANE 2-I
LANE 3-U                         LANE 3-J
LANE 4-V                         LANE 4-K
LANE 5-W                         LANE 5-H
LANE 6-X                         LANE 6-L

           POINT #5              POINT #3
           LANE 1-Y              LANE 1-M
           LANE 2-Z              LANE 2-N
           LANE 3-AA             LANE 3-R
           LANE 4-BB             LANE 4-O
           LANE 5-CC             LANE 5-P
           LANE 6-KK             LANE 6-Q
```

```
|                    LANE GRADERS ANSWER & SCORE SHEET                    |
|         LANE #1        |       LANE #2         |       LANE #3          |
| GRID COORD/PT/?        | GRID COORD/PT/?       | GRID COORD/PT/?        |
|  12345678  -1-A        |  12345678  -1-F       |  12345678  -1-C        |
|  12345678  -2-G        |  12345678  -2-I       |  12345678  -2-J        |
|  12345678  -3-M        |  12345678  -3-N       |  12345678  -3-R        |
|  12345678  -4-S        |  12345678  -4-T       |  12345678  -4-U        |
|  12345678  -5-Y        |  12345678  -5-Z       |  12345678  -5-AA       |
|  12345678  -6-FF       |  12345678  -6-GG      |  12345678  -6-HH       |
|         LANE #4        |       LANE #5         |       LANE #6          |
| GRID COORD/PT/?        | GRID COORD/PT/?       | GRID COORD/PT/?        |
|  12345678  -1-D        |  12345678  -1-B       |  12345678  -1-E        |
|  12345678  -2-K        |  12345678  -2-H       |  12345678  -2-L        |
|  12345678  -3-O        |  12345678  -3-P       |  12345678  -3-Q        |
|  12345678  -4-V        |  12345678  -4-W       |  12345678  -4-X        |
|  12345678  -5-BB       |  12345678  -5-CC      |  12345678  -5-EE       |
|  12345678  -6-II       |  12345678  -6-JJ      |  12345678  -6-KK       |
```

So what do you use for the points? Well, preferably a tall wooden brightly painted 4x4 pole. But if you don't have or can't acquire them, try using empty plastic water bottles or 'plastic* throw-away picnic plates.

So what do you attach or write down on these plastic water bottles or picnic plates? Well, you gotta somehow let the navigators know that they've found the right point. And if it's not the right point, then they should know where they're located so they can get right back on course.

Rick F. Tscherne

 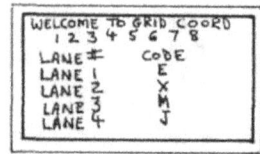

So for starters, write down on the point "WELCOME TO GRID COORDINATE 12345678." This will let them know that they have either found the right point or so they can " self-correct" themselves to get back on course.

If the point is going to be used for more than one lane, then write down all the lanes that will be passing through it and a code word or letter beside it that lane. This will prove that they have successfully made it to the point and reduce the chances of anyone cheating for another team.

If you use empty plastic water bottles for the points, place some bright colored paper inside of it before hanging them up in a tree so they can be spotted or seen much easier.

Before I went to Bosnia I purchased some magic markers, a bag of 550 paracord, and several packages of "yellow" plastic picnic plates. So if I was ever tasked to set up one of these quickie land navigation courses, I only had to pull out my things and begin setting it up, presto!

NOTE: Before attaching any tie-down string to the plastic picnic plates. Take a nail and hold it over a fire and then melt a small hole through the plastic plate to prevent the string from tearing apart the plate during windy days.

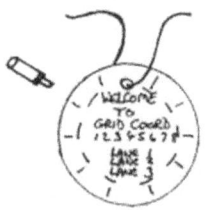

# SMART INVESTMENT STRATEGIES FOR THOSE WHO WANT TO RETIRE EARLY

Listen up guys & gals, if I told you once, I've told you about a half dozen times. (Or at least in every other book of mine..)

*The Complete Ranger Digest: Vol. VII*

That if you don't start saving and putting away some of your money, you ain't never going to be able to retire .Nor are you ever going to have any money for emergencies, retired or not. You think I'm bull shitting?

Well, let me tell you something Mr./Ms."Know-it-all," I know of many "retired" E-7s, E-8s, and E-9s who still have to work to make ends meet to survive. Why? Because they didn't save and invest for their retirement, that's why. They thought (like you) that they would worry about it when the time comes. Well guess what? Surprise, surprise, surprise. The time came sooner than expected and now they gotta work the rest of their lives to survive, financially that is...

Well, I hate to keep sounding like a broken record, so this will be the last time that I'm gonna tell ya all. Either you get off your butts and start saving some of your money each month by investing in one or more of these mutual funds. Or you ain't never gonna see a day of retirement. And the longer you keep delaying it, the less you'll have, it's that simple.

NOTE- INFO BELOW AS OF JANUARY 1998

## BEST RETIREMENT FUNDS

| | RISK | | 3 YR AVG RTN % GAIN (LOSS) | | | | 5 YR AVG RTN EXPENSES | | PORTFOLIO COMPOSITION | | | |
|---|---|---|---|---|---|---|---|---|---|---|---|---|
| | LOW   HIGH  1 2 3 4 5 6 7 8 9 10 11 12 | STYLE | RISK LEVEL | FOUR WEEKS | ONE YEAR | THREE YEARS (ANN. AVG.) | FIVE YEARS (ANN. AVG.) | FIVE-YEAR TOTAL COST | MINIMUM INITIAL PURCHASE | % STOCKS | % BONDS | % CASH | TELEPHONE (800) |
| **LARGE-COMPANY STOCK FUNDS** | | | | | | | | | | | | | |
| Selected American | LG/VAL | 8 | 3.0 | 38.5 | 34.9 | 19.7 | $55 | $1,000 | 95 | 0 | 4 | 243-1575 |
| Clipper | LG/VAL | 7 | 2.2 | 29.4 | 31.0 | 19.8 | 60 | 5,000 | 61 | 11 | 28 | 776-5033 |
| Vanguard/Windsor II | LG/VAL | 7 | 4.0 | 32.4 | 30.9 | 20.3 | 22 | 3,000 | 93 | 0 | 7 | 851-4999 |
| Westwood Equity Return | LG/BL | 7 | 1.2 | 27.7 | 30.7 | 21.6 | 82 | 1,000 | 90 | 9 | 1 | 937-8966 |
| Vanguard Index 500 | LG/BL | 8 | 2.8 | 33.1 | 30.3 | 19.4 | 61 | 3,000 | 98 | 0 | 2 | 851-4999 |
| T. Rowe Price Blue Chip Growth | LG/BL | 7 | 1.8 | 26.5 | 30.2 | N.A. | 69¹ | 2,500 | 91 | 0 | 9 | 638-5660 |
| Vanguard/Primecap | LG/VAL | 10 | -(2.1) | 35.0 | 29.9 | 23.2 | 32 | 3,000 | 90 | 0 | 10 | 851-4999 |
| Baron Asset | MED/BL | 10 | 1.5 | 33.0 | 29.8 | 23.5 | 71 | 2,000 | 93 | 0 | 6 | 992-2766 |
| T. Rowe Price Dividend Growth | LG/VAL | 5 | 3.3 | 30.7 | 28.6 | N.A. | 61¹ | 2,500 | 83 | 3 | 10 | 638-5660 |
| Harbor Capital Appreciation | LG/GRO | 11 | -(0.1) | 27.3 | 28.4 | 19.7 | 42 | 2,000 | 99 | 0 | 1 | 422-1050 |
| Strong Schafer Value | MED/VAL | 7 | 0.8 | 31.2 | 28.2 | 20.4 | 69 | 2,500 | 99 | 0 | 1 | 368-1030 |
| Oppenheimer Quest Opportunity A | LG/VAL | 8 | 2.3 | 21.0 | 27.6 | 18.8 | 148 | 1,000² | 82 | 0 | 18 | 525-7048 |
| T. Rowe Price Mid-Cap Growth | MED/BL | 9 | 1.0 | 15.8 | 27.6 | 20.7 | 57 | 2,500 | 90 | 0 | 10 | 638-5660 |
| Dodge & Cox Stock | LG/VAL | 8 | 1.2 | 27.9 | 27.2 | 20.7 | 33 | 2,500 | 89 | 0 | 10 | 621-3979 |
| Babson Value | LG/VAL | 8 | 1.9 | 27.9 | 27.1 | 20.8 | 53 | 1,000 | 95 | 0 | 4 | 422-2766 |
| Columbia Growth | LG/BL | 9 | 0.8 | 25.4 | 26.7 | 17.8 | 40 | 1,000 | 98 | 0 | 2 | 547-1707 |
| Brandywine | MED/BL | 11 | -(5.8) | 14.1 | 25.6 | 19.6 | 58 | 25,000 | 97 | 0 | 3 | 656-3017 |
| Third Avenue Value | SM/VAL | 7 | -(1.1) | 23.0 | 25.0 | 19.6 | 67 | 1,000 | 55 | 6 | 38 | 443-1021 |
| Yacktman | MED/VAL | 7 | 0.6 | 17.3 | 24.4 | 14.1 | 55 | 2,500 | 84 | 0 | 16 | 525-8258 |
| Vanguard Wellington | LG/VAL | 7 | 2.5 | 23.2 | 23.5 | 16.2 | 17 | 3,000 | 65 | 34 | 1 | 851-4999 |
| Robertson Stephens Value Plus Growth | LG/BL | 12 | -(5.2) | 11.6 | 23.2 | 22.5 | 79 | 5,000 | 100 | 0 | 0 | 766-3863 |
| Mutual Beacon Z | MED/VAL | 7 | 1.6 | 24.1 | 23.0 | 19.5 | 42 | 1,000⁵ | 79 | 4 | 17 | 368-1030 |
| Cohen & Steers Realty Shares | MED/VAL | 8 | 2.4 | 24.5 | 22.7 | 18.9 | 60 | 10,000 | 98 | 0 | 2 | 437-9912 |
| Neuberger & Berman Guardian | LG/VAL | 10 | 1.0 | 17.7 | 22.1 | 15.9 | 46 | 1,000 | 89 | 2 | 8 | 877-9700 |
| Strong Opportunity | LG/VAL | 8 | -(1.3) | 22.3 | 22.1 | 18.0 | 72 | 1,000 | 91 | 0 | 9 | 368-1030 |
| Invesco Total Return | LG/VAL | 6 | 2.0 | 24.4 | 21.6 | 15.6 | 53 | 1,000 | 64 | 29 | 7 | 525-8085 |
| Strong American Utilities | LG/VAL | 7 | 4.1 | 27.8 | 21.3 | N.A. | 63¹ | 1,000 | 96 | 0 | 4 | 368-1030 |
| Dodge & Cox Balanced | LG/VAL | 6 | 1.2 | 20.8 | 20.7 | 15.9 | 31 | 2,500 | 57 | 38 | 4 | 621-3979 |

NOTICE: PAST RESULTS DOES NOT GUARANTEE FUTURE RESULTS.
(BUT IT BEATS THE HELL OUT OF SAVINGS BONDS,SAVINGS ACCT AND BANK CDs).

167

Rick F. Tscherne

# HAVE YA EVER HEARD OF HAPPY SOCKS?

This next topic is a real "touchy one," it's a subject that no one in the military dares to openly talk about. But anyone who has ever been on a long deployment or a hardship tour. Well, they have probably done it at least one time or more. But, no one in their right mind is ever going to openly admit that they have, especially to their buddies. And that's OK, because it's a private matter, and what you do in your spare time is your business and no one elses.

Now originally, I wasn't going talk about this subject. But after receiving quite a few anonymous letters, well, I thought, "Hey what the hell, why not?" It's better to read about it in the *Ranger Digest* than in Reader's Digest, right?

Instead of going into detail in how to do it, (I'm no expert, ya know..), you'll have to use your own imagination and read between the lines. And if you don't know what this subject is all about by now, or if ya do and you're disgusted, then I'd suggest you just skip the rest of this page.

Now the term "happy socks" I learned from a couple of SF MPRI instructors in Bosnia. But the technique is nothing new, hell, convicts do this all the time in prison. First you take a pair of socks and place one inside the other. Then you start rolling the open end until it becomes smaller and tighter and fits snugly over your.....well, you know what.

Once it's on, crawl inside your fart sack, lay on your stomach, think of some beautiful honey, wait until it gets hard, raise your butt up a little and start rocking. As soon as you relieve yourself, remove it immediately before it leaks through the socks and place it inside your laundry bag. (WARNING: To avoid being detected, practice late at night.)

In all my years as a squad leader, platoon sergeant, and drill instructor, I'.ve never seen a jar nor tube of vaseline inside anyone's wall locker during a room inspection. If I had, I would have thought and said what anyone else would have said, "Hey dickhead, what do ya use this stuff for???"

During the Persian Gulf War I once received a letter from a soldier who said he uses his liquid hair shampoo or bar of soap in a private toilet or shower stall to relieve himself of his "sexual fantasies." (OK, I guess that's a technique..)

Listen fellas, let's clear the air once and for all. If ya wanna do it - fine, do it. If ya don't - then don't it. It's your business. But if you think about it, if you can't wait until you get back home to be with your honey. Well, maybe this ain't such a bad idea after all. Check out the next page....

HEY, CHECK OUT THIS RECENT STUDY...! GUESS THIS MEANS I'M GOING TO LIVE FOREVER! (NOT)

WORLD NEWS — THE STARS AND STRIPES — Page 9

# Sexually active men live longer, study says

### But findings dismissed in commentary calling proof flawed, doubtful

LONDON (AP) – Men who have frequent sex are less likely to die young, according to a study published – along with some skeptical commentary – in the latest issue of the British Medical Journal.

During a decade of follow-up, the researchers found that men who had sex less than once a month had twice the death rate of those who had sex at least twice a week.

A commentary also published in the journal dismissed the findings as flawed and possibly a matter of chance.

Such studies are not intended to prove cause and effect. However, George Davey Smith of the University of Bristol and Stephen Frankel of Queen's University, Belfast, argued that the evidence appeared strong.

"The association between frequency of orgasm and mortality in the present study is at least – if not more – convincing than many of the associations reported in other studies and deserves further investigation to the same extent," they wrote.

The authors studied 918 men drawn from a pool of 2,512 men aged 45-59 and living in or near Caerphilly, Wales.

They were divided into three groups: those who had sex twice or more a week, an intermediate group and those who reported having sex less than monthly.

A decade later, researchers found that the death rate among the least sexually active men was twice as high as that of the most active. The death rate in the intermediate group was 1.6 times greater than for the active group.

The death rate for coronary heart disease was 2.2 times greater for the least active compared with the most active.

The authors said they tried to account for differences in age, social class, smoking and general health.

The study was dismissed in a separate commentary by Matthew Hotopf and Simon Wessely of the Department of Psychological Medicine at King's College, London. They said the researchers failed to consider, for instance, whether less frequent sexual activity was caused by depression that in turn was a symptom of disease.

"Although the authors claim to have accounted for coronary heart disease by using baseline reporting of chest pain, this is a blunt instrument," Hotopf and Wessely wrote. "It would not take many cases of early undetected heart disease to give the results reported here."

They said they expected their doubts to be ignored.

"As you read this, we confidently predict that 'Sex makes you live longer' will occupy more newspaper inches over the holiday period than the queen's speech," they wrote.

"The public will hear what they want to hear, and they will be deaf to the problems of bias, confounding, reverse causality or chance," they wrote.

---

## Masturbation no sin – just a sane answer to sex drive

DEAR ANN LANDERS:
Please rerun the column about masturbation as a safe alternative to unprotected sex. It is very important.
— NO NAME, NO STATE

DEAR NO NAME:
The column you are referring to ran on Oct. 24, 1993. It was an edited version of an article by Dr. Steven Sainsbury of San Luis Obispo, Calif., which appeared in the Los Angeles Times. Here it is again:

"My 13-year-old patient lay quietly on the gurney as I asked the standard questions 'Are you sexually active?' She said, 'Yes.' Next question: 'Are you using any form of birth control?' The response was 'No.' Next question: 'What about condoms?' Response, 'No.'

"Her answers didn't surprise me. She had a rip-roaring case of gonorrhea. It could easily have been AIDS. I treat teen-agers like this one every day. Most are sexually active. Condoms are used rarely and sporadically.

"Yet in the midst of the AIDS epidemic, I continue to hear condoms touted as the solution to HIV transmission. Condoms are being passed out in high schools, sold in college restroom dispensers and promoted on TV. The message is: Condoms equal safe sex.

"As a physician, I wish it were true. It isn't. It is a dangerous lie.

"Fact No. 1: In 1989, a survey among college women, a group we presume to be well-informed on the risks of herpes, genital warts, cervical cancer and AIDS, showed that only 41 percent insisted on condom use. If educated women can't be persuaded to use condoms, how can we expect teen-agers to do so?

"Fact No. 2: Condoms fail frequently due to improper storage, handling and usage. The breakage rate during vaginal intercourse is 14 percent. For a person who averages sex three times a week, a 14 percent breakage rate equates to a failure nearly every two weeks.

"For condoms to be the answer to AIDS, they must be used every time, and they can never break or leak. So what's the answer? The only answer is no sex until one is ready to commit to a monogamous relationship. The key words are abstinence and monogamy.

"I can hear the masses. Condom fans murmur words like unrealistic, naive and old-fashioned. Well, perhaps

Ann LANDERS

what is needed to stem the tide of AIDS and unwanted pregnancies is a return to those old-fashioned concepts.

"To quote Dr. Robert C. Noble, a University of Kentucky infectious disease expert, 'We should stop kidding ourselves. There is no safe sex. If the condom breaks, you may die.'"

DEAR READERS:
Powerful piece, isn't it? Well, now I am going to suggest a far more realistic solution than abstinence.

The sex drive is the strongest human drive after hunger. It is nature's way of perpetuating the human race. Males reach their sexual peak as early as 17. There must be an outlet. I am recommending self-gratification or mutual masturbation, whatever it takes to release the sexual energy. This is a sane and safe alternative to intercourse, not only for teen-agers but also for older men and women who have lost their partners.

I do not want to hear from clergymen telling me it's a sin. The sin is making people feel guilty about responding to this basic, fundamental human drive. I love my readers, and my mission is to be of service.

*Rick F. Tscherne*

# ATTENTION COFFEE LOVERS!

Hey soldier! Are ya tired of drinking that shitty instant coffee that comes with your MRE? Huh? Are ya? Huh? Then why don't you carry your own favorite brand of coffee to the field?

You don't need to carry a coffee pot, you just need a container for your coffee and a product called "Coffee Quick."

What is it? It's a mesh plastic basket that holds several spoons of coffee. Your choice, regular or drip,

How do you use it? Oh, it's pretty simple.

Step #1: Heat up a canteen cup of hot water.
Step #2: Fill the "coffee quick" basket full of coffee grinds
Step #3: Place the "coffee quick" basket inside the canteen cup and stir, stir.
Step #4: When it reaches the color blend that you want, remove the "coffee quick" basket from the canteen cup and taste it.

NOTE: If the coffee is a bit too weak for your taste, just add another scoop of coffee to the basket and stir. If it's too strong, then reduce the amount of coffee the next time around.

Where can you order this "Coffee Quick" basket? From: Miles Kimball, 41 West Eighth Avenue, Oshkosh, Wisconsin 54906.
Cost: $2.98 ea. Item#: 576884
(NOTE - Item is currently not listed on their website..?)

*The Complete Ranger Digest: Vol. VII*

# A FEW MORE HAND & ARM SIGNALS

*Submitted By: Sgt. Stephen M. Mich (USAF)*

Cover Me - I'm Fucked!

Fuck You - I'm Covered

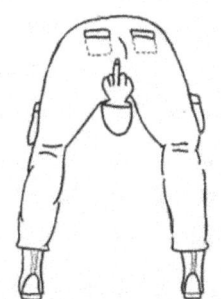

Fuck You, Asshole
(Submitted By Ranger Rick)

Rick F. Tscherne

# WELL EXCUSE ME LADIES..

Well, I wasn't sure if I was going to offend any Muslim soldiers or not, but I went ahead and did it anyway. And if you know anything about the Muslim religion, you'll understand why some may find this next training tip offensive.

Now some of you female soldiers will probably not like this next tip, and I'm sure I'll probably hear from you too. But hey, it turned out the Bosnians liked it and they also learned how to read a map from it too.

Now I've always tried my best to make all my classes as interesting and enjoyable as possible, but sometimes it's been kinda hard to do, ya know what I mean? Especially if you gotta give a class on how to "identify terrain features."

So, while looking through the ol' US Army Common Task Test (CTT) manual, I saw in it how you can use your fist in explaining how and what the different types of terrain features look like. But, I found this training technique to be a bit BOOOOOOORING!

So I decided to come up with my own teaching technique in how they can identify terrain features much more easily. Look at the next page > > > and tell me what ya honestly think?

Did I do good or what? I made this into a study guide and a viewgraph slide, and the Bosnians LOVED IT!

PS: I apologize to any female soldiers who finds this teaching technique offensive, as it was strictly made for training purposes only. And for those of you who don't accept my apology, well then... tough @*#%!

Note: Match these to the drawing on the next page.

| GRID | HILLTOP | SADDLE | RIDGE | VALLEY | SPUR | DRAW | CLIFF | DEPRES. |
|---|---|---|---|---|---|---|---|---|
| 58952965 | | | | | | | | |
| 58853025 | | | | | | | | |
| 58802560 | | | | | | | | |
| 58752760 | | | | | | | | |
| 58802365 | | | | | | | | |
| 58202980 | | | | | | | | |
| 58802980 | | | | | | | | |
| 59602760 | | | | | | | | |

*The Complete Ranger Digest: Vol. VII*

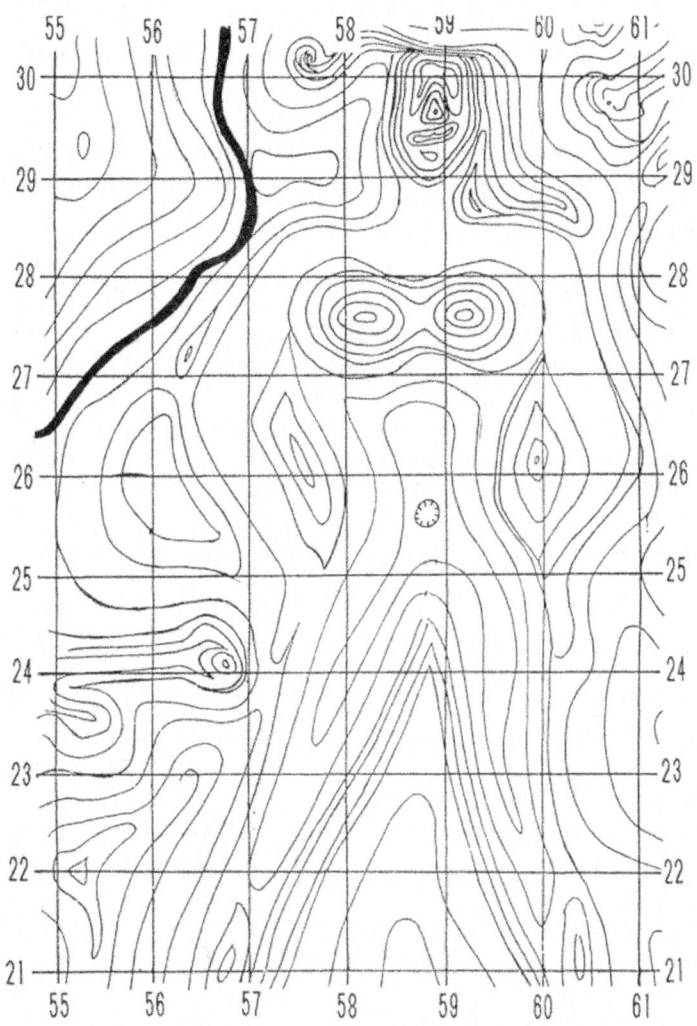

*Rick F. Tscherne*

# MAKING YOUR OWN SIGNAL MIRROR

Oh man, I wish I would have thought of this years ago, I could have saved me some $$$$$ in replacing all those military signal mirrors that I either broke or lost while in the Army. Well, it might be too late for me, but not for you.

In your local PX near the stationery section they sell a very inexpensive wall locker mirror made of 'plastic' Well, if you buy one of these suckers, you can easily cut it into 4 x rectangle shape mirrors with a straight-edge razor blade. In fact, when you're done cutting it up, they'll look a lot like mini military style signal mirrors.

And if you drill a hole near one of the corners, you can then attach a lanyard (dummy cord) to it so you won't lose it. Now if you wanna make a cover for it so it won't get scratched up. Just wrap a layer of toilet paper around the mirror and then lightly wrap a layer of 100 mph tape around the mirror and toilet paper.

The purpose of putting toilet paper around the mirror is to prevent the 100 mph tape from sticking to the mirror thus forming a homemade cover for it. Now all ya gotta do is trim up the tape and remove the mirror.

If you're a fire team or squad leader, you can buy a couple of these plastic wall locker mirrors and make a signal mirror for each of your unit members.

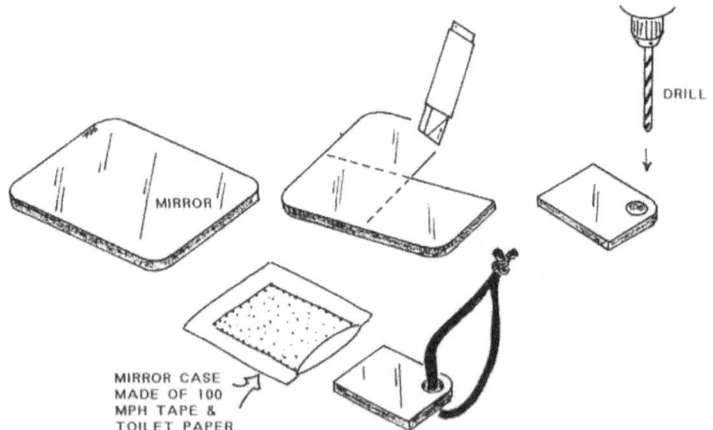

Or, if you got the time, and patience, you can cut or saw this plastic wall locker mirror into six or more round mini mirrors and then glue or silicone it to the inside portion of a military lensatic compass (see drawing). Of course this will take a bit more work and patience to do, but then you'll be able to use it either as a signal mirror or for putting on camouflage paint.

When cutting or sawing this plastic mirror, do it slowly and take your time or you'll mess it up. Do all your cutting on the mirror side and NOT on the reverse side, then sand the edges smooth with a piece of sand paper.

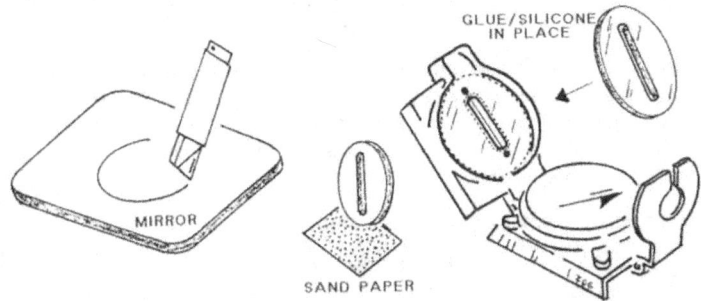

Here's another idea for a compact signal mirror that came from SGT FRANK GILLILAND. He cut out a piece of plastic mirror the same size and shape as his ID Tags, placed a rubber ID tag silencer around it and wears it along with his dog tags. Not only can he use it as an emergency signaling mirror, but for putting on camouflage paint too. Neat, huh?

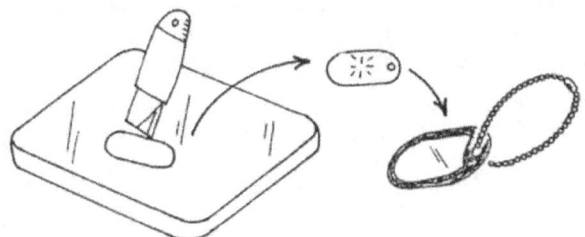

## WATER SEAL YOUR BOOTS, HAT...

Before I went off to Bosnia, I bought me an expensive pair of "Matterhorn Boots." Not because they looked pretty, but because I've got delicate feet and I wanted to make sure my little tootsies stayed nice & warm during the Balkan winter.

And as I was walking out of the AAFES Military Clothing Sales Store here in Vicenza, Italy. I saw a container of *SNO-SEAL* and grabbed one just in case I needed it in Bosnia.

Now to be honest with ya, I've never used this stuff before, nor has anyone ever told me if this stuff really works or not. So I said to myself,"Hey, what the hell, I'll give it a try."

Well, to be honest with ya, I forgot I had it. And it wasn't until mid winter while standing in a large puddle of Bosnian slush that I remembered I bought it. And even though my boots were suppose to be waterproof to certain degree, some of that ice cold slush leaked in.

So after I finished drying out my boots and removing as much of the shoe polish as possible with some gasoline. I then followed the instructions on the label and smeared some on and then used a hot hair blowdryer to melted it into the pores of the leather. And guess what? Yep, this sh— really does work. Not bad, not bad at all.

Now the only drawbacks to using this stuff is...

(1) Because it's something like a Vaseline or a soft wax, if you happen to walk through some dry or wet dirt or mud. It'll stick to your boots like glue and be a pain-in-the-ass to remove.

So what can you do about it? Well, if you're in the field, hold the boots over a small fire to loosen up the Sno-Seal and wipe off the dirt & mud with some MRE toilet tissue.

Now if you're back in the rear, heat'em up with a hot hair blowdryer and then wipe'em off with an old rag or piece of paper towel. If you should remove too much of the Sno-Seal, you'll have to reapply some more if you want them to stay snowproof and waterproof.

Or another method that I discovered that really works well. Is to place the boots in a bathtub or kitchen sink and blast'em clean with some very, very hot water from a shower or sink "spray hose."

Whichever method you decide to use, brush'em with a shoe brush afterwards to even out the coat of Sno-Seal that's left on the boots. And if necessary, yes, reapply some more.

(2) When applying sno-seal to your boots, no matter how hard you try, you're gonna get some on your hands & clothes.

The Solution? Put it on with a small paint brush and wear a pair of throw-away medical gloves to protect your hands. If ya wanna carry and use some of it in the field, I'd highly recommend you place some of it inside a plastic zip-lock sandwich bag. Then you'll be able to store and place it on your boots much more easier without making a mess.

(3) When applying sno-seal to boots, it's definitely easier to smear and smooth it on with a brush and then use a hot hair blowdryer to melt it into the pores of the leather. But what if you're in the field?

The Solution? Though you can smear it on with a tissue and leave it on like so, it repels water much better when it's melted into leather. But if you're in the field, the easiest way to melt it into the pores of the leather is by holding it over a small fire, a candle, or over a hot exhaust pipe of a vehicle or track. Once you see it starting to melt, smear and wipe away any excess build-up.

HOT IDEA! Hey guys & gals, wanna make your black military gloves waterproof? Just place a small amount of sno-Seal on the gloves and rub it in evenly and then blast'em with a hot hair blowdryer.

How about your BDU field cap, wanna make it waterproof too? Hey, no problem! Again, just place a small amount of Sno-Seal onto the outside portion of the cap, rub and smear it in evenly and hold it over a small fire to melt it into the pores of the fabric.

Now your hat and head should stay waterproof for quite awhile or until the Sno-Seal wears out or until you decide to wash it out in a washing machine. Smart, huh? But don't add too much Sno-Seal to your hat or gloves or you'll pick up a lot of dirt and mud. (Note: Works best in a cold weather winter environments.)

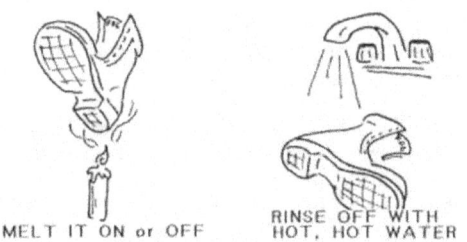

MELT IT ON or OFF    RINSE OFF WITH HOT, HOT WATER

## A CANTEEN CUP COVER

While out in the field, how many times have you used your canteen cup to make or to get yourself a cup of coffee from the chow line? A whole bunch of times, right?

OK, now how many times has a buddy asked you to bring him back a canteen cup of coffee too? A bunch, right?

OK, now how many times have you tried to walk back to your tent or vehicle without spilling it? A whole bunch of times, right? Did you spill any? I'll bet ya have. Did ya cuss and scream when ya spilled some? I'll bet ya did. Was the coffee already cold before you made it back? I'll bet it was.

Well, this use to happen to me too. So let me tell you what I did to solve the problem.

Because the canteen cup is an odd shape, it was kinda hard to find something solid to place over it. The only solution was to use something flexible or rubbery so it would "stretch" over it.

So then I began trying an assortment of different things to see which one would work best. First I tried using a plastic bag and wrapping some rubber bands around it, but it didn't keep the coffee in the cup very well. Bad idea.

Then I tried stretching a balloon over it, but before I could get one successfully on, I broke a dozen of them. Plus, I couldn't get the same balloon back on twice once I removed it. Again, another bad idea.

So then I found something more durable, a CONDOM, the non-lubricated type. And yes, it stretched quite easily over the top of the canteen cup. And yes, it kept the coffee inside the cup too. But, it looked like a condom on top of canteen cup.

So then I tied a knot in it as far down and as close to the rim of the canteen cup as possible and cut off the rest. And yea, it worked, and it was also re-useable too.

But one day I went to my unit aid station to see if I could get a package of disposable rubber surgical gloves. Well, they gave'em to me, and yes, it fit nice and snug over the top of the cup too. And yes, it kept the coffee inside the canteen cup too during movements. (Vehicle fit foot.)

So it wouldn't look like a surgical glove covering a canteen cup, I took a strand of nylon string from the inside portion of some 550 paracord and tied a knot as close to the rim of the cup as possible. Then I folded it over, tied another knot in it and cut off the rest of the glove. Did it work? Yep, it sure did, problem solved.

But then there was another problem, every time I wanted to take a sip from it, I had to remove the entire rubber cover. So then I began experimenting again.

I decided to puncture a small hole through the rubber covering to see if I could sip through it, but it didn't work very well. So then I bought me one of those plastic durable kiddie drinking straws, cut it down to size, made a "pin hole" in the rubber covering and stuck the straw through it. And yep, this worked a lot better.

It was not only spill proof, but the rubber covering also kept the coffee warmer too. I later tried using a rubber "dishwashing glove" because it was much more durable than the surgical glove. I tied a knot into it the same way as I did the surgical glove, cut off the rest and made a pin hole into it as well.

After I placed the rubber covering on top of the canteen cup, I then carefully cut a small short slit (with a razor blade) near (but not on) the edge of the cup. Then I tried drinking from it. And yea, it worked somewhat, but 1 preferred to drink from it through the pin hole and straw.

IMPORTANT: If the rubber dishwashing glove has a synthetic inner lining, just turn the glove inside-out so that the "rubber portion" covers and touches the coffee and NOT the synthetic lining.

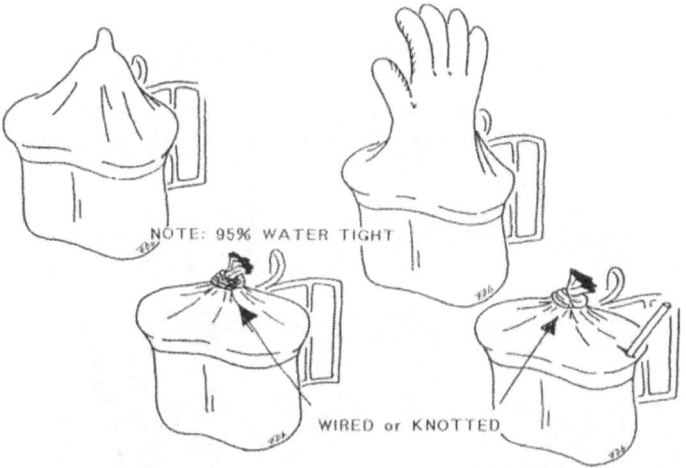

## GOT AN IDEA FOR A NEW PRODUCT?

Hey Guys & Gals! Check out this handy-dandy product, it' called an IMPS-NET. (Individual Multi-Purpose Survival Net)

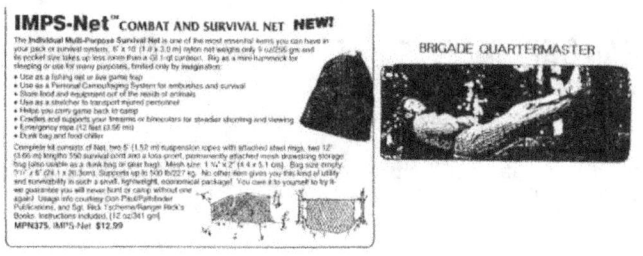

*Rick F. Tscherne*

If you'll remember in my last *Ranger Digest Handbook (VI)*, I showed you how to convert a mini hammock into a multi-purpose tactical net. Well, Brigade Quartermaster liked a few of my ideas and came up with their own product.

Which goes to prove, if you think you've got a good idea for a new product, don't sit and dwell on it. Either market it yourself or share it with someone who can do it for you. And if you haven't bought one of these IMPS-NET yet, then get off yo'ass and buy one, because it will no doubt pay for itself in only one day of use. (And it's pretty inexpensive too.)

Another idea that I sent to Brigade Quartermaster, is the *Ranger Rick Poncho Liner Conversion Kit*. Though this idea was originally sent to US Cavalry back in the late 1980s, they failed to listen to my advice on how it should be marketed.

I suggested they sell it as a do-it-yourself kit to reduce manufacturing costs and to keep the retail selling price down to a minimum. But instead, they choose to market the poncho liner complete and called it a Delta Liner. Which they didn't sell very many of and discontinued it a few months later.

Why? Well, primarily because of the price, who in the hell wants to pay 50 bucks for a $19.99 poncho liner with a couple of zippers attached to it? Obviously Brigade Quartermaster agreed with me and went along with my suggestion and so far they've been selling a whole bunch of these kits too.

# RANGER RICK SAYS, THINK ABOUT THIS...

To all you soldiers who see something wrong with your unit or with some of the leaders who run it and are afraid to speak up in letting your chain of command know about it. Remember what Thomas Jefferson once said, "BE REBELLIOUS!"

To all you soldiers who someday may find yourselves being deployed on a real world mission to a far away exotic place. Remember this: It's better to return home carrying the flag than to be carried home "underneath one" - THINK SAFETY!

There are two types of orders that you hope you will never hear from the mouth of your commander that mean you're in deep shit, "FIX BAYONETS" and "..AT ALL COSTS."

Married service members who are thinking of cheating on their spouses should consider the "financial consequences" if you get caught. Will your spouse DIVORCE YOU for the alimony or KILL YOU for the SGLI insurance money?

### THE TOP 10 SIMILARITIES BETWEEN A PENIS AND A COMPUTER.
As David Letterman would say, "Heeeeere we go..."

#10 Those who own one would be devastated if they ever broke it.

# 9 Those who don't own one think it must be pretty darn nice to have, but can't figure out why those who do own one make such a big darn fuss out of it.

# 8 Those who own one think that those who don't - are inferior.

# 7 Those who don't own one would sure like to get their hands on one and try it out.

# 6 Those who own one know that it's not only a lot of fun to use when it's up, but it's hard as hell to concentrate on doing anything else.

# 5 Those who own one know that if you're not very careful

with it and take some necessary precautions, it could easily spread a virus.

# 4 Those who own one and use it on a regular basis will eventually become more and more addicted to using it.

# 3 Those who own one not only like to tell others about how often they use it, but exaggerate and brag about what they can do with it.

# 2 In the past, it's only purpose was to transmit information from person-to-person, which some people today still believe that's it's only purpose. But those who own one know that it's not only a lot of fun to use, but using it to get to know other people too.

And the #1 similarity between a penis and a computer is....

Those who own one will agree that if you're not very careful with it, it can get you in a "whooooole lot of trouble."

*Rick F. Tscherne*

# PEPSI CAN LANTERN & STOVE

While working for MPRI in Bosnia I lived in a small town (Jelah) near Doboj that had "frequent" power outages. A few of my fellow instructors and I use to joke as to why these power outages occurred.

We figured the Serbs probably controlled the electricity and the Muslims controlled the water and they liked playing games with one another.

"Hey Muslims, want power for your TVs? Tough! Suffer!"And then the Muslims would not get their power turned back on until they retaliated by shutting off the Serb's water.

"Oh yea, Serbs, let's see how much shit your toilets can hold without water to flush 'em." (Of course, we were only joking about this....)

Anyway, one night while sitting next to a candle waiting for our power to come back on. I came up with an idea for an improvised lantern, I call it my "Pepsi Can lantern." It's pretty simple to make, all ya gotta do is find a couple of empty soda cans and....

a) Cut the two soda cans in half and throw away the top parts.
b) Take one of the soda can halves and puncture a hole into the top center part.
c) Same soda can, make a few slits along the sides with a
pair of scissors. (Note: This will enable the soda cans to be placed one-inside-the-other )
d) Same soda can, find a piece of narrow cloth and run it through the small hole.
e) Grab the other soda can half and fill it half full of "Kerosene."
f) Connect the two soda can halves together by placing one-inside-the-other and secure it with some tape.
g) Wait about 10 minutes to allow the cloth to become soaked with kerosene and then light it with a match.

WARNING - DANGER - WARNING - DANGER

1. Though you can use gasoline or diesel as an alternative
fuel, kerosene burns much slower and is a lot safer to use.
2. The shorter the wick, the smaller the flame and the less fuel it will consume and burn up.
3. Always place the soda can lantern in a safe place where it won't get accidentally knocked over by others or yourself.

NOTE: I dedicate this pepsi can lantern tip to my MPRI roommate, Dan Bateman. Who I enjoyed the hell out of annoying everyday with my daily inventions and innovations.

Here's another innovative idea for a lantern. Find an empty 50 Cal. brass shell casing, fill it with some fuel, place piece of cloth inside, bend the top part closed with a pair of pliers and... PRESTO! "A .50 Cal. Lantern."

## CAMOUFLAGING TIPS & TRICKS

When it came to teaching the Bosnians why it's important to know how to properly camouflage themselves, they weren't really interested in learning this skill. The only thing that they were interested in learning from us, was how to attack. Why? So they could learn how to kick ass and get back some of the land that they lost during the war.

Almost all the Bosnian units that I trained appeared to have plenty of ammo and weapons on hand, but they lacked the basic stuff like camouflage sticks, camouflage covers, camouflage bands, etc. Though I was able to do some scrounging and acquire some of these items from my American IFOR/SFOR buddies, I had to rely mostly on my Ranger Rick "ingenuity" in showing them how to improvise. For example;

How to make an elastic camouflage band out of a rubber "inner tube" tire by cutting it up and placing it over the edge of their helmet.

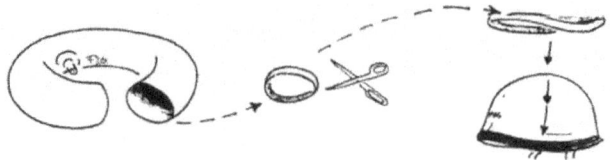

How to use a white bed sheet as a winter (white) camouflage coverall. Or, if a white bed sheet wasn't available, how to smear and spread white flour or foot powder on a (BDU) uniform to blend in with the winter (snow covered) terrain.

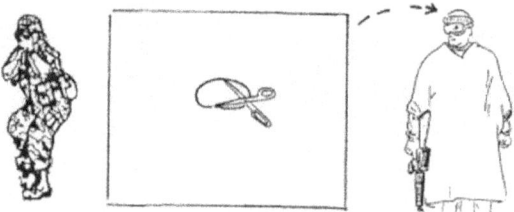

How to camouflage their face and hands with a "burnt" piece of cork or wood. And or how to mix baby oil and shoe polish together to make their own homemade camouflage "war paint".

These were just a few simple tips, tricks, and ideas that I gave them in how they could improvise the things they lacked.

Well, while preparing for my camouflage class one day, I came up with an idea in how to modify and camouflage a BDU patrol hat. All you need is some o.d. 550 paracord, a sewing needle, and then....

STEP 1: Take some 550 paracord and remove all the inner nylon strands.

STEP 2: Take one of the nylon strands and thread it through the needle.

STEP 3: Now you have a choice, you can either sew the empty 550 paracord shell entirely around the top & bottom of the BDU patrol hat. Or you can cut the 550 paracord shell into smaller pieces and then sew them onto the side of the hat. NOTE: TO prevent the ends of 550 paracord shell from unraveling, melt the ends with a match or zippo lighter.

To camouflage the hat, simply take some foliage and place it in between the 550 paracord shell and the hat. (Note: Due to this modification, the BDU patrol hat can never be worn in garrison again, only out in the field.)

## SLEEPING BAG TIPS & MAINTENANCE

If you go to the field as often as I did when I was on active duty. I'll bet many of you guys & gals don't even bother to air out your sleeping bags in the mornings. Do ya?

You probably just crawl in, snooze, and then crawl right back out and pack it up, right? Though this is NOT the proper way to take care of your sleeping bag, it's probably the most common way ya all use 'em. Right? I'll bet ya do? Don't lie.'

Now the correct way to use a sleeping bag, is to "shake it out" before you get into it at night. And in the mornings you should "air it out" before packing & rolling it up. Why?

Well, the purpose of shaking it out before you get into it, is to allow the feathers inside the bag to become "fluffy" so they'll retain your body heat better. And the purpose of shaking it out after use, is to help remove any odor, sweat, dirt, and or little critters that may have accumulated inside

Got a problem opening and closing your sleeping bag? If there appears to be nothing broken or blocking the teeth, then it may only need to be lubricated. No, NOT with oil, but with a bar of soap or some candle wax.

Take the bar of soap or candle and rub it directly onto the teeth and then move the slider up and down several times to see how it works. If it still continues to jam or catch onto something, inspect the slider and teeth more closely for defects. If the slider is defected, then more than likely the entire zipper (slider and teeth) will have to be replaced.

But if only a few teeth appear to be bent, grab a needle and bend them back into shape and position. If a few teeth are missing, move the slider slowly and carefully over this portion of the zipper when opening/closing it.(See drawing A)

If the missing or badly bent teeth are located at the bottom portion of the zipper, instead of replacing the entire zipper, you can sew this portion closed. (See drawing B)

If your slider keeps coming off at the TOP or BOTTOM of the zipper, then sew some thread (or wire) around the set teeth where it keeps, sliding off. (See. drawing C)

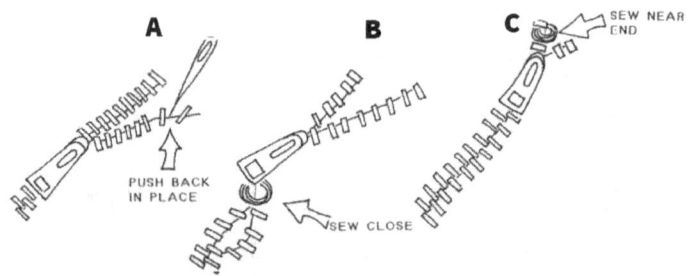

## CANTEEN CUP STOVE

There's a product on the market called a G.I. Canteen Cup Stove. If you turn it over on one side, it can be used as a stove for heating up a cup of coffee or some MRE meals. And if ya turn it over on the other end, it can be slid snugly over the bottom half of the canteen or cup so you can carry it easily inside your canteen pouch. Cool, huh?

Well, I'm not trying to steal any business away from the manufacturer who made these nor the stores who sell them. But if you want to save yourself a few bucks, here's how you can make one out of some chicken wire.

Get yourself some "square mesh" chicken wire and cut off a piece about 15 X 5 inches in length. The square mesh type, NOT the pentagon nor the hexagon type, just the square type.

Then wrap, bend, and mold it around the bottom half of the canteen cup (as shown in the drawing) until it conforms to the size and shape of the canteen cup.

Remove the chicken wire from the canteen cup, turn it over and then try it out by resting the canteen cup on top of it. To prevent or reduce the chances of the chicken wire from snagging and or grabbing hold of the inside portion of the canteen cover, trim the sharp edges with a pair of scissors or wire cutters.

To use, just place a heat tab or candle underneath it and place the canteen cup on top of it. And there you have it.

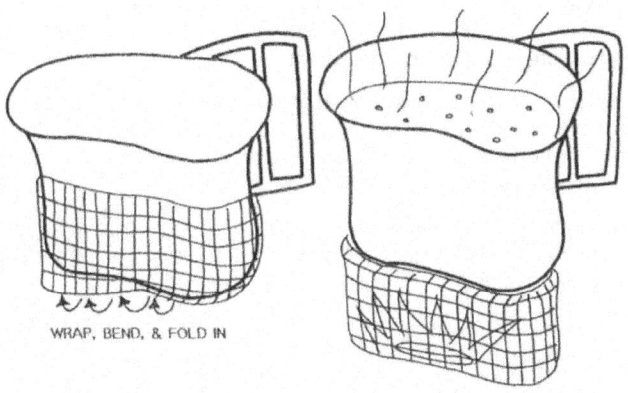

WRAP, BEND, & FOLD IN

## WARNING - DANGER - MINES!

I have no doubt that Bosnia is one of the most mine infested countries in the world. Why? Because I've been there and I've seen 'em. And believe me, I wasn't out souvenir hunting for 'em neither.

Now let me tell you what happen to one of our MPRI instructors in the field one time.....

Every time we wanted to go off post to train, we first had to get written permission from SFOR/IFOR Headquarters. We weren't allowed to train wherever we wanted to, only in designated and approved areas. Basically speaking, SFOR/IFOR didn't trust the Muslims, the Serbs, or us.

Now at one of these approved training areas, even though we were told it was safe and cleared of mines, one of our instructors almost stepped on one. He was walking with a Bosnian unit in the field when his translator spotted a partially exposed PMA2 anti-personnel mine in front of him.

Now common (military) sense would tell you that if you find one partially exposed land mine, you are probably in or near a mined area. Right? And wouldn't you think there are probably a few more of these critters laying around? At least I hoped an experienced leader would, especially a retired US Army Viet Nam vet.

Well, when this instructor contacted our boss, MPRI 3rd Brigade Senior Trainer "Alzheimer Joe" (a retired Colonel/O-6 and Viet Nam Vet) via his two-way radio. Our boss instructed him to "...just go around it!"

When I heard this, I said to myself, "Smart, Joe, real smart. Why don't you just tell him to keep going until he hears an explosion...."

Because this instructor was a retired US Army Sergeant Major, he knew better and didn't follow Alzheimer Joe's instructions. Instead, he back stepped and immediately withdrew out of the area.

We later found out that a Bosnian Army Engineer unit "supposedly" cleared the area of mines. (Yea, right, huh-huh, sure they did...). So from that day on whenever we heard that a training area was safe and cleared of all mines, the first question we asked was, "Oh really, by who???"

We also found out later that good 'ol Alzheimer Joe supposedly told one of our trainers that the land mine "was only a small toe-popper and it would've only taken off his foot."

Well, whenever I went out in the field with a Bosnian unit, I took a few extra safety precautions:

\* When traveling off roads, if a Bosnian leader didn't feel comfortable about a route that I choose for him to follow. I went with his instincts and allowed him to choose the route.

\* When possible, I stayed on hardball and paved roads and only on dirt roads that showed a lot of vehicle tracks and use. And when I encountered trails and paths, unless it also showed a lot of use, foot prints, etc. I tried to avoid traveling on them altogether, (which wasn't always possible).

\* Though we were advised to walk at least 10 meters behind a Bosnian soldier and to never lead their unit in the field. I normally walked directly behind the lead "element leader" and in his tracks.

\* I always wore several pre-made "adjustable slip knot tourniquets" on the upper part of my legs and arms. (See drawing.) So if I (God forbid) stepped on a land mine and survived the blast, hopefully I would be able to slide these pre-made tourniquet down to the wound before bleeding to death. Plus I also carried my ol' dog tags & some blank casualty cards too.

ATTENTION READERS: Check out this homemade improvised mine prober that I made out of an ol' magic marker and a steel rod and see how and where I wore my pre-made tourniquets...

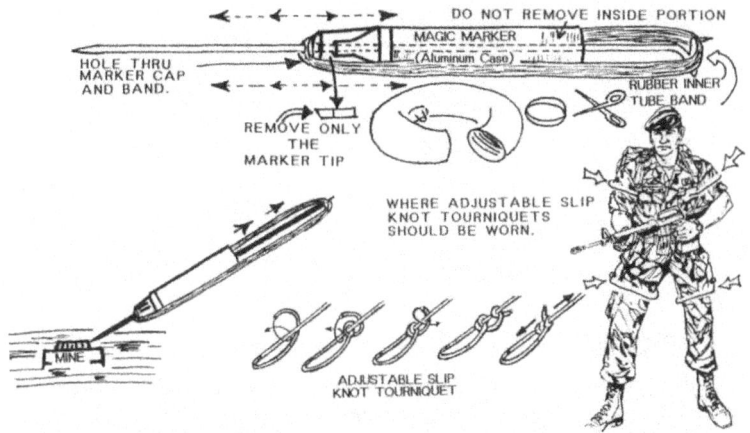

## UNDERSTANDING YOUR BODY TEMPERATURE

Most soldiers could care less as to why their body temperature fluctuate so much in different environments, that is until they become sick, hot, or cold. But once you understand as to how your body thermostat works, you'll be able to control the amount of heat it produces or releases.

Body temperature is a measurement of heat in one's body due to consumed foods and body activity. Although the average temperature for a healthy body is 98.6 degrees Fahrenheit. It will fluctuate up and down depending on the types of foods you have eaten and the type of activities you are doing. The average body temperature is at it's lowest point in the mornings and steadily raises until about late afternoon, then it begins to steadily fall again until you are sound asleep.

*Rick F. Tscherne*

In a cold weather environment the temperature of the skin and certain parts of the body may drop far below the temperature "deep within" the body. And in hot weather environments, these same body parts may also raise far above the temperature as well.

Overall, the body temperature is controlled by the brain which sends various signals to certain nerves & glands throughout the body to help it maintain a normal temperature. It's only when you fail to recognize certain body symptoms & warning signs is when you become overheated, extremely cold, and or seriously ill. So remember these basic things...

5 Major Heating Points

Now that you know where the five major heating points are located, try to keep them as dry as possible. The more they heat up, the more you'll sweat and become "wet."

When Out In The Field:

1. Eat regularly and exercise.

2. Drink plenty of liquids/water.

3. When it's hot - open clothing and try to stay in the shade.

4. When it's cold - button up and try to stay in the sun.

5. When you begin to sweat, wipe it off with a handkerchief or tie a cloth around the forehead and or neck to catch it.

6. When tired and sweaty, rest and change your underclothes.

7. If you start to feel dizzy or sick, don't try to downplay or ignore these symptoms because your body is trying to warn you that something is wrong. Sit down, rest, drink plenty of water & wait until you're feeling better before moving on.

# IMPROVISED BATTERY POWERED HANDWARMER

Hey boys & girls, here's a field expedient method in how to make an improvised battery powered handwarmer. I discovered it by accident one day while playing around with a couple of used batteries.

All ya need to do is find yourself a "C" or "D" size battery and a short piece of thick wire and attach it to the + & - ends of the battery. To control the heat setting, connect and disconnect the wire off and on to the battery.

WARNING: Never keep the wire continuously connected to the battery for long periods of time. Once the battery heats up to a comfortable warm temperature and NOT too hot. Then disconnect one end of the wire and wait until the battery temperature has cooled down a little before reconnecting it. Ya gotta keep connecting and disconnecting the wire off & on to avoid causing damage to the battery and yourself.

SPECIAL NOTE: Batteries used in this mode shouldn't be bought for this purpose only, as they don't last very long. Use old batteries instead of new ones. (Pssssst! Wanta save some $$$$$? Use Uncle Sam's Gov't issued batteries. It doesn't matter if they're new or used as long as they are NOT dead batteries. Get my drift, Rambo?)

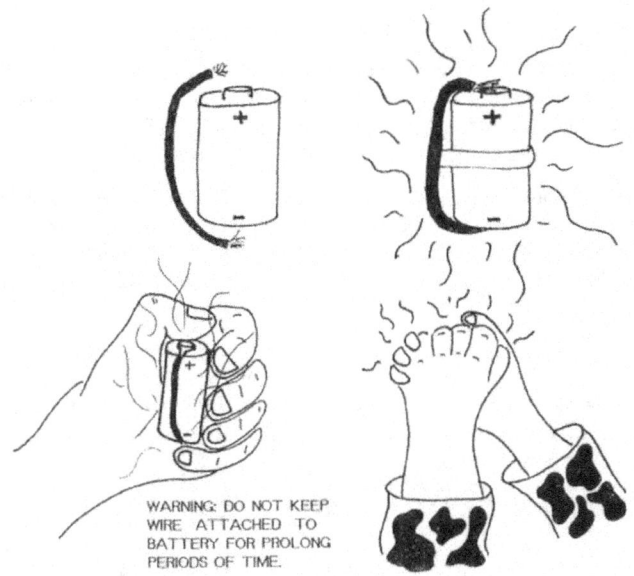

## MILITARY HUMOR

A young man walked into an Army Recruiting station one day to see what kind of job opportunities were available to him. After meeting with the recruiter and being told what he could enlist for, he asked if he could also go Airborne too. "Of course," the recruiter replied.

The young man then says, "Great, when can I sign up?" The recruiter says, "right away" and then he quickly pulled out all the enlistment papers and began filling them out. After the young man signed all the documents, he then asks the recruiter, "What can you tell me about Airborne School?"

"Well," the recruiter replied, "the school is located at Fort Benning, Georgia and it's three weeks long. The first week is where they separate the men from the boys, and the second week is where they separate the men from the fools."

Excited, the young man eagerly asks, "and the third week, what happens during the third week?" The recruiter turns to him and says, "the third week is when the damn fools jump."

\* \* \* \* \* \* \* \* \* \* \* \* \*

One day a very religious national guard soldier was sitting in his home along a river front when suddenly a police car pulls up to warn him of raising flood waters. The soldier tells the policemen, "I'm not afraid of a little water, if it gets bad, I know the Lord will save me."

A few hours later the river begins to raise and soon reaches the top of his porch, and a short time there after a river patrol boat arrives to evacuate him. But the soldier turns down their offer and tells them, "I'm not worried about the water, if it gets bad, I know the Lord will save me."

A few more hours later the water raises even higher and forces the soldier out of his home and onto his roof. Soon a helicopter arrives, lowers down a rope and tells him to grab onto it. But the soldier refuses to grab the rope and yells to the pilot, "I'm not worried about the water, if it gets bad, I know the Lord will save me."

Soon, the flood waters quickly rise and the soldier drowns. When he arrives in heaven he asks the Lord, "I was praying and waiting for you to save me, Lord, why didn't you come?" The Lord turns to the soldier and says, "Hey listen, I sent you a police car, a patrol boat, and a helicopter, what more do you want from me?"

## AN EMPTY SHELL CASING TIP

*Submitted By: Sgt. Stephen M. Mich (USAF)*

Dear Ranger Rick,
Enclosed you'll find some drawings and ideas that I hope you'll find amusing and publish in your next *Ranger Digest*. Keep those Ranger Digests coming!

According to Sgt. Mich, German snipers in World War II were known to climb trees with empty ammo shell casings. They pounded them into a tree with their E-tool and then climbed up using them as steps.

Ranger Rick's Comments: Because I've never heard of this technique before, I decided to try it out myself.

Well, you can forget about trying to use 5.56mm brass shell casings, they're too short, too weak, and they keep doubling over and bending.

The 7.62mm shell casings will work to a "certain degree'," but only in "soft wood trees." Provided you grab hold of the tree and at the same time distribute some (but not all) of your weight from one foot hold to the other as you're climbing.

The 50 Cal. shell casings work the best, provided you can pound them in deep enough into the tree with your E-tool. Though it's much easier with a hammer or a hatchet than an E-tool. But if neither are available, try using a BFR (Big F———n Rock), it's a lot quieter and less awkward to use.

NOTE: Soft wood trees are pine and birch.

READERS BEWARE: Unless you're desperate to climb a tree without any limbs, I suggest you avoid using this technique.

# MAKING A PARA-GRIP FOR YOUR KNIFE

While breezing through some military supply catalogs I couldn't help notice that some of the knives come with 550 paracord already pre-wrapped around the handles. Though this is a great idea, I just wonder how much it increases the price of the knife. (Hmmmm....)

The purpose of having paracord wrapped around a knife handle is not only for emergency survival needs, (snares, fishing line, etc). But so that you can hold it more securely in your hand. Provided of course, it's wrapped firmly around it.

It's not all that difficult to wrap, really? In fact, it's pretty damn easy if you got the right knife (handle) and some 550 od parachute cord.

Now most of the "Wanna Be Killers" and "Rambo Cowboys" that I've known usually tied the cord to the handle first and then began wrapping it either from top-to-bottom or vice-versa. And after a short period of time, it would eventually slide off unexpectedly.

Though I only know of one smart way to wrap it, you have a choice of having a knife para-grip handle "with" or "without" a lanyard. The only purpose of a lanyard is so that you can place it around your wrist while using it and so that you don't drop it. Like when you're fighting a couple of bad guys, (yea, right, huh-uh, sure, dream on...).

The first thing you need to do before wrapping the paracord around the handle, is melt the ends of the cord so it won't unravel. Then allow it to soak in some water for about 15-30 minutes. The reason why you soak it in water, is so that it will stretch as you're wrapping it around the knife handle. Then as it dries it will tighten-up and become more firmly secure in place around the handle.

Now instead of me trying to explain to you in so many words how to wrap it around a knife handle, just follow the drawings on the next page. When you've finished wrapping 550 paracord around the handle, if you want to speed up the drying process, just use a hair blow dryer. But don't turn up the blow dryer too high or you might melt some of the cord and then you'll have to start all over again.

Before wrapping the paracord around the handle, if you so desire, you can wrap some survival fishing line, a couple of small hooks, and some od green booby trap wire around it. But don't add too much around it or you'll make the handle a bit too bulky.

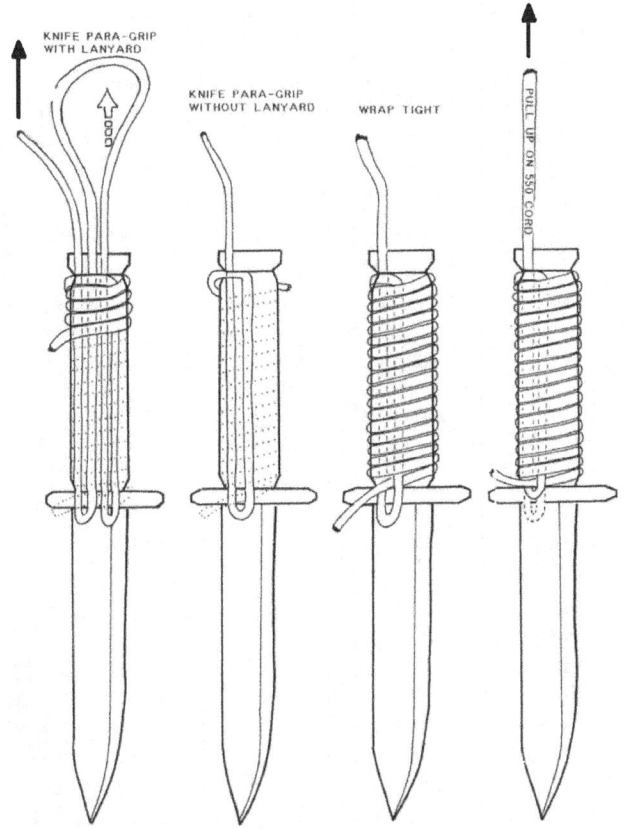

# HOW TO MAKE TINY LIGHT MARKERS FROM REGULAR ONES

I'm sure most of you have seen the small packages of "ITTY BITTY LIGHTSTICKS." They come in packages of 10 and will glow for about 4-6 hours, or so they say. And they can be used for an assortment of things, such as for marking trails, equipment positions, and so on. Just like regular size chem-lights.

Well, I don't know if they're available through the military supply system, nor do I seriously doubt anyone would want to waste their money on them as they are kind of "expensive." But if you ever need a bunch of tiny lightsticks, you can always make 'em yourself. How?

First, visit your unit aid station and ask a medic if you can have an IV (intravenous) tube and either a nose spray or an eye dropper. Take the nose spray/eye dropper, empty it out and then wash it thoroughly, it's got to be real clean.

Take the IV tubing and cut it up into smaller pieces about 2 inches in length so they'll fit nice and snug inside a 35mm plastic film container. You should be able to produce a whole bunch of these.

Next, visit your friendly supply sergeant or military clothing sales store and get yourself a couple of standard regular size chem-light sticks. Once you have them, DO NOT break nor tear open the air-tight wrapper until you're ready to use it. The purpose of this air-tight sealed packaging is to protect it from direct sunlight exposure. If the air-tight seal is broken, and depending on how long it's been exposed, it may not work the next time you need to use it.

When you're in the field and you see a need to use smaller chem-lights instead of regular size ones. Such as for marking some trails, positions, individual equipment, etc, here's what you do......

STEP 1: Break, shake, and activate the chem-light stick and very carefully cut off one of the ends without spilling it.
STEP 2: Take one of the 2 inch tubes and bend and or tie one of the ends closed while leaving the other end open.
STEP 3: Take the eye/nose dropper and fill it with the glow liquid from the chem-light stick.
STEP 4: Take the eye/nose dropper and squirt a few drops of glow juice inside the open end of the short IV tube.
STEP 5: Bend and seal the open end closed with either a small stick, piece of wire or string and PRESTO! Now you're ready to attach it to whatever you need to use it for.

If you can't acquire a medical IV tube, you can always substitute it with a McDonald's or Burger King straw. And if you can't get a nose or eye dropper, again, you can always use a straw instead. Just place the straw inside the chem-light tube, place your thumb over the other end to trap the glow juice inside and lift out. Then very carefully release it over or inside the short tube or straw.

NOTE: When removing the "glow juice" from a light stick, regardless if the packaging states it will glow continuously for 6, 8, or 12 hours, it will lose 50% or more of it's glowing time strength when exposed to the air. So keep this in mind when figuring out when you intend to use it or it may burn out sooner and lose it's glowing power way before you get to use it.

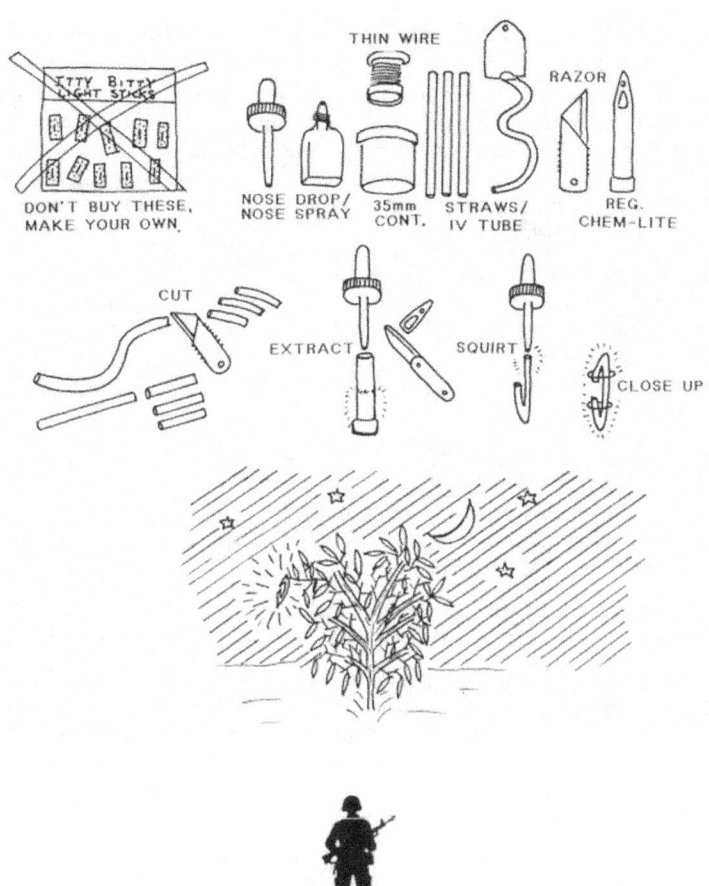

# WHAT DO I THINK OF THE ARMY'S SUGGESTION PROGRAM?

Now some people think I enjoy "slinging mud" at the Army, but I don't. I've always loved being a soldier and being in the Army, but I haven't always agreed with the leaders who run it.

In my 20+ years in the Army I've sent no less than a half dozen suggestions into the Army's Suggestion Program. And I either did not receive any replies back, or a reply stating "..sorry, not feasible for combat related field conditions.."

Among several of my suggestions was the *Do-It-Yourself Warning & Operation Order Handbook* that I developed and submitted to the Army back in the early 1980s. But it got rejected and sent back to me. Today, this book is widely used among almost every small unit leader from fire team to platoon size element. Thanks to my determination in getting it published and no thanks nor assistance from the US Army.

A few years ago there was a well known Infantry Colonel from Fort Benning who (I won't name or embarrass) tried to copy my book and market his own patrol order handbook, but his book wasn't very successful. Why? Because it was NOT compact in size, inexpensive, and most importantly - it wasn't "user friendly" written. In other words, it wasn't written in plain simple english.

Anyway, while in Bosnia in 1997 I received a couple of letters from some friends asking me if I saw the latest "Army Times" newspaper. In it they claimed were some of my tips and tricks from my *Ranger Digest V* handbook. Which was how to convert an MRE cardboard box into a disposable water container (P.66) and shit box (P.68).

Well, at first it didn't bother me that they published my ideas, I only wished they would have given me the credit on where they acquired the information.

In fact, when I first came up with this idea back in 1994, I sent it to the Commanders of the 82d Airborne Division, 101st Airborne Division, and the US Army Infantry Center at Fort Benning, Ga. And a few of them replied back to me too. They said, "..thank you for sending us your ideas, your techniques have merit and will be forwarded to our unit commanders for further evaluation..." (OK, I believed that they would.)

But now, what really pisses me off, is that someone stole my idea, submitted it to the Army and took credit for it too. Hell, whenever I receive any ideas from a Ranger Digest reader, I don't steal 'em, I give them the full credit for it.

Anyway, back in September 1997 I wrote a letter to the "Army Times" editor to ask where they got this information from, and not just once, but twice! And today, January 15, 1998, I still haven't received a reply back.(Hmmmm, I wonder why?)

## RAMBO CHESS & CHECKER SET

Hey you Rambo Cowboys! Here's a couple of neat novelty ideas that you might wanna try.

Buy yourself one of those cheapo $ 1.99 checker sets and modify it by gluing some toy soldiers onto the top of the checkers. It'll make the game much more interesting to play, militarily that is...

Oops, that's right, what ya gonna do when you gotta king one of them? Hey, no problem, number the checkers from 1 to 12 somewhere on the top or along the side. Then, when one or more of your soldiers reaches the other side of the board, you write down that number on a piece of paper.

Now the goal is for your opponent to remember which of your soldiers has been crowned "Little Rambo" and can move freely in any direction. If your opponent forgets, that's his tough luck. If he challenges you and asks which of your soldiers have been crowned "Little Rambos," you just hold up the piece of paper with the number written down on it. But don't forget, you gotta remember these numbers too.

Wanna make a neat looking tactical chess set? First, you need to acquire some of these ammo rounds.

| QTY | TYPE of Rnd. | USED FOR | QTY | TYPE OF Rnd. | USED FOR |
|---|---|---|---|---|---|
| 2 | .50Cal | Kings | 4 | 45.Cal | Bishops |
| 2 | 7.62mm | Queens | 4 | 9mm | Knights |
| 4 | 5.56mm | Rooks | 16 | 5.56mm blk | Pawns |

Unless you're in a combat zone, and for safety reasons too, I suggest you use only empty shell casings. If you can't acquire some of these military rounds, visit your local gun shop and buy some civilian rounds. To make the rounds a bit more stable so they'll stand up better, glue 'em to some bingo or poker chips.

Wanta make the playing board a bit more tactical looking too? Change the red squares to OD green, but leave the black squares the same color.

Here's another novelty idea. Find yourself 24 non-lubricated condoms and a red and black magic marker. Paint 12 of the condoms red and 12 of the condoms black and your ready to play some serious checkers. No doubt you and your friends will find this game of condoms unusual and hilarious.

*Rick F. Tscherne*

# RANGER RICK'S COMMENTARY

Because of the bad publicity the Army has been receiving lately, thanks to ex-SMA McKinney, today's topic is on "Sergeant Majors."

Do I believe Sgm McKinney is innocent of making unwanted sexual advances towards his aid, Sgm Brenda Hostner? Well, let me put it this way, I once had a lot of respect for the man, but not anymore. Not after five other females came forward to make the same accusations against him.

And what made matters worse, he threaten to expose other leaders in the United States Army who were not prosecuted for the same offense. Not to mention, trying to play the "race card" as to why he was being prosecuted by the Army. Well, after all this, he definitely lost my respect and support.

Now I can understand if one female soldier made these accusation against him, and maybe even two, then it would be only his word against theirs. But at last count there were six female soldiers making the same accusations against him. Now that's an awful lot of accusers, and I wonder how many more others are not speaking up to avoid the publicity.

Of course he's going to deny all these accusations, he's fighting for his retirement pay and benefits. But he's either going to be charged and prosecuted, or he's going to be allowed to retire with full military benefits. And I'll bet he'll probably be allowed to retire to save face, both, his and the Army's.

Now the one who started all this trouble was a retired Sergeant Major by the name of Brenda Hostner. And to be honest, I have "less respect" for her than for the accused, ex-SMA McKinney. She should have made these accusations and filed charges against him while she was still on active duty, which she didn't do.

Like everyone else, I wonder why she waited until she was retired before denouncing and accusing him of sexual harassment? She was a Sergeant Major in the United States Army, what could he (or they) possibly do to her? As far as I'm concerned, if she didn't have the " intestinal fortitude" to speak up when he made these unwanted sexual advances towards her. Then it's obvious she should never have been selected and promoted to the rank of Sergeant Major.

I guess she must have been absent and or asleep during the Sergeant Majors Academy's sexual harassment class. Because she definitely didn't know how to handle the situation.

Now I've known a lot of Sergeant Majors in the Army and I gotta admit, most of them were very professional and very dedicated leaders. But no matter where you are stationed in the Army, every installation has it's own fair share of poor leaders, and that includes Sergeant Majors.

Now because I've spent most of my career here in Italy, I've seen a lot of Sergeant Majors come and go, and not all of them pcs'd nor retired under favorable conditions.

I knew of a Command Sergeant Major who had an 18 year old son who was on first name basis with the MPs. He was always in trouble and being picked up and put on the MP blotter report. His wife was also known for bouncing checks too.

Now if you or I had this same problem, what do you think the Army would have done to us? By regulations, the Army would have probably shipped our families back home to CONUS. Right?

I knew of another Command Sergeant Major who had an extramarital affair with an enlisted female soldier. (I guess she was trying to get promoted.) He was relieved and forced to retire with full military benefits.

Now if you or I had done this, what do you think the Army would have done to us? By regulations, we probably would have gotten an Article 15 and barred from re-enlistment. Right?

I knew of another Command Sergeant Major who was known to be a heavy drinker, especially on Friday and Saturday nights. It wasn't bad enough that he drank too much, but he use to drive himself home too. And the MPs at the gate knew when he was drinking (or drunk) but they wouldn't dare stop him.

Now if you or I had this same problem, or had done the same thing and got caught, what do you think the Army would have done to us? By regulations, we would have been put into an alcohol abuse program, had our driver's

*Rick F. Tscherne*

license taken away, and barred and or given an Article 15. Right? You betcha!

These were just a few of the incidents that I knew about that involved Sergeant Majors. And I'm sure there were probably a lot more other incidents that I never heard about.

The bottom line is this, guys & gals. The more rank you have, the more leniency the Army will probably be towards you. Is this fair? Not in my book it's not. As far as I'm concerned, the more rank you have, the more severe you should be punished because you should know better. Especially if you're a leader.

Yours Truly,

*Ranger Rick Tscherne*

# BOOK EIGHT

# FOREWORD

Here ya go, another *Ranger Digest - VIII*. More tips, tricks, and ideas than ever before. *Hooah!*

As long as you guys & gals keep writing, then I'll keep on publishing these handbooks. I might be retired from active duty, but I'm still doing what I like best - teaching & training soldiers.

And not only do I get mail from soldiers, but also from hunters, campers, survivalists, and outdoor enthusiasts too.

Now I know my books are not professionally written. But as I've told ya in previous Ranger Digests, I market only my tips, tricks, and ideas, NOT my writing skills. (But ya gotta admit, I am getting better at this, huh?)

My goal is to teach soldiers what the military doesn't necessarily teach and explain it in layman terms or "soldier talk." But sometimes it's kinda hard to do, which is why I rely heavily on drawings and illustrations.

And although most senior (active and retired) NCOs and Officers disapprove of my books. Because they consider my style of writing "provocative & unprofessional," the lower enlisted, NCOs, and officers LOVE 'EM! At least that's what they keep telling me in their letters, no BS.

Personally, I believe the only reason why a senior NCO or Officer would not like my books, is because they're envious and jealous of my accomplishments. And to those of you who are...."Eat Ya Hearts Out! "

Anyway, I hope ya all like this latest edition. And if ya got some free time on your hands, then why don't you drop me a line and let me know what you think of it. Because your tips, tricks, comments, and letters DO COUNT.

Well, that's about it for now, til next time kiddies...

*—Ranger Rick*

P.S.: Now don't forget, if someone asks where you learned these tips & tricks from, ya tell 'em..."from my buddy Ranger Rick."

**AUTHOR'S DISCLAIMER:** *The Ranger Digest* is a series of training handbooks strictly designed for US military personnel. The author and his contributors cannot be held liable for injuries or deaths caused by the use of any of these tips, tricks, or ideas. You are advised to use them at your own risk.

*The Complete Ranger Digest: Vol. VIII*

# SPECIAL THANKS

As always, I dedicate a special page to all those who took the time to write in wanting to share their favorite tip, trick, and or idea with me and my readers. And if it wasn't for these caring Ranger Digest contributors, there wouldn't be a *Ranger Digest VIII* today. Thanks fellas!

## RANGER DIGEST VIII CONTRIBUTORS

| | | |
|---|---|---|
| Maj David Oaks | Cpt Robert C. Fraser Jr. | 1Lt Karl L. Mims |
| 1Lt Jimmy Deak | Ssg Steve Brittian | SP4 Greg A. Banker |
| PFC Ben Donaldson | PFC Chris Watson | Lee Churchwell |
| Luis R. Ramos | Dr. David A. Williams | Mike Giles |
| Chris Ayers | David M. Handa | Daniel Garcia |
| F.W. Eickelen | Michael Chase | Paul Gromkowski |

AND AGAIN, AN EXTRA SPECIAL THANKS TO MY RANGER DIGEST BUDDY & ARTIST:

### AIRBORNE SGT FRANK D. GILLILAND

**ATTENTION RANGERS, SF DUDES, DELTA FORCE, & NAVY SEALS!**

Hey, where the hell are you guys? Rarely do I ever get any mail, tips, tricks, and or ideas from any of you bad asses. What's the matter, don't ya all know how to write? Com 'on, get off your butts and start sharing with us some of your tips, tricks, and ideas. I'm a challenging ya, ya know?

# WATCH CADDY
### BY: RAINE INC.

*Black Poly-Pro Web
*2 1/8in. Universal Clip
*Full Length Velcro Watch Mount
*Double Stitch Construction
*Made in the U.S.A.

The Watch Caddy was designed for the Military & Law Enforcement communities. It allows the conventional time piece to be positioned Anywhere on the person that will support the use of the Universal clip, such as belt, pocket, collar, etc. The Watch Caddy also allows the watch or watch type accessory (compass/ watch beeper/ calculator/ etc.) to be additionally positioned on any personal piece of equipment, such as back packs, web gear, duffel bags, fanny packs, purses, etc. Providing a Safe, Convenient, Out of the way placement of the conventional time piece or similar item. Available through any Raine Inc. product distributor, or contact by the internet at WRKN4@AOL.COM

use in
Military / Tactical / Law Enforcement/ Security / EMT / Outdoor / Field / Sports / Extreme / Industrial / Work / Amputees / Disabled / Everyday

ATTENTION READERS: This nifty device was designed & developed by my buddy "Airborne Frank" D. Gilliland and forwarded to Raine Inc for review. And...well hell, you can see what they thought of it.

### RANGER DIGEST VII COMMENTARY UPDATE
Reference to ex-Sergeant Major of the Army Gene McKinney's sexual misconduct chargers. As I stated in my *Ranger Digest VII* - "Ranger Rick Commentary:" (quote)

"He's either going to be charged and prosecuted, or he's going to be allowed to retire with full military benefits. And I'll bet he'll probably be allowed to retire to save face, both, his and the Army's." (See below.)

Well, did I call this, shot right or not? You didn't need to be familiar with the Uniform Code of Military Justice (UCMJ) to see this verdict coming. Not only was this a slap in the face to all the women who serve in the armed forces, but to our NCO Corp too.

Do you really believe sexual misconduct and harassment will now start to disappear in the military? In my opinion, not until the Pentagon starts enforcing the regulations and prosecuting everyone "fair & square" regardless of their rank and or position.

It seems we always hear about the "little fish" getting caught and prosecuted, but rarely do we ever hear about the "bigger fish" (Sergeant Majors, Colonels, & Generals) getting caught, charged and prosecuted for doing something wrong or inappropriate. Ya know?

Why? As I stated in my *Ranger Digest VII*, "the more rank you have, the more leniency the Army, and also your chain of command, will probably be towards you." Is this fair? Nope! Not by my standards.

But according to Pentagon sources, the services are now revising their retirement procedures for the "bigger fish" so they can't retire early (or unexpectedly) to avoid (or escape) prosecution when they have been suspected of (or done) something inappropriate.

Good news? You betcha! In the past senior officers and NCOs who did something wrong or inappropriate were given a choice. Either retire early with full military honor and benefits, or stand trial and try to clear their names and risk losing everything. Now if you knew you were guilty of doing something wrong or inappropriate, which way out would you take? Get my drift? I knew you would.

**AFTER THE VERDICT:** Army Sgt. Maj. Gene McKinney, left, Friday leaves a Fort Belvoir, Va., courtroom with members of his legal team after he was cleared of all but one of 19 charges in a military sex scandal.

**Army's former top enlisted man cleared of sexual misconduct**

FORT BELVOIR, Va. (AP) — Sgt. Maj. Gene McKinney was cleared Friday of all sexual misconduct charges made against him by six military women who said he pressured them for sex. He was convicted on only one charge: obstruction of justice for urging one of the women to lie.

McKinney, the former top enlisted man in the Army, stood ramrod straight as the verdicts were read; he showed no emotion. His wife sobbed and mouthed, "Yes, yes, yes," as each of the 18 not guilty verdicts were read.

McKinney faces a maximum of five years in prison, dishonorable discharge or reduction in rank. Had he been convicted on all 19 charges, he could have faced 55½ years in prison.

Rick F. Tscherne

# LATE & OVERDUE RECOGNITION

Just a few of the many, many, African-Americans who served proudly in the United States Army are finally getting the recognition that they deserve. And these two soldiers, STAFF SERGEANT EDWARD A. CARTER and LIEUTENANT HENRY O. FLIPPER definitely deserve it.

### A soldier's story of heroism

Edward A. Carter II, born on May 26, 1916, in Los Angeles, was the son of missionaries. He and his family left the States when he was 5, and traveled throughout India and the Far East before settling in Shanghai, China.

There, he studied both traditional and military studies. He learned three languages - Mandarin, German and Hindi - and also became adept at military strategy. Carter, who is featured in the book, African American Book of Values, joined the Chinese army and rose quickly to the rank of lieutenant before army officials found out he was only a teenager and discharged him.

His family returned to the United States, and he quickly returned to battle, fighting as a volunteer on the side of the Loyalists in the Spanish Civil War.

He returned to Los Angeles in 1940, where he met his future wife Mildred. The next year, he enlisted in the United States Army.

After basic training he was assigned to Fort Benning, Ga., as a cook, and moved there with wife and son, Edward III. Before moving, Mildred said, her husband instructed her on how to behave in Georgia in order for them to survive. His relatives said that "years later he talked about the segregated units, and the fact that they (white soldiers) felt that the black soldiers should have a mop and a bucket, and that they thought they were only fit to clean up."

But Carter wanted to stay in the Army, and he knew what to do and how to get along. He made sergeant in less than a year, family members said and became mess sergeant of the officers' club.

In 1944, he was called to service in Europe. After three months of working supply for the troops and volunteering daily for front-line duty, Carter finally got his chance at some real action. However, there was one condition. At the time, no black NCO could command white enlisted troops. He accepted this condition and was demoted to private.

As a member of the 12 Armored Division, 56th Armored Battalion, Company D in March 1945, Edward and his fellow troops were attached to Gen. George S. Patton's 3rd Army, and joined in the push to the Rhine River.

On March 23, Carter's unit was ordered to attack Speyer, Germany, a Nazi stronghold. His tank took enemy fire around 8:30 in the morning. He volunteered to lead three men towards the offensive — about 150 yards away. He was shot five times. After two hours, Carter killed six of the eight enemy troops that surrounded him. He then used the remaining two as human shields and then as informants.

Less than a month after being wounded, he returned to the battlefield. Carter's sergeant's stripes were reinstated and he was immediately put in charge of training his platoon. He received multiple awards, including the Distinguished Service Cross, the Bronze Star and the Purple Heart, but never the Medal of Honor, although his name was submitted.

In 1993 Daniel Gibron headed a panel that studied the nomination of Staff Sgt. Carter and other black servicemen. They determined that the lack of awards to these nominees resulted from the contemporary climate of racism.

So on Jan 13, 1997, President Clinton awarded Carter - who died in 1963 — and six other black American servicemen the Medal of Honor for their distinguished service in World War II. The next day his body, originally buried in Sawtelle National Cemetery in Los Angeles was transported to Arlington National Cemetery and re-interred among the rows of our nation's war heroes.

# A FLASHLITE AS A FIRE STARTER?

Do you know how many ways there are to start a fire? Let's see, you can start a fire with matches, a lighter, a magnifying glass, a flashlight reflector.... Huh? What? You've never heard of starting a fire with a flashlight reflector?

Well, ya can't start a fire with any ol' flashlight reflector. It's gotta be a good one and not some worn out, scratched up, or f——— up reflector neither. Provided you also got some really dry tinder and some strong sun rays too.

First, remove the reflector from the flashlight

Then go ahead and remove the bulb from the reflector

Then place some really dry tinder in the center where the bulb use to be, pushed up from the back side

Face the reflector in the direction of the sun to get the strongest and hottest sun light/rays on the end of the tinder.

And when it starts to smoke, blow on it very lightly to increase the burning of the tinder. Then remove it, place some more dry tinder around it and continue to lightly blow on it until it burns into a flame.

HOT TIP: Got a few extra rounds? Remove the lead bullet from the shell casing, pour a few grams of gun powder on the dry tinder and then place it in the reflector - helps get it burning fastar.

## RANGER BAND WEAPON HAND GUARDS

Not too long ago I saw an advertisement (see below) in a military mail order supply catalog. What is it? It's a set of specially designed rubber hand guards for an M-16 or AR-15 rifle, they're called "Rub Kits."

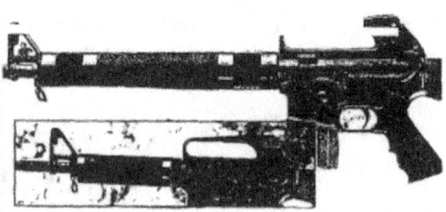

NEW! Highly dependable products for military and law enforcement personnel, these kits allow users to achieve a comfortable, non-slip shooting grip. Both kits (CAR-15 and AR-15) are specifically designed to retain the important lines and aesthetics of the weapon. Each kit features a grip and a forend tube. The grip is highlighted by compound palm swells and proportioned finger grooves that position the hand naturally. OverMolded™ with modern durable rubber, it is unaffected by oils and solvents often found around firearms. The accuracy-enhancing, free-floating, two-piece forend is also OverMolded™ with a rubber grasping area that insulates the hand from the heat and shock associated with rapid fire. Includes installation instructions, however essential tools are not included. USA Wt: 2 lbs. **Black.**® 77.⁰⁰

According to the ad (quote), "allows the user to achieve a comfortable, non-slip shooting grip." Big f——— deal!

Now who's gonna be that damn dumb to pay $99 for a pair of custom made non-slip rubber hand guards when you can purchase a bag of "ranger bands" for a buck or two and slide 'em over the hand guards to achieve the same thing - a comfortable, non-slip shooting grip.

Duuuuh, you don't need to be a rocket scientist to figure this out, bozo.

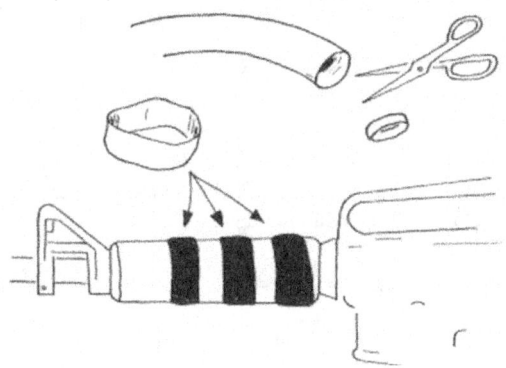

## COILED TELEPHONE CORD USES

*Submitted By: Mr. Mike Gilles*

Here's a neat trick I'm sure you're gonna like. Especially if you carry a sidearm, compass, strobe, maglite, small two-way radio, or some other expensive item on your belt or lbe.

Instead of wearing a regular nylon "dummy cord" attached to the items that I just mentioned so you won't lose 'em in the field, try attaching something a bit more flexible and attractable.

Go to your local electronic supply store (like Radio Shack) and purchase some "coiled" telephone cord. You know, the type that's attached to a telephone handset? They're not very expensive and they come in different lengths and colors too.

Then visit your local hardware store and purchase a couple of "mini" snaphooks/snaplinks and a roll of thin wire.

To determine how much cord you'll need, hold one end of the coiled cord to your pistol belt/lbe and with your other hand "stretch" the cord to where you'll be holding and using the item that you're going to be attaching it to.

EXAMPLE: If you're going to attach the coiled cord to your pistol, then hold the cord in your pistol hand at the "ready/aim position" and stretch it to your belt or lbe where you intend to attach it.

Six (6) inches of "unstretched" coiled telephone cord will "stretch" to an approximate length of 36 inches. Which is about the right amount needed for the average person for attaching a pistol, maglite, or knife to their belt or lbe. And for a compass, strobe light, and or a two-way radio, you'll need about four (4) inches of unstretched coiled cord.

NOTE: The stretch ratio of coiled telephone cord is 1/6, which means one inch of "unstretched" cord will "stretch" to a max length of six inches.

Once you've determined how much coiled cord you'll need, cut it down to size and attach the two mini snaphooks/snaplinks to both ends of the cord with the thin wire. (See drawing)

IMPORTANT: Insure you cut the "exact amount" of cord you need, no more - no less. If it's too long, like cotton and nylon dummy cords, it will also get caught and snagged on trees, bushes, and wait-a-minute vines too.

Because these coiled cords come in many different colors, for tactical purposes, purchase only green, sand, or black cords.

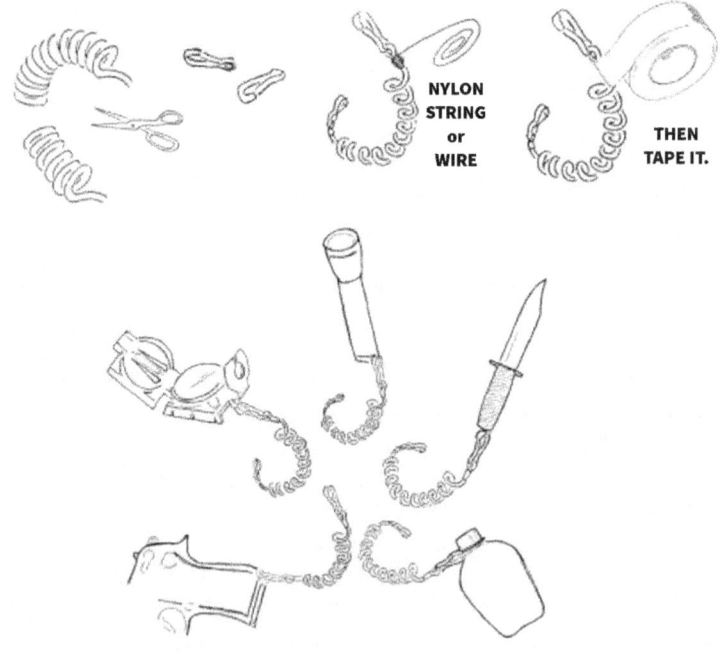

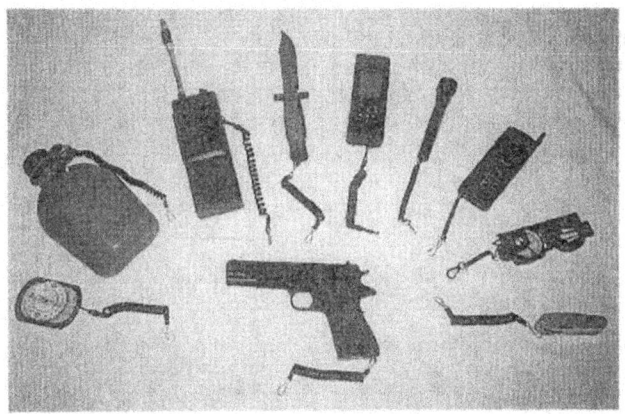

## MAGLITE MOUNTING BANDS

Here's a tip I'm sure you're gonna appreciate. How many of you troopies been attaching your maglites to your weapon with some electrical or 100 mph tape? A whole bunch of ya, right? Just like the way I showed ya in one of my earlier Ranger Digests, right? I'll betcha have.

Well, I discovered an easier method to mount a maglite to a weapon, especially if you don't carry tape to the field.

Now you have a choice, you can either buy a package of "ranger bands," or you can make your own from an old bicycle inner tube. If you make your own, you just need to cut the bicycle tube into the sizes that you want'em, which will probably work out better for you.

Then take these rubber bands and slide 'em over and around the handguard of your weapon, but not just one, several of'em. Then grab your maglite, place it where you want it, and slide the rubber bands over it.

Hooaah! Now you'll be able to mount and dismount your maglite to your weapon much easier. Cool, huh? I knew you'd like this tip.

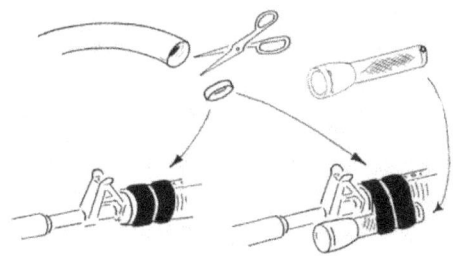

*Rick F. Tscherne*

UPDATE: Great overview information (plus great pics too!) of many weapon light mounting systems and mounts:
http://www.militarymorons.com/weapons/wlights.html
http://www.militarymorons.com/weapons/wlights2.html
http://www.militarymorons.com/weapons/wlights3.html

# YOU KNOW YOU'VE BEEN IN THE ARMY TOO LONG

* You make your newborn child attend the base "Newcomer's Orientation Class" within the first 30 days of his life.

* You go to a barbecue party and make your family eat "tactically dispersed."

* You make your mechanic "replace the sandbags" in the floorboard of your car as part of the yearly tune-up.

* Your kids volunteer to pull "air guard" on the roof of their school bus everyday.

* Your oldest kids keep calling the youngest in the family "Cherry, Trainee, or Recruit."

* Your Christmas tree is decorated every year with "chem-lites and engineer tape."

* Your wife gives the command "fix bayonets" every year at the Thanksgiving Dinner table.

* You make your kids show their "meal cards" at the kitchen door before letting 'em in to eat. Except the oldest, who's on separate rations, you make him pay cash for his meal.

* Your wife takes a "one knee" security position behind her

shopping cart while waiting her turn at the store check-out lines

* You do all your "back to school" shopping at Army & Navy Surplus Stores, Brigade Quartermaster, and or the base Military Clothing Sales Store (MCSS).

* Your son fails third grade and tells everyone he's just a "phase three recycle."

* Your kids "salute their grandparents" every time they come over to visit.

* Your kids get an "LES" (Leave & Earning Statement) along with their allowance.

* You make your kids clear the base "Housing Office" before going away to college.

* Your kids take turns pulling "security patrol" around your house every night.
* Your kids have all their "toys and clothes" hand receipted to them.
* Your wife always conducts an "AAR" (After Action Report) after sex.
* Your wife calls foreplay before sex "Prepping the Target."
* Your wife's 3 favorite lipstick colors are "green, loam & sand."
* Your house has "sectors of fire" sketches posted at every window
* Your kids call their mother "HOUSEHOLD 2."
* Your daughter must "sign out on pass" before going on a date.
* Your kindergartner calls his school recess "Smoke Break."
* Your kids call the tooth fairy "Slicky Boy."
* Your kids first two initials begin with "AR, FM, TM, or DA."
* Your pickup/car has "your name stenciled" on the windshield.
* Your kids recite their ABCs "phonetically."
* Your wife keeps "mermit feeding cans" in the china cabinet.
* Your wife left you and you hold a "Change of Command" ceremony.
* Your dog's name is "Ranger" and you call your cat "Airborne."
* Your kids call their sandbox the "Warfare Training Center" box.
* Your newborn's first three words were.... "ALL OK JUMPMASTER!"

## AN MRE PLASTIC BAG SEALER

*Submitted By: Michael Chase*

Dear Ranger Rick,
There's a battery powered device on the market called a EUROSEALER. It's used for sealing up food so they'll stay fresh while they're in the refrigerator.

Well, I've been using my Eurosealer for waterproofing some of the things that I take to field, such as socks, underwears, t-shirts, etc.

But instead of using regular plastic food bags, I've been cutting up and using the plastic MRE package, they're a lot more durable. The cost? $13.69 Available from: Amazon.com

Rick F. Tscherne

http://www.amazon.com/SCG-TS-100-Sealer-Battery-Operated/dp/B000E8OKQ8

Battery Operated! Take it anywhere! Seal in the fresh flavors, textures, and nutrients of leftovers instantly in the original bag when... You're camping Out on a boat After a picnic or cookout At the beach Locks in freshness with patented micro-thermal technology, so even liquids won't leak! Uses 2 AA batteries - not included.

# HMMM, INTERESTING...

Now here's a few Army Times articles that I found interesting, and I'm sure you will too.

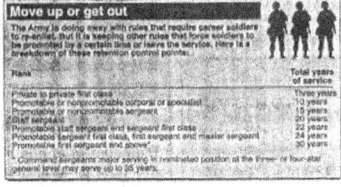

*The Complete Ranger Digest: Vol. VIII*

# AMMO CAN TIPS & TRICKS

Once upon a time I needed a few ammo cans, so I went to my nearby off-post military supply store to look for 'em. But after seeing how much they wanted, well, they were a bit too damn expensive for me.

So then I looked through some military mail order supply catalogs, and again, they were still a bit too damn expensive for me. (Yep, that's right, I'm a tightwad.)

So I called up one of those Gov't/Military Property Disposal Offices, which I think today they're called Reutilization Offices or something like that... And I asked'em when they were gonna have their next sale. And boy was I lucky, they were gonna have one in just a few days - hot dog!

So a few days later off I went to the PDO sale. And as soon as I saw the head MFIC I asked him, "Hey buddy, have ya got some ammo cans for sale?" And he said, "Sure do."

"Great," I said, "I'll take about five of them."

"Uh-uh, no sir," he said, "sorry, but we sell 'em by the pallet load, and it's either a pallet load or nothing at all."

So I said to the man, "Well, how much is a pallet load?" And he said, "For 5.56 ammo cans, it's $20 bucks a pallet."

And I says to the man, "well, OK, then I guess I'll take me a pallet load." (Dummy me, I forget to ask how many came on a pallet).

Well, after I paid the man and he handed me my receipt, he then lead me to the back of the warehouse. And boy, was I shocked to find out that a pallet load contained 100 ammo cans.

So I says to the man, "Hey wait a minute, buddy, I only need a few, not a 100 f—— ammo cans!"

And he turns to me and says, "Hey, you asked for'em, you paid for 'em, there's no refunds, now get'em the f—— outta here."

Well, I guess it really wasn't such a bad deal after all, they only cost me 20 cents a piece. But I had to make a dozen trips with my 1982 VW Rabbit to get 'em all home. Not to mention, my wife wasn't very happy to see half our garage filled with ammo cans.

Well, after giving away as many as I could, I still have about 50 left. And so far here's what my buddies and I have used them for:

## SECURITY CONTAINER

Now before you can convert it into a security container, you will need to get yourself one of those "loop bolts" with 2-3 nuts and two washers. Measure the "loop" portion of the bolt to the hole that is already in the ammo can latch (see drawing) to make sure it will fit through and over it, which it probably won't. No problem, just take a hammer and whack the "loop" a couple of times to make it narrower so it'll fit through the hole.

IMPORTANT: Before whacking and making the loop narrower, place your lock through it so you won't make it too narrower that your lock won't fit inside of it.

Then take your drill and make a hole directly in the center of this latch hole and place the loop bolt inside and fasten it in place with the nuts and washers and then test it out. If it doesn't close snugly, just bend the loop bolt whichever way you need until it closes comfortably over the ammo can latch.

# The Complete Ranger Digest: Vol. VIII

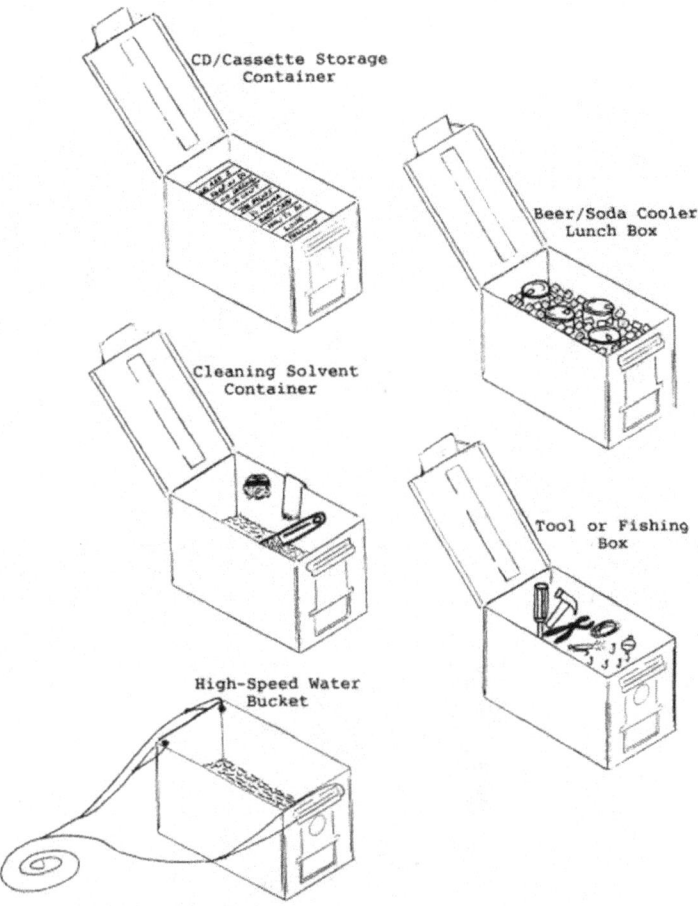

*Rick F. Tscherne*

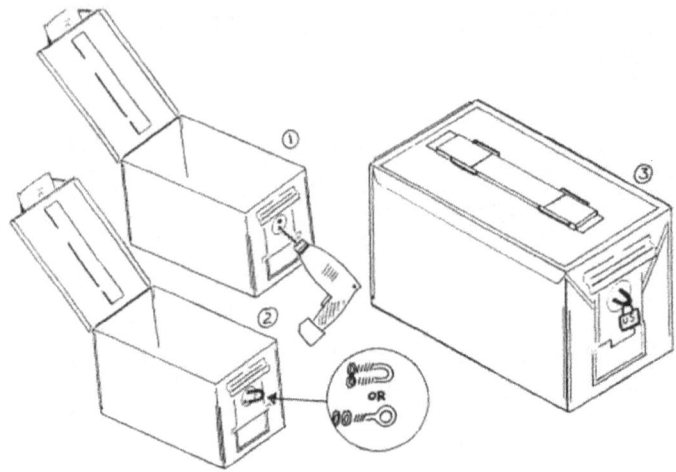

## AMMO CAN ORGANIZERS

Hey! Wanna max out the storage space inside your handy-dandy multi-purpose 5.56mm or 50 Cal metal ammo can? Then purchase a couple of these "Ammo Can Storage Trays". They're great for storing things like nuts, bolts, tools, ammunition, fishing gear, electronic parts, and more.

They're made of durable plastic and won't shrink, crack, or warp, and the tray handles fold down so you can stack'em one-on-top-of-the-other. Neat! The cost? **NOW $6.97 for a set of three.** Order from:

CHEAPER THAN DIRT
2536 NE Loop 820
Ft Worth, TX 76106-1809
https://www.cheaperthandirt.com/product/mtm-case-gard-50-caliber-ammo-can-organizer-tray-3-pack-black-aco-026057362380.do?sortby=ourPicks&refType=&from=Search

*The Complete Ranger Digest: Vol. VIII*

## AMMO CAN FIELD STOVES, HEATERS, & B-B-Qs

One of my favorites and mother-of-all-uses, is the "Mini BBQ / Field Stove Ammo Can," or the Infantryman's Poor Man Field Stove.

I came up with this idea after reading about a company buying a bunch of old 20mm Ammo Cans, converting 'em into portable field stoves and selling 'em at a real expensive price.

To modify and convert a 5.56 ammo can into a Mini BBQ / Field Stove, all you need is a drill, a small drill bit, pliers, wire cutters, a pencil, ruler, and five (5) "wire" coat hangers.

Take your pencil and ruler and make two straight lines along the two wide sides of the ammo can about 2 and 4 inches below the top edge. Then measure and make a small visible mark about every 2 1/2 inches along these lines on both sides of the can. Then take your drill and make a series of holes where you've made these marks.

IMPORTANT: As you're drilling the holes, try to make them as straight and parallel with one another as you possibly can so you can easily insert & remove the coat hanger wires (when they're cut, bent, & shaped) from the side of the can.

Now take your wire cutters and snip off all the "hooks" from the coat hangers and straighten 'em all out the best you can with a pair of pliers. Then take one of the wires and run one end through both of the holes (across from each other) in the ammo can and bend the wire over to the next hole beside it.

Place a mark on the wire where it touches the next hole and remove it. Where you've made this mark, now bend the wire over and make it the same length as the other side of the wire. Now test it out for size by inserting it in "all" the holes in the ammo can.

Continue making these rectangle shaped wires until you have filled up all the holes in the ammo can. If they fail to slide easily in & out, then you screwed up (Bozo) and either didn't bend the wires correctly, or the holes were not drilled evenly spaced apart and or straight across from one another, (Dummy!)

When you have finished measuring, cutting, and bending the wires, bend slightly up or down only the tips/ends of these wires so they'll stay firmly in place without moving and sliding around. Now all ya gotta do is get some wood, charcoal, paper, etc and you're ready to do some serious cooking.

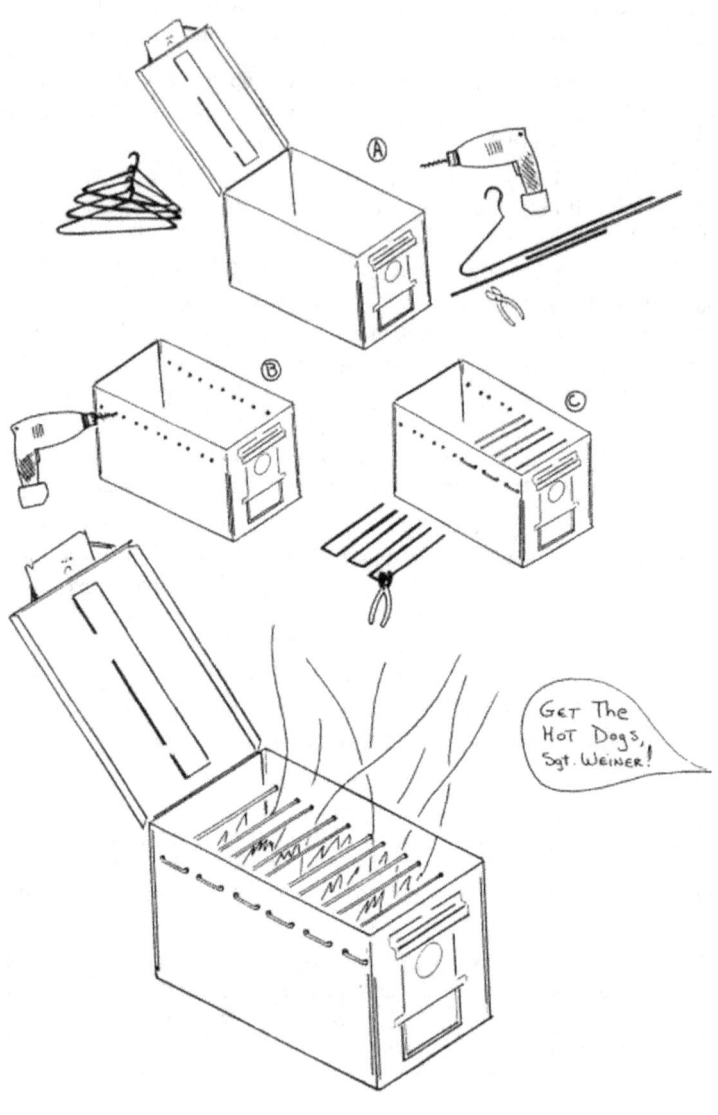

# A NEW WONDER WEAPON?

Hey guys & gals, have ya heard about this new weapon system the Pentagon plans on purchasing and issuing to all the light infantry units in the year 2006? No? Well read on.....

In the next century American soldiers will be going into battle with a weapon so advanced that it will make today's assault rifle seem as primitive as a "musket ball & rifle."

According to military analysts, tomorrow's wars will be low-intensity conflicts and fought in the streets, alleys, and slums of third world countries. Therefore, they had to developed a new type of weapon system that would not only be capable of winning urban warfare battles, but also in familiar terrain such as forests, deserts, and jungles.

The answer? A weapon system the Pentagon dubbed the "Objective Individual Combat Weapon," or the OICW.

It's a double-barrel weapon system that fires both, NATO standard 5.56mm ammunition and 20mm high-explosive fragmentation rounds. It's capable of firing 20mm air-bursting rounds over the heads of hidden targets as far away as a 1000 meters. A single trigger is linked to both barrels by way of a laser guided electronic firing system as sophisticated as what you will find on today's US tanks.

"On tomorrow's battlefield, the OICW will leave no place for the enemy to hide," says the developer of the weapon system. The key to the OICW'S success is in it's electronic fire-control system, which enables the rifle to determine when it's "smart" 20mm rounds need to be detonated.

Like today's conventional explosive rounds, these shells detonate on impact. However, they can also be set to explode after passing through a cement or wooden wall and or a sheet of metal. This capability could be especially useful in third world shantytown shacks where many abandoned or pirated cargo containers serve as common living quarters.

The OICW will be able to take out targets that an M16 rifle and or M203 grenade launcher can do today except with "pinpoint accuracy." Which will be important on tomorrow's battlefields as the enemy will be hiding among civilians - a popular survival tactic used in urban warfare. And with the OICW'S range of 1000 meters, five times greater than today's M16/M203, it will be able to do just that.

**Ranger Rick's Comments**: With a price tag of $10,000 (+) each verses $500 for an M16A2 rifle, let's hope it works and we get our money's worth out of it.

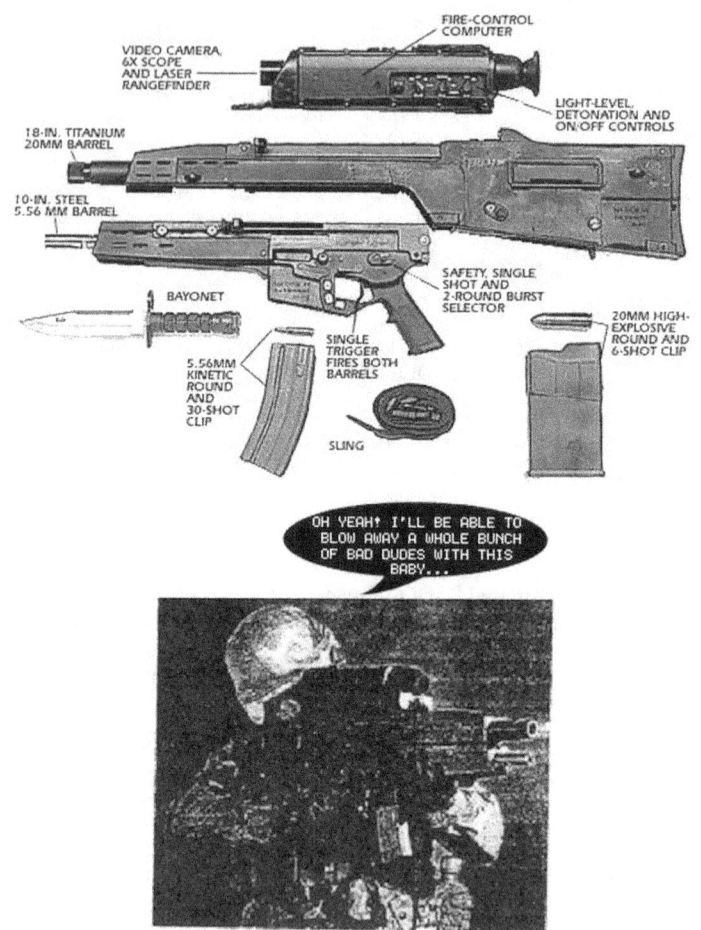

UPDATE: This weapon was not ultimately adopted, but the technology did morph into a different dedicated air-bursting 25mm smart grenade launcher system, called the XM-25 CDTE (Counter Defilade Target Engagement System).
Five of the weapons were deployed with the 101st Airborne Division in Afghanistan in October 2010, along with 1,000 hand-made air-burst rounds. The soldiers reported that the weapon was extremely effective at killing or neutralizing enemy combatants firing on US troops from covered positions. The US troops have nicknamed the weapon, "The Punisher."

http://en.wikipedia.org/wiki/XM25_CDTE

Video: http://www.military.com/video/guns/grenade-launchers/xm-25-demo-video-from-atk/666202460001/

# A TOILET PAPER FIELD TIP

*Submitted By: Luis R. Ramos (CAP)*

Dear Ranger Rick,
I've become a fan of your Ranger Digest handbooks, they not only teach soldiers a lot of useful tips, tricks, & ideas, but outdoorsmen and survivalists too.

For example in one of your Ranger Digest handbooks, the tip on how to convert a BDU patrol cap into a survival cap was a terrific idea. But in one of your other books, I kinda disagree with you on removing the cardboard center piece from a roll of toilet paper before packing it away.

Though it will no doubt make it smaller, compact, and easier to pack away, you can always use this little bit of space to store some of your toilet articles inside of it. Like your toothpaste, toothbrush, razor, deodorant, etc.

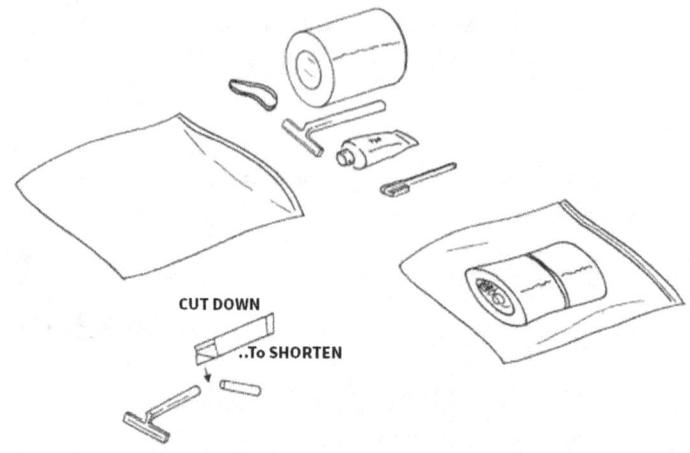

CUT DOWN ..To SHORTEN

## REPAIRING PONCHOS

*Submitted By Chris Ayers*

How many times have you turned in a rain poncho for a serviceable one just because it leaked a few drops of water when you used it in the field as a shelter? A bunch of times, right? I'll bet ya have.

Did you have to wait in a line a long time to turn it in? If not, I'll bet when you handed it over to the supply clerk he or she probably asked ya, "So what's f——— wrong with it?" And when you told 'em, "It f——— leaks." They probably said, "I don't see any f——— holes." Right?

Unless, of course, you did what everyone else does today. You took out your trusty ol' pocket knife and made a few more holes so the supply clerk could obviously see what's wrong with it, right? Yep.

Well, it's not that damn difficult to repair a rain poncho today. All ya need is a sewing needle, some thread, and a small tube of "rubber" or "silicone" glue. Yep, that's all.

To locate every single hole or tear in your poncho, hang it on a wash line outside in the direction of the sun. And with the poncho fully open and stretched between you and the sun, you shouldn't have any problems seeing all the pin holes and tears.

To repair small and tiny holes, wipe the area clean and place a small amount of rubber/silicone glue directly on the hole and smear in with your finger. Then go to the other side of the poncho and repeat the same procedure. NOTE: It doesn't do any good to do just one side, you need to do "both sides" to insure the hole is completely closed.

To repair small rips and tears, take a needle and thread and sew it up the best you can. Then wipe the area clean, place some rubber or silicone glue directly on the stitches and smear it in. Then go to the other side of the poncho and repeat this same procedure.

Because rubber and silicone glue can be purchased in very small tubes, always carry some in your rucksack in case you need to repair your rain poncho out in the field.

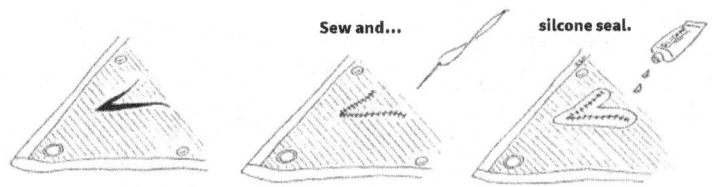

## TAC LITE PRESSURE SWITCH TIP

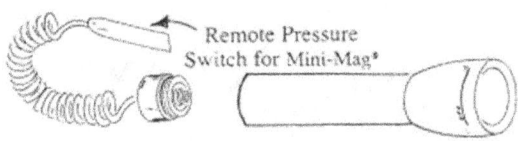

*Submitted By: David M. Handa*

Dear Ranger Rick,

If you've ever used a Sure-Fire or Mini-MagLite Mounting Kit attached to your weapon, I'm sure many times you've been frustrated with the pressure switch always falling off. As the velcro that's provided with these damn kits ain't enough to keep them in place. Especially when moving through brush, weeds, wait-a-minute vines, and dark buildings too.

I got tired of my pressure switch coming off and solved the problem. And so far it's not only worked well me, but also for my buddies too. Here's what you do;

a) Go down to your local bicycle repair shop and get some various size inner tubes that they're throwing away.
b) Take a pair of scissors and cut off several short pieces so they look like

flat short rubber bands. Then slide them over the hand guards of your rifle or shotgun and or the grip of a pistol.

c) Position your Sure-Fire/Mini-MagLite "pressure switch" where you want it and then slide the rubber bands over the entire pressure switch itself or just on the edges of it.

I know you probably think the rubber bands will keep the pressure switch constantly on, but it doesn't. Unless, of course, you're using some real tight and narrow bands or tire tube. And if you are, then just mount the bands on the edges of the switch.

My buddies and I like the smooth, tactical, and professional look, so we cover the entire switch with a 4-5 inch piece of inner tube.

If you want to save the hassles of going down to your local bicycle repair shop and making your own mounting bands. You can always buy a bag of "ranger bands" from your local military supply store or mail order catalog.

WARNING: Should you decide to use this mounting technique on pistols with "grip safeties." Due to the emplacement of the rubber band around the grip, the grip safety may constantly be in the off and disengaged position. Caution is advised.

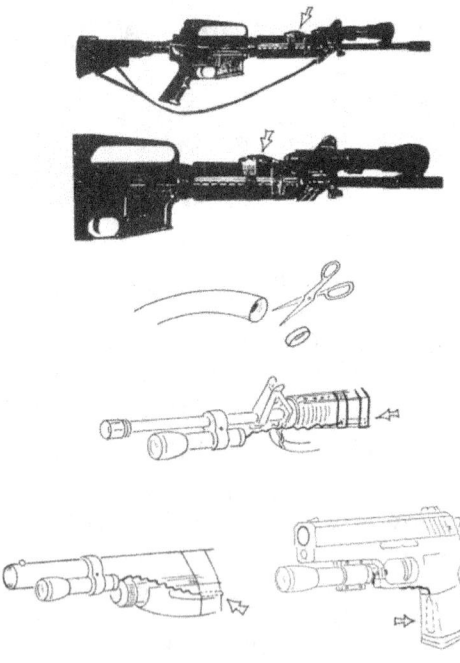

*The Complete Ranger Digest: Vol. VIII*

AND ANOTHER ONE OF MY ARTICLES MAKES IT INTO THE....

It seems everytime I pick up a monthly Southern European Task Force Outlook newspaper here in Vicenza, Italy there's a column listing the command's recent DUI offenders. So it's obvious, like many other military communities here in Europe, we have a growing problem with drunk drivers.

A long time ago, back in the 1980s, my 'ol Airborne unit had a similar problem - not only with DUIs, but drugs too.

Our battalion commander, Lt. Col. Needham called for a "loddy-doddy everybody" battalion formation. I was a platoon sergeant at the time and remember that day like it was yesterday. Good 'ol Needham climbed on top of a wooden physical training platform, took a long silent look at the battalion and said,"Starting today, regardless of your rank, race, religion, married or single, if you get picked up and test positive for drugs or for driving under the influence of alcohol, you will be given a battalion Article 15. And when you come before my desk, be prepared for punishment. I WILL MAX YOU OUT! And if you want to know what's the maximum punishment I can impose on you under the UCMJ, just ask your company commander or first sergeant. Are there any questions?"

Well needless to say, from that day on the troops nicknamed him "Nuke 'em Needham." Because when you get caught and stood before his desk, you knew you were gonna get "nuked" and lose your rank and/or a whole lot of money.

Though I was not a strong supporter of Needham's new policy, I gotta admit, he did get rid of and/or drastically reduce the battalion's DIM and drug problems. He believed that if you hit a soldier where it hurt the most, in the wallet, he won't have any money left for booze or drugs. (Come to think of it, didn't Gen. Ration have a similar theory?)

229

Now back in the late 1970s, when I was assigned to a Ranger unit, we had a battalion commander who didn't believe in Article 15s. He said, "It's better to take away a soldier's free time than to ruin his career with an Article 15."

Though we didn't have any serious DUI in the battalion, we did have problems with soldiers getting themselves thrown in jail for drunk and disorderly conduct and bar fighting. But this only happened on the weekends, as we were usually training out in the field Monday through Friday.

That is, when we weren't away on a long deployment somewhere else. So naturally, it was just a tradition for us Rangers to cut loose and raise hell on the weekends after spending so much time out in the field.

Well, the battalion commander got tired of getting late night phone calls on the weekends from the local police department about his Rangers tearing up the town. So he passed word down through the chain of command that if we didn't start behaving ourselves, then he'll start taking the weekends away from us too.

Needless to say, we didn't heed his warnings right away. And we didn't learn our lesson until he started calling battalion alert formations on the weekends and taking us on 12-mile road marches in the middle of the night.

Now if you showed up drunk you were either sober by the time you completed it or you simply passed out somewhere along the way. And if you passed out or quit before successfully completing the march, you had to do it all over again the next day or the following weekend, with your chain of command. Talk about being in double trouble...

In another unit, a fellow platoon sergeant and friend of mine, used to call in his entire platoon whenever one of his men got picked up by the military police or the local police. His theory was, "If I gotta get up in the middle of the night to retrieve one of my soldiers, I'm gonna make the whole platoon come with me until they start behaving and policing themselves."

Yes, all this may have happened more than 20 years ago, but what worked for us old-timers back then I guarantee will still work today.

## ATTENTION TERRORISTS

Here in US Army Southern European Task Force (USASETAF) - Italy, you can always tell when someone calls in a bomb scare. As the MPs at the gate are the first to put on their helmets, flak vests, and conduct random vehicle searches.

Now I may not be a highly trained anti-terrorist security officer. But I know if a talented, intelligent, and determined terrorist wanted to place a bomb on a US military installation overseas -he's going to succeed. And only the dumb, stupid, and amateur terrorists are gonna f—— up and get caught.

Have ya ever wondered how many places you can hide a bomb in a car? I have, and after reviewing a security & search manual, here are the most common hiding places;

*Question*: How many different types of vehicle searches are there? Answer: Three, (1) a quick search, (2) a thorough search, (3) and a "detail" workshop search.

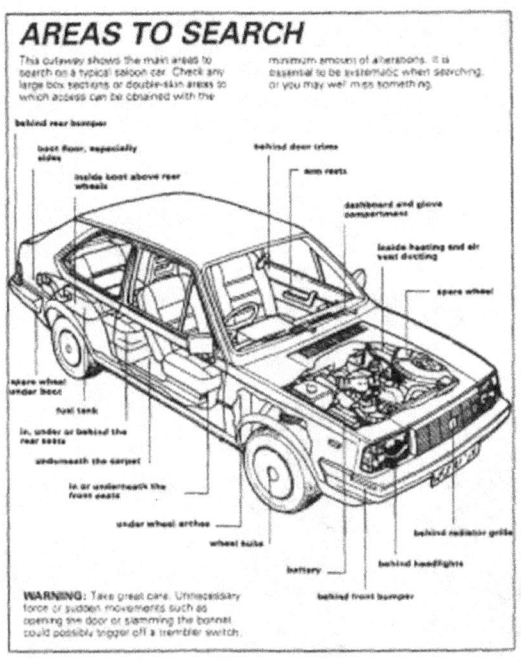

# A "RANGER" RADIO RIG

*Submitted By: Lee Churchwell*

Dear Ranger Rick,
I'm writing for two reason. First, to tell you how much I appreciate you putting together these *Ranger Digest* handbooks. And second, to offer you a tip of my own for your next edition.

Whenever I go outdoors I always like to carry my portable hand held CB radio and police/fire/emergency radio scanner just so I can keep up in what's going on in my area.

*Rick F. Tscherne*

But I got tired of carrying these radios in my pockets, so I thought about buying one or two of those high-speed radio pouches that the police and fire departments carry their radios in. But when I saw how much they cost, oh man, I almost choked on the price. (Arrrrrrrrgh!)

Being a little short of cash, I decided to make my own. And what I came up with looks pretty much like the real thing, except it only cost me a $1.

I bought a used M16 30-round magazine pouch from my local army & navy surplus store and modified it by cutting off the top flap and removing the plastic latch clip. Then I added some 550 parachute cord to both, the plastic latch and the nylon pouch and presto! A homemade improvised "Ranger Radio Rig." Check it out.

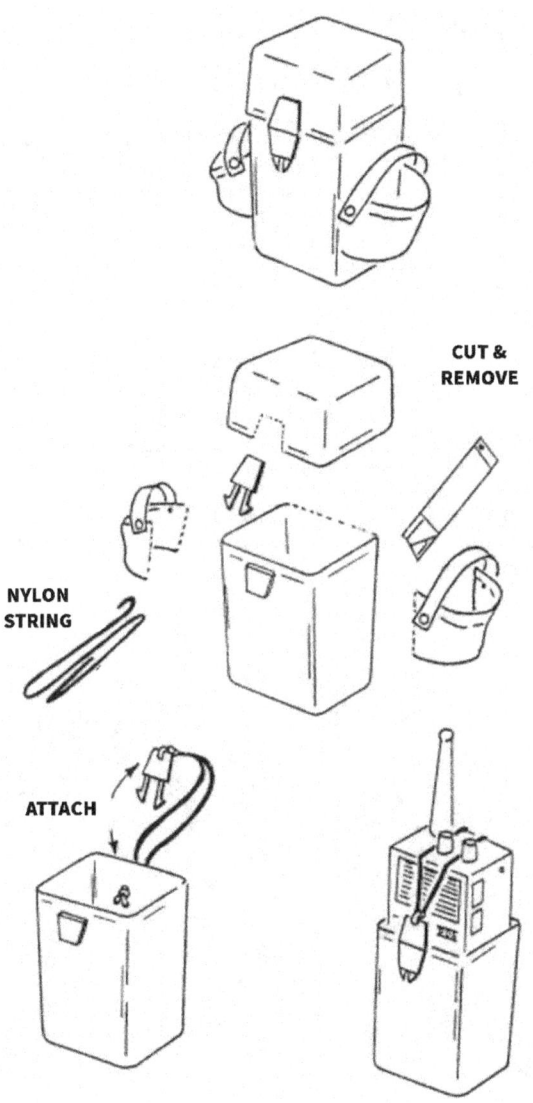

*Rick F. Tscherne*

# AN IMPROVISED CALL-FOR-FIRE TRAINING BOARD

While serving in Bosnia & Herzegovina with MPRI, one of the many classes we were tasked to give, was how to call in artillery and mortar fire. Unfortunately, due to IFOR/SFOR restrictions, we couldn't practice calling in the real McCoy on serb positions, (darn it).

So what many MPRI instructors did, was use viewgraph slides to teach their class. Except me, I thought it would be too B-O-R-I-N-G and put all the Bosnian leaders to sleep. (Zzzzzzzzz)

So I grabbed me a couple of toy soldiers, tanks, & vehicles, a green, black, & blue thick marker, some 100 mph tape, cotton balls, nails, an empty plastic water bottle, one blank viewgraph slide, and a piece of 4 x 4 regular ol' cardboard and made my own Call-For-Fire Training Board. (Yea buddy, it's called the "Art of Improvising, Modifying, and Adapting.)

Well, it worked so well, not only did other MPRI instructors ask me to make them one too, but so did many of the Bosnian unit commanders. Which I was happy to do, and here's what it looked like.

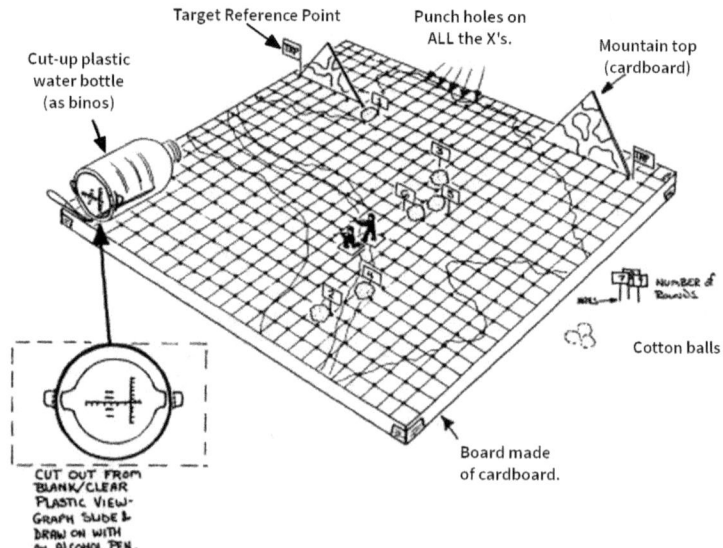

*The Complete Ranger Digest: Vol. VIII*

# A FEW MORE TIGHT 360'S

When operating at night as a small 3-5 man patrol, depending on the terrain and enemy situation, it may not always be feasible to set up a well dispersed 360 security perimeter. Instead, it might be wise to go "tight."

Here's a few "TIGHT 360s" that are commonly used by the British SAS and our American Special Forces and Rangers. (See *Ranger Digest V* - Pages 84-85 for more tight 360s.)

What's great about these techniques? Well, it enables you to quickly alert and communicate with the person on your left and right in complete silence. One tap/jab - get up, two taps/jabs - stay down, three taps/jabs - enemy approaching.

What's bad about 'em? All it takes is one grenade and or several well placed shots from a weapon with a night device attached to it and "you're history."

Rick F. Tscherne

# THE SPEEDHOOK

Here's a neat little survival gadget that's available through the DOD supply channels, it's called a SPEEDHOOK. (Ha! I'll bet ya never knew it existed, did ya?)

A fellow by the name of Mr. Andrew J. Pratscher came up with this handy device after he was in a car accident and couldn't enjoy fishing the same way he use to.

And according to an article published in the American Survival Guide. Every time he tried to set a fishing hook, he was just too slow at setting it.

He says, "I just couldn't move as fast as I use to due to an old car injury. My right shoulder was injured and I had to have some surgery done on it. But some things just can't be repaired to work the same way as they use to."

"I learned how to do a lot of things, but I just couldn't do them very fast, to include catching fish. So out of desperation I started messing around with a few ideas. And through trial and error I came up with a device that's helped me to catch fish the way I use before the car accident, I call it a Speedhook."

The Speedhook works best at catching crappie, bluegills, perch, carp, rock bass, largemouth bass, smallmouth bass, wallege, white fish, and catfish.

And Mr.Pratscher also says, "all my Speedhooks are assembled by hand and come with a life time money back guarantee." NOTE: There are two types, the Regular Speedhook - $ 4.95 each, and the Military Issued Speedhook - $ 9.95 each. To order, contact:

SPEEDHOOK
Dept. ASG, PO Box 11215 Merrillville, IN 46411
Ph: 1-877-SPDHOOK
http://www.speedhook.com
Email: admin@speedhook.com

YouTube video link: http://www.youtube.com/watch?v=96DhW2gDsRY

Attention Military Service Members: To order this device through the DOD supply channels, the NSN is #4220-01-3795598.

*The Complete Ranger Digest: Vol. VIII*

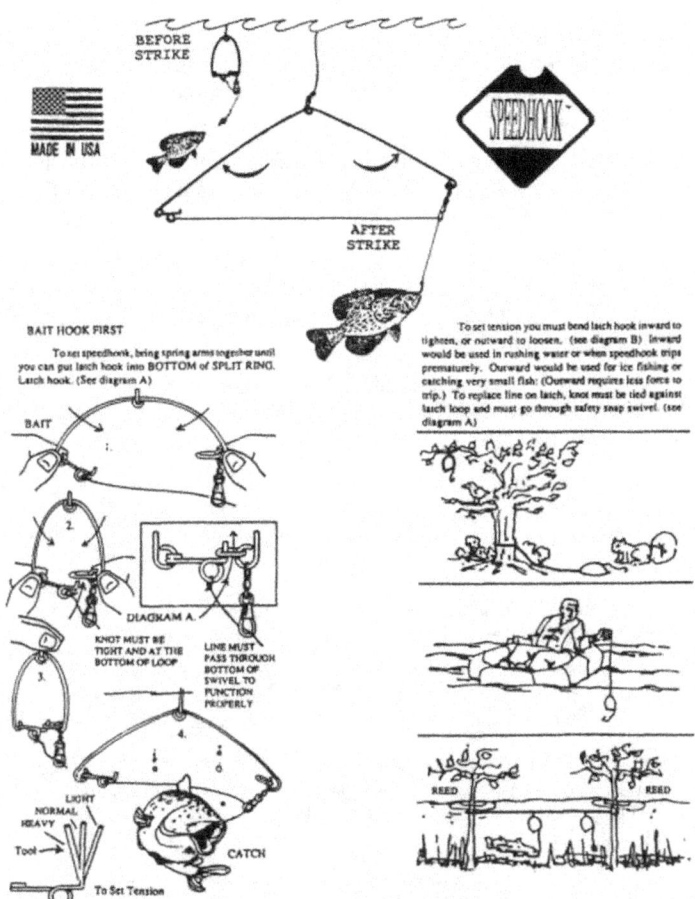

FAILURE TO FOLLOW INSTRUCTIONS MAY RESULT IN BODILY INJURY. NEVER PICK UP SPEEDHOOK FISHING RIG BY THE FISH HOOK. SPEEDHOOKS ARE TO BE USED TO CATCH FISH, ANY OTHER USE MAY RESULT IN BODILY INJURY. NEVER TOUCH LEADER OR FISH HOOK WHEN SPEEDHOOK IS IN THE SET POSITION.

Purchaser assumes all liability with the use of SPEEDHOOK. Not recommended for children without adult supervision.

Speedhook is illegal in Minnesota

*Rick F. Tscherne*

# NEW COMBAT BOOTS

Hallelujah! Well it's about time the services started to wise up and buy a decent pair of boots for the troops. And one that's been used for years by civilian hikers, mountain climbers, and outdoor enthusiasts, the "MATTERHORN."

It's not only stylish, lighter, cooler, and much more water-resistant than the current military issued leather boot, but it's made of Gore-Tex and Teflon too. Except the soles.

The military VIPS finally realized the ol' leather combat boot was just slightly better than going barefooted. As all the stress of running, walking, marching, etc, was being transmitted to the foot & leg with very little cushioning and support. (Duh, no shit!)

And the daily blisters, raw heels, fallen arches, ingrown toenails, trench foot, shinsplints, etc, all the traditional infantryman's foot problems...may now be of the past. (We hope.)

According to the boot manufacturer, they told the Pentagon; "You guys are still stuck back in the 1940's when it comes to boot technology, why don't you take advantage of all the research and development that has gone into the civilian boot market."

Well, so far only the Marines have agreed to purchase these new boots. And the Army? Well, they're still trying to decide if they want them or not. (As usual, "a day late & a dollar short" in deciding what's best for the troops.) But I'm pretty sure they'll probably choose the same boots, but then again, maybe not. God forbid should they choose something the Marines are already using.

The cost? Uncle Sam is gonna pay $89 for these Matterhorns. Which ain't bad compared to what they're already paying for the current cowhide leather boots - $69. And the civilians? Well, if they want to own the same pair, it's gonna cost 'em a whopping $133. (Ouch!)

Cove Shoe Co.'s Matterhorn boot

*The Complete Ranger Digest: Vol. VIII*

# COMMANDO'S GET NEW ROLE

According to military sources, the Pentagon is prepared to mount commando strikes against terror groups or rogue nations that threaten to use weapons of mass destruction against the US and or other friendly nations.

Units of the multi-service U. S. Special Operations Command are engaged in intensive training missions to seize nuclear weapons and to destroy chemical or biological factories and storage sites. The commandos would be used when conventional military firepower, such as the cruise missiles attack in 1998 against Afghanistan and Sudan, would be ineffective.

The 47,000-person U. S. Special Operation Command, headquartered at MacDill Air Force Base, Florida took on the new mission about three years ago (1995).

Though they've been practicing the "loose nukes" scenarios for some time now, they have just begun to receive a major infusion of additional funding for "direct action" operations to find, neutralize, and retrieve nuclear weapons in hardened or underground sites that missiles or bombs can not destroy.

### READY FOR ACTION

Four units from the multi-service U.S. Special Operations Command would have responsibility for attacking hardened underground sites to destroy or seize nuclear, biological and chemical weapons in the hands of terror groups or rogue nations. They are:

■ Joint Special Operations Command: Highly classified headquarters unit under the U.S. Special Operations Command that supervises counter-terrorist commando units. The staff of several hundred is based at Fort Bragg, N.C.

■ Special Forces Operational Detachment-Delta: Activated as a primarily counterterrorist unit in 1979, Delta Force is believed to number about 800 commandos and support personnel. It led the ill-fated 1980 embassy hostage rescue mission in Iran. Delta Force commandos took part in a successful hostage rescue mission during the 1989 intervention in Panama, and operated behind the Iraqi lines during Operation Desert Storm, hunting for Scud missiles in 1991. Delta commandos also have participated in several operations responding to terrorism, including the attempt to capture the Palestinian hijackers of the Italian cruise liner Achille Lauro in 1986. The unit is based at Fort Bragg.

■ Naval Special Warfare Development Group: The commando unit replaced the original Navy counterterrorist unit Seal Team 6, formed in 1980 to fight terrorists in a maritime environment. It took part in the 1983 Grenada intervention, where several members were killed in action. In 1992, members of the unit made a rescue in Haiti of several former Haitian government officials whose lives were in danger. The unit is based in Hampton Roads, Va.

■ 160th Special Operations Aviation Regiment: Known as the "Night Stalkers," the helicopter unit serves as a helicopter delivery arm for Delta Force commandos. The regiment was in combat in Panama and Operation Desert Storm. It is based at Fort Campbell, Ky.

Other Special Operations units: For counterproliferation missions overseas, the nonclassified branch of the U.S. Special Operations Command would likely provide intelligence and combat reinforcements for a commando mission. Units include:

■ Air Force Special Operations Command: Headquartered at Hurlburt Field, Fla., the organization flies both planes and helicopters configured for covert, long-range flights and penetration of hostile areas. The command also operates AC-130 gunships that can deliver artillery fire.

■ Army Rangers: Members of the 75th Ranger Regiment, including the Fort Lewis, Wash-based 2nd Ranger Battalion, are trained to support special operations missions. Their role would be to seize strategic areas such as airfields, then provide firepower support for commandos.

■ Army Special Forces: These five, battalion-sized units, including the 1st Special Forces Group at Fort Lewis, would provide expertise in a particular geographic area, including local contacts and language assistance, if needed.

■ Navy Seal Teams: Six conventional Navy Seal teams — three on each coast — could provide seaborne combat power to a counterproliferation operation. Other Navy special warfare units could provide transportation on high-speed patrol craft or mini-subs launched from submarines.

— Seattle Post-Intelligencer

"These are very tough targets," stated a Pentagon spokesman, "and we are prepared to use human resources to take out any nuclear or chemical weapons before they are used against the US or another friendly nation."

And according to military sources, the handpicked personnel and units that would be performing these new missions, would come from...(see above)

Rick F. Tscherne

# CAMMIE POLE LADDER

Have ya ever practiced entering a fortified building from a second or third floor window? You know, such as tying a rope around a log and tossing it through a window? Or maybe you lifted a squad member up into the window with the log. Did ya? Uh? Did ya?

Now let's be honest, would you really want to use these techniques in real combat? As the kid from the movie "Home Alone" would say, "I don't think so."

No doubt it would be a lot easier to gain entry into a second or third floor window with the use of a grappling hook or a ladder. That is, if you're fortunate enough to have any of these items in your unit, which not many do.

But one item there seems to be no shortage of in every unit, and that's vehicle camouflage net poles. And believe it or not, you can make a nice portable field ladder out of them. How? All ya need is a drill, a drill bit, some thick nuts & bolts, a measuring tape, string, a marker, and the metal stakes that come with the camouflage net kit. Really, I'm not BS-ing you.

Just take the number of poles that you want to use, run a piece of string straight up and down the center of the pole (from top to bottom) and tape it (the string) in place. The purpose of this string is to help you to drill all the holes straight and even down the center of the pole.

Then grab a measuring tape and magic marker and mark down alongside this string how far apart you want the steps to be and start drilling the holes. But not just through one side, but through both sides of the pole.

IMPORTANT: The distance between each step must be exactly the same for all the poles so they can be interchangeable with one another.

When you've finished marking and drilling the holes in this first pole, you can either repeat the procedure for the other poles. Or place this first pole on top of other poles and drill right through these same holes and into the next pole.

To insure they get drilled straight and even just like the first one, take some 100 mph tape and fasten 'em (the two poles) firmly together before drilling.

When you have finished drilling all the holes in all the poles that you intend to use, lay the poles side-by-side in pairs of two to determine how far apart you want them to be.

Then grab one of the metal stakes that come with the camouflage kit and lay it across the poles and mark it where you need to drill the holes so it can be used as the first step of the ladder.

Then use this metal stake/step as an "example" to assist in drilling the holes in all the other stakes. NOTE: Carefully lay and align this stake on top of the other stakes one-at-a-time and drill right through these same holes.

Lay your poles on the ground side-by-side with the holes facing up, take your stakes/steps and place them across the poles so they (both, the stakes and poles) are aligned one over the other and then begin fastening them together with the nuts & bolts.

NOTE: When drilling the holes, use a large drill bit and large nuts & bolts so you can "hand tighten" the steps to the poles without the use of any special tools (wrenches, pliers, etc). This will assist in assembling and disassembling the ladder much more quickly. And always carry a few extra in case you lose some of the pieces.

Regardless of how many poles you decide to use, this ladder will be easy to transport and set up & take down in a matter of minutes. And not only will you be able to use it for entering building windows, but for climbing trees to emplace snipers, observers, and or for rescuing paratroopers caught in trees. Not to mention, they can also be used as portable foot bridges for crossing small creeks and streams.

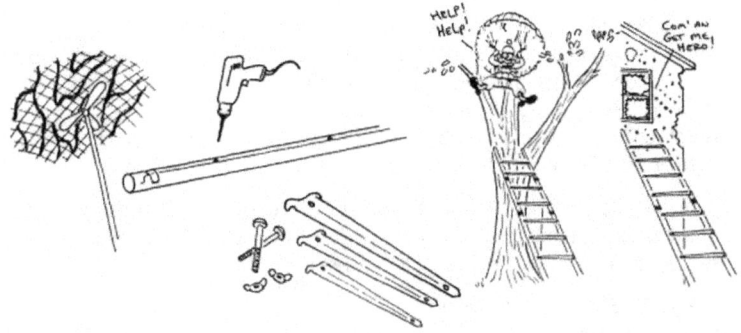

## GLOBAL POSITIONING SYSTEM

For years I've been putting off buying one of these, at least until they've gotten a bit more "down to earth" cheaper. And when they did, I bought one, a cheapo Magellan GPS 2000 XL.

What's a GPS? It's a self-contained handheld navigational device that uses a series of orbiting satellites to compute your location on earth, whether you're at sea, in the air, or on dry land.

How does it work? Well, each navigational satellite transmits it's own precise location. And when your handheld GPS receives at least three different signals from the 24 + navigational satellites that are orbiting the earth, it computes your location and transmit it to your GPS receiver.

How many different types of handheld GPS are there? Though there are several models to choose from, if you're in the military and you're planning on buying one, make sure it comes with MGRS. Which stands for Military Grid Reference System. Otherwise you're gonna get stuck with a GPS that you won't be able to use with a US or NATO issued military map.

How accurate are they? The cheap ones are accurate to within 10 (+/-) meters, but the more you pay, the more accurate they are.

And the short falls? Well, if you think you're gonna be able to move, stop, turn on your GPS to get a "quick fix" on your position and then move on again within a "few minutes," you're wrong. Once it's programed or initialized, depending on the terrain and weather conditions, it could take anywhere from 5, 10, 15 minutes and even longer for the satellites to compute your location.

Are they worth the money? You betcha! Whether you're an experienced navigator or not, these handheld GPS are pretty handy out in the field, regardless if you're deployed to the desert, jungle, or even the arctic circle.

Some other features? A track plotter, which is used for tracking and recording up to 5 different routes,20 legs, and 200 landmarks, weighs 10 oz (+/-), runs on 3-4 AA batteries for 20-30 hours of continuous use, and comes with 3-5 different navigational viewing screens. Plus a lot more other features too. And they're pretty easy to operate, you don't need a college education or a degree in computers to figure out how to work one.

Where do you buy them? From almost any outdoor or military supply store or mail order catalog. How much are they? Anywhere from $150 for the cheapos, to as much as $400 for the expensive ones.

UPDATE: GPS unit reviews online:
www.gpsreview.net
www.geocaching.com/reviews/gps
reviews.cnet.com/best-gps/

*The Complete Ranger Digest: Vol. VIII*

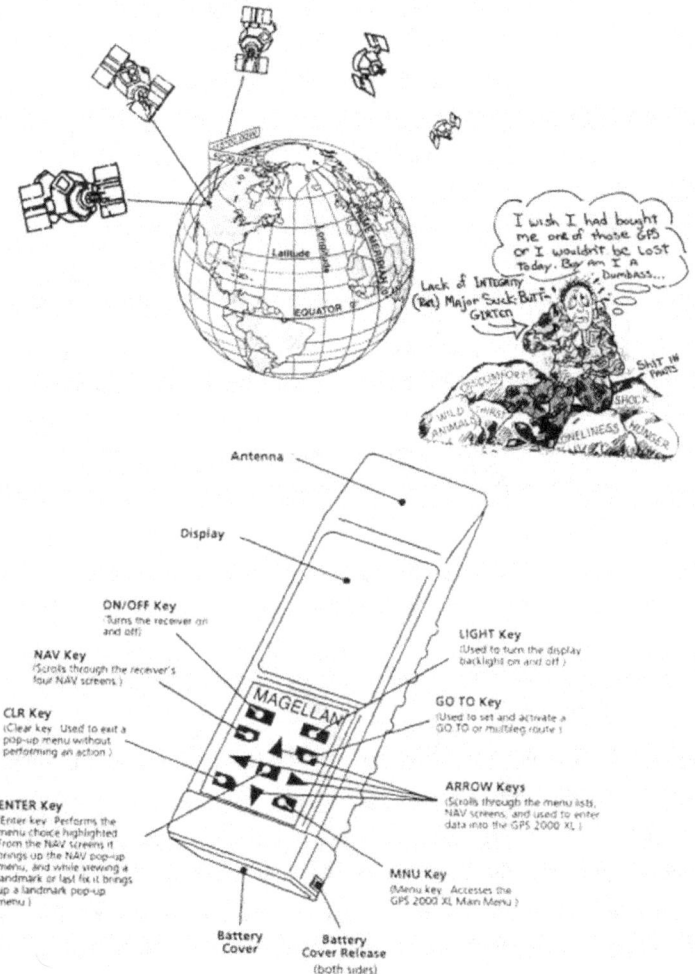

Rick F. Tscherne

# VELCRO ON KEVLARS

*Submitted By: PFC Ben Donaldson*

Dear Ranger Rick,

I think I've got a couple of good ideas that you can use in your next Ranger Digest. But before I tell you about them, I'd like to pass along to your readers some advice.

If you're serious about your military job, you should invest some of your own money into buying some of the equipment that the Army doesn't provide you.

Here in the Army National Guard we're often asked to do more with less and we're not always issued the latest nor the best military equipment. But if you have your own, well, you'll be much better prepared. Plus, you'll be able to modify it without getting in trouble, as it belongs to YOU and NOT the Army or the state.

Working at night can sometimes be a royal pain in the butt, especially if you only have one free hand to work with because you have to hold a flashlight in the other.

So I bought some velcro and glued a piece of it to my MagLite and sewed the other piece to the side of my kevlar camouflage cover. Now whenever I work at night and need to use both my hands, I just attach my MagLite to my kevlar helmet.

Ranger Rick's Comments: Don't worry about being out of uniform just because you've got a piece of velcro attached to your kevlar camouflage cover. If you're like most soldiers, you probably own two, one for garrison and the other for the field. Naturally, you only need to sew it to the one you wear to the field.

When attaching...

a) the velcro to your MagLite, make sure you glue it firmly in place.

b) the velcro to your kevlar cover, sew it either on the left, right, or both sides of the camouflage cover.

IMPORTANT: Before sewing the velcro into the camouflage cover, hold your MagLite up to the helmet with one hand and a map in the other to determine where exactly the velcro needs to be sewn.

ALSO: Extra Heavy Duty adhesive Velcro tape:
http://www.kk.org/cooltools/archives/001014.php

*The Complete Ranger Digest: Vol. VIII*

# M16 PISTOL GRIP STORAGE COMPARTMENT

*Submitted By Daniel Garcia*

Hey Ranger Rick!
Your Ranger Digests are absolutely great, they've helped me many times in the field, and now I got a tip of my own that I'd like to pass along to your readers.

Soldiers can easily convert the hollow portion of their M16 "pistol grip" into a small storage compartment by ordering an old World War II/Korean War Ml Carbine accessory piece called a "Magazine Dust Cover."

Though these dust covers were originally designed to help keep dust, dirt, and sand out of a M1 Carbine magazine. They'll stretch and fit snuggly over the bottom portion of an M16 pistol grip so you can store a few little extra items inside of it. Such as ear plugs, cleaning patches, a mini bic lighter, and even some survival things, (string, fishing hooks, wire, etc.).

TO USE: Hold the M16 rifle upside down and place the items inside the pistol grip and "stretch & slide" the dust cover over the front and back portion of the grip. You can order it from:

Sherluk Marketing & Trading
P.O. Box 6991, Toledo, Ohio 43612
(Cost: $1.00 each)

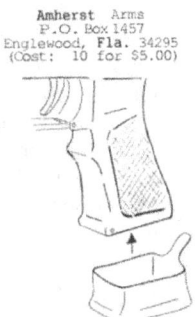

Amherst Arms
P.O. Box 1457
Englewood, Fla. 34295
(Cost: 10 for $5.00)

AR Stowaway Pistol Plug

PMS Inc. is introducing a new AR Stowaway Pistol Plug that allows you to convert an existing AR-15 or A2 pistol grip into a storage space for extra firing pins, springs and other accessories. The plug is intended to be an affordable alternative to the Stowaway Pistol Grip. The plug is made of rubber and is friction fitted into the bottom of your existing A2 pistol grip.
For further information, call (800) 878-3767.

UPDATE: Numerous newer/better options for aftermarket storage grips are now available for the M16/M4 rifle:
- Magpul
- Tango Down
- US Palm
- Hogue, and the Hogue grip w/ integrated Samson Field Survivor multi-tool

Rick F. Tscherne

# HMMMMM, WHERE HAVE I HEARD THIS BEFORE?

Not too long ago I received a letter from one of my Ranger Digest readers, and attached to his letter was an interesting article that he tore out of a military magazine.

I couldn't tell which magazine it came from, but it was about an S-4 captain who claimed "he developed" a unique and convenient way to resupply water to the troops out in the field. Check it out below:

## Water Resupply in the Light Infantry

CAPTAIN W\_\_\_ C _____, JR.

One of the most difficult logistical missions in light infantry is water resupply. These soldiers must have water to survive, but they must also carry what they drink. In cool weather, six quarts will last 24 hours. in hot weather, soldiers will drink more than eight quarts in 24 hours, which means they will have to be resupplied every 12 hours. From a battalion S-4's perspective, the difficulty is in making sure water gets to every soldier in a usable package.

When I was a battalion S-4 in the 2d Battalion, 27th Infantry, during a rotation at the Joint Readiness Training Center and all of the training for it, I learned a lot about water resupply.

There are various ways to resupply water in light infantry: One way is to deliver water cans to line companies with the logistical package (LOGPAC). The problem with this method is that the platoons and squads are usually spread out and performing missions. There is not time enough to distribute five-gallon cans and collect the empty cans during the short LOGPAC window. Soldiers have to carry them around until the next LOGPAC. Supply sergeants have to bring along at least 80 water cans so they can keep 40 with the company between LOGPACs (not counting cans that will be lost).

A second method is to use 50-gallon blivets during LOGPAC. But it is unrealistic for a company to use them, because all its soldiers must be brought to one location to fill their canteens.

The solution we came up with was to use six-gallon plastic milk containers, the milk bags used in the mess hall. We bought them empty from a milk company. More than 1,000 bags cost less than $800 in Hawaii and should be even less expensive in other areas. The 1,000 bags, which came with a sealed white tube attached, took up the space of a footlocker. To fill a bag with water, a soldier pops the tube off, puts water in, and replaces the tube. To fill a canteen from the bag, he cuts the end of the tube and water streams into the canteen.

We used the water bags for the first time during a brigade field training exercise. The one problem we had to solve was carrying the bags once they were filled. If they were not packaged, they were difficult to carry around and load. We wanted a package that was already part of the supply system and one that could be thrown away.

MRE (meals, ready to eat) boxes fit both of these needs. The support platoon put the MREs in trash bags in the brigade support area before bringing them out at LOGPAC, and then they put the full water bags in the MRE boxes. This worked very well. The boxes are easier to load and are in-tended to be thrown away when they're empty. When the LOGPAC was delivered, all the supply sergeant had to do was kick out the MRE boxes and the trash bags, which reduced our LOG-PAC time.

Once the soldiers had been resupplied, they were able to treat everything delivered at LOGPAC like trash. They left it for pick-up and moved out.

There are some other benefits to water-bag resupply. The bag's two-ply plastic will not burst unless it is punctured by a sharp object, and it can be reused. A soldier can put any amount he wants in the bag and carry it in his rucksack like a five-quart blivet.

When the bags are in MRE boxes, they can easily be slingloaded. They can also be stacked inside aircraft. (We conducted five battalion air assaults in preparing for and conducting our JRTC rotation. We slingloaded or slacked water boxes with almost every air assault.) The same is not true of water blivets or cans. With water-bag resupply, it is easier to preposition or cache water. When pre-positioning cans or blivets, there is always a concern that they will be left behind. With the water bags (at less than 80 cents each), there is no worry about leaving them behind. A unit can preposition bags in two different sites, knowing that only one of them will be used

My recommendation is that the Army make water-bag resupply the standard for light units. If water bags were made to fit light infantry unit specifications, the resupply process would be easier for everyone and also save money. Anyone who has been to the JRTC knows that many water cans are lost or left behind in the boxes. During unit training, it's the same story. Water bags cost far less and can be reused if necessary, and soldiers can carry empty bags around if they have to. Another saving, both in dollars and in unit effectiveness, is in heat casualties, most of which occur because individual soldiers do not have access to enough water.

Water-bag resupply is the cheapest, most efficient way of getting water to the people who need it most—the light infantrymen at company level.

### Rick F. Tscherne

*Captain W C_____Jr.*, is assigned to the\_\_\_ Battalion \_\_\_Infantry, in \_\_\_\_. He has served as a platoon leader, a rifle company executive officer, and a battalion S-4. He is a 1990 ROTC graduate of The Citadel.

OK, so what do you think? Does it sound familiar? *Check out my Ranger Digest VI* - Page 66-67.

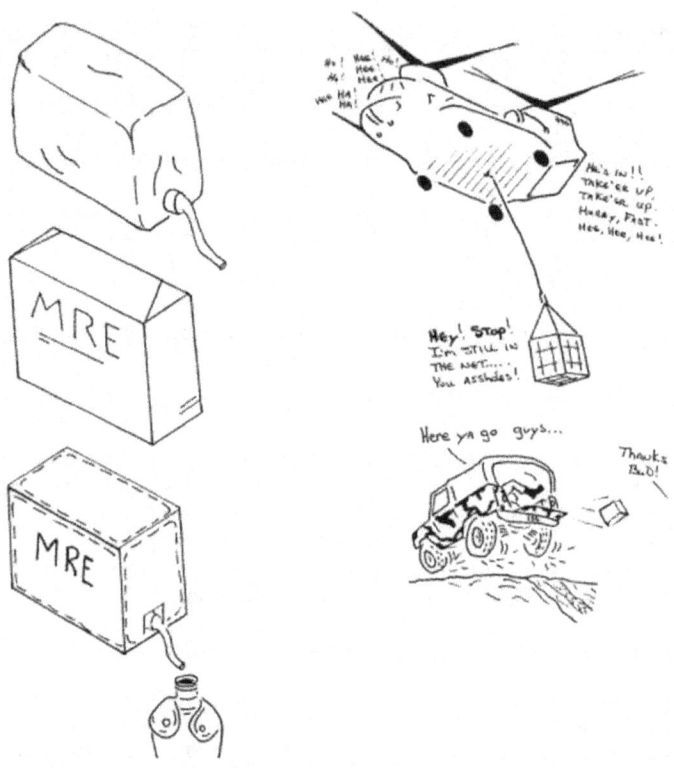

*The Complete Ranger Digest: Vol. VIII*

# COMMON WEAPONS YOU'LL SEE IN PEACE & IN WAR

The SOPMOD M-4 Accessory Kit adds a range of specialized devices to the already capable M-4 carbine.

*Rick F. Tscherne*

# TIPS & TRICKS FROM A NAVY SEABEE

### Submitted By: PO2 Paul Gromkowski

Dear Ranger Rick,
I have all your Ranger Digest handbooks and enjoy reading all the ingenuity ways in how to live comfortably out in the field. I have spent the last 8+ years in the Navy Reserve as a Seabee, and like our slogan says, "We Build - We Fight." We spend half our time in the field sharpening our combat training skills and the other half constructing and building things.

Anyway, here's a few of my own tips & tricks that I would like to contribute and share with your readers.

1# - I've always hated to feel wet grass, dirt or mud under my feet whenever I got up out of my cot or sleeping bag. So I use the cardboard sleeves that hold MRE boxes together as a floor mat. I just open them up and lay them on the ground next to and or under my cot or sleeping bag and no more annoying wet grass or dirt.

2# - Whenever I deploy to the field laundry time is always a pain in the butt. Especially when you've got to share 1 x washer and 1 x dryer with a hundred or so other guys. So to cut down on the wash cycle and the amount of clothes to wash and dry, I only bring my dark colored clothes to the field. One load of clothes to wash and one load of clothes to dry and I'm good for about a week.

#3 - In the field and at home clean hands lead to good health. Most colds, viruses, and diseases are spread when it moves from your hands to your mouth, nose, eyes, and or other opening parts of your body. So I "always" wash my hands before I touch or eat any food. But as you know when you're deployed to the field or to a remote country, it's not always possible to find clean running water for washing. And sometimes there's no water at all except for drinking.

Which is why I always carry in my rucksack or tool box a few small bottles of "gelled alcohol." They're pretty handy to use because you don't need a towel after putting some on your hands, as it evaporates very quickly. And as an added bonus, this stuff is flammable just like Sterno and Trioxane and makes an excellent fire starter too. CAN DO!

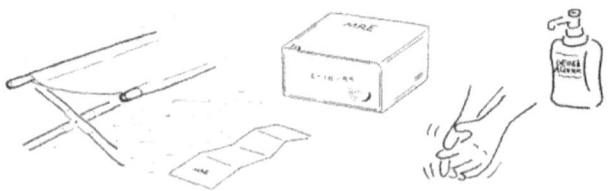

*The Complete Ranger Digest: Vol. VIII*

# SIGNAL MIRROR ADDITIVES

*Submitted By Ssg Steve Brittian*

There's only one side of a signal mirror that is useful, and that's the side with the mirror. And if there isn't any sunlight out, well, you ain't gonna be doing very much signaling with it, are ya?

Now the backside of a signaling mirror has no use whatsoever, so to make this "worthless side" a bit more useful, here's what you can do to it.

a) Get yourself some luminous (glow) tape and lay it down on a flat surface and place your signaling mirror on top of it. Then cut out a piece of illumines tape about the same exact size and glue it to the "worthless side" of the mirror.

Or...

b) Get yourself some bright orange paper, acetate both sides, place the signaling mirror on top it and cut out a piece about the same exact size and glue it to the "worthless side" of the signaling mirror.

Whichever method you decide to use, at least the "worthless side" of the mirror will come in handy for something, either for short-range "day" or "night time" signaling.

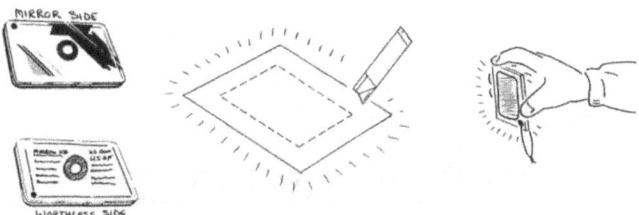

# COMBAT EYE GLASS HEADSTRAP

*Submitted By: Major David Oaks*

Hey Ranger Rick,
This tip comes from a buddy of mine, MAJOR MIKE STEELE, who's assigned to the 505th Infantry / 82d Airborne Division. Hooaah!

Like a lot of Airborne troops, my buddy wears a pair of those high-speed nylon combat eye glass frames whenever he goes to the field and or makes a parachute jump.

But the cheap black rubber headbands that come with these frames seem to quickly break, melt, and or fall apart whenever they come in contact with military issued bug juice. Well, over a period of time, it can cost you a lot of $$$$ to keep replacing them.

So what my buddy did to solve the problem was replace his cheapo headband with a parachutist's kevlar helmet "retention strap." This strap is already o.d. green, comes with velcro, and can be adjusted to fit on almost any size shape head.

But best of all, it's inexpensive and available in all the AAFES Military Clothing Sales Stores (MCSS). Check it out!

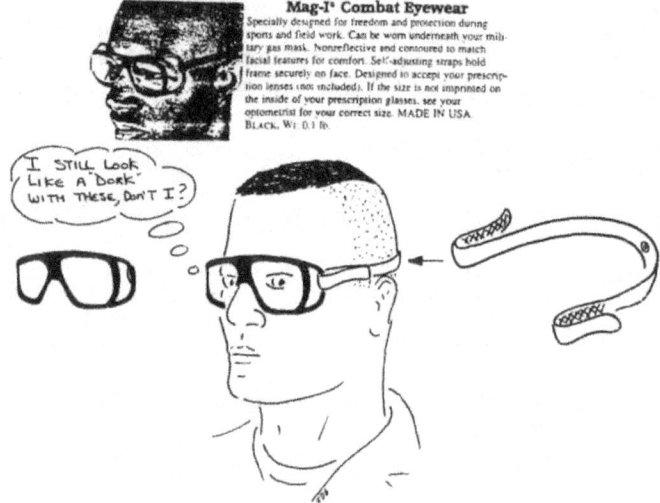

## CANTEEN MODIFICATIONS

Not very long ago one of my Ranger Digest readers sent me an interesting letter. He said, "Hey Ranger Rick, wouldn't it be nice if our canteens came with a "window slit" on the side so you could see how much water you actually have remaining without guessing? It would also make it easier to determine how many water purification pills you need to place inside of it too."

Hmmmm, you know something? He's right, it would be kinda nice.

Well, I put on my ol' McGyver "thinking cap" and this is what I did to one of my own military plastic canteens.

I took a pencil and ruler and drew a straight line down the center of the canteen from top to bottom. Grabbed me a heavy duty sharp razor and made a cut on the left and right side of this penciled line and then cut out this thin piece of plastic.

Then I filled the slit with some "clear silicone," waited until it dried, then tested it out by filling it full of water. And guess what? Yep, it worked. As I drank and emptied some of the water I could easily see right through the silicone the amount of water remaining. Neat! Cool! Hooah!

Another way you can make this window slit is with a bandsaw.

WARNING: If you've never used a bandsaw machine before, I strongly advise you to find someone who has so they can give you a hand and save you from losing one.

Hold your canteen in the upright position while standing on the opposite side of the machine where you would normally be standing. This is so you can see exactly where you're cutting, which means you will be pulling "towards yourself" rather than "pushing away" from yourself, you know what I mean? Uh?

Now when you're done making one cut/slit, carefully make another cut/slit right along side the first one and then remove this piece of plastic with a sharp knife or razor.

Then take some "clear silicone" and carefully squirt it all along and inside this narrow slit. And don't be stingy with it neither, it's better to squirt a lot of it inside than very little at all or it will eventually leak on you. Then take a flat long narrow stick and place it inside and flatten the silicone along the walls of the inside portion of the canteen.

Then wet one of your fingers with some "saliva" (mouth spit) and smooth out the silicone on the outside portion of the canteen so it's just about flat, but NOT completely. If the silicone begins to stick to your finger, this means your finger is too dry and does not have enough "mouth spit" on it, just add some more saliva and you'll be good to go.

Once the silicone is completely dry, fill the canteen full of water to see if it leaks. And if it does, then just add silicone to the part of the slit where it leaks.

Now you'll be able to see and not guess anymore how much water you actually have inside your canteen(s).

Rick F. Tscherne

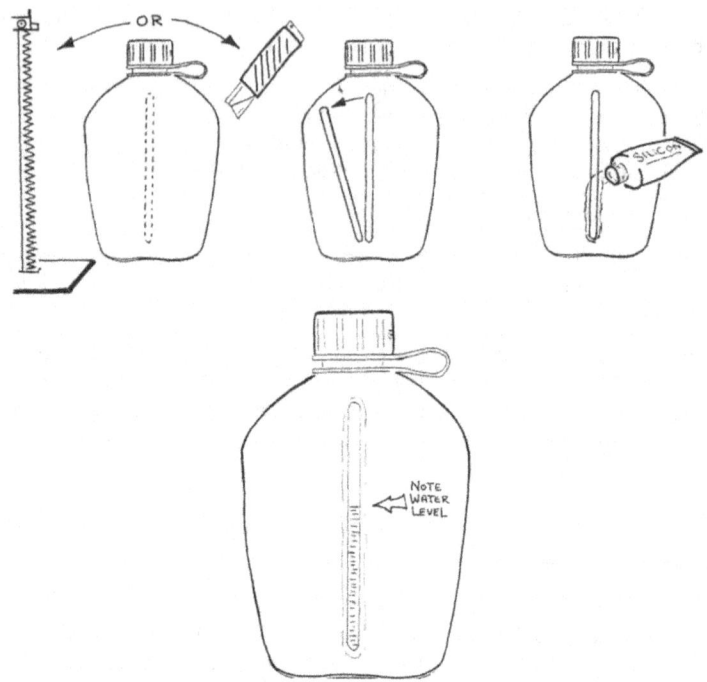

## WATERPROOFING MAPS

*Submitted By: CPT. Robert C. Fraser, Jr.*

When regular military acetate is not available, you can always use clear shelving paper made by RUBBERMAID. It's not expensive and it's available in all of the Home Depot, K-Mart, and Walmart department stores.

But before placing it on a map, it's important you remove as many wrinkles as possible. To remove most of (but not all) the wrinkles, take an iron (without any water inside of it) and turn the temperature setting to MEDIUM HOT. Then lay the map on an ironing board and constantly keep the iron moving over it until most of the wrinkles are gone.

Then lay the map down on a large table and place some scotch tape only on the corners to keep it from moving and sliding around.

Take your shelving paper, roll a little bit out, peel back some (but not all) of the wax paper from the back side and place it slightly beyond the edge of the map. Then slowly, very slowly begin peeling and rolling it out and placing it on the map.

As you peel and roll, press down hard on the shelving paper. If bubbles start to appear underneath, simply flatten them with a plastic credit card. And if this doesn't work, then take a straight pin and poke a few tiny holes into the bubbles to release the trapped air.

When the front of the map is completely covered, repeat this same procedure for the back side. Yep, that's right, ya gotta do the back side too, because it doesn't do any good to just waterproof one side and not the other. So take your time, do it right the first time and you won't have to worry about doing it a second time.

When you're finished with both sides, take a pair of scissors and cut off any access shelving paper along the sides of the map.

Three 2d Lieutenants died and arrived at the pearly gates of heaven. Saint Peter tells 'em, "If you can answer one simple question, I'll open up these gates and let you in."

Saint Peter turns to the first 2d lieutenant (an OCS graduate) and asks, "What is Easter?"

The lieutenant smiles and says, "Oh, that's an easy question, it's that holiday in November where everyone gets together, eats turkey, and gives thanks for..."

"Wrong" interrupts Saint Peter. And he turns to the second lieutenant (an ROTC graduate) and asks, "Tell me, what is Easter?"

The lieutenant says, "I think Easter is that holiday in December when we put up nice trees, decorations, and exchange gifts to celebrate..."

"Wrong" replies Saint Peter angrily shaking his head in disgust. And he then turns to the third 2d lieutenant (a West Point graduate) and asks, "Do you know what Easter is?"

The West Pointer smiles confidently at the other two lieutenants and says, "Yes sir, I most certainly do."

"Oh?" says Saint Peter with a surprise look on his face. "Really? Then please tell me what is Easter?"

The West Point lieutenant says,

"Easter is a Christian holiday that coincides with the Jewish celebration of Passover. Jesus and his disciples were eating at the last supper and Jesus was later deceived and turned over to the Romans by one of them. The Romans took him to be crucified and he was stabbed in the side, made to wear a crown of thorns, and was hung on a cross with nails through his hands and feet. He was then buried in a nearby cave which was sealed off with a large boulder."

Saint Peter, now smiling from ear to ear, is shaking his head up and down in approval when the West Pointer continues on and says..

"And every year this boulder is moved off to the side so that Jesus can come out, and if he sees his shadow, we'll have six more weeks of winter."

## AN IMPROVISED CPR PRACTICE DUMMY

While training one of the many Muslim army units in Bosnia & Herzegovina, I became good friends with a Bosnian medical officer. And to assist him to teach his soldiers in how to properly administer CPR and Cardiopulmonary resuscitation, I built a couple of practice CPR dummies.

They were made of the following items:

| | |
|---|---|
| 1 x Plastic Water Bottle | 1 x Clorex Bleach Container |
| 1 x Flat Wooden Board | 1 x 50 cm Rubber Hose |
| 2 x Plastic Trash Bags | 1 x Roll of 100mph Tape |
| 1 x Tube of Silicone/Glue | 1 x Black Magic Marker |
| 1 x Heavy Thick Book | |

Though it was very primitive looking, it definitely worked the same way as a real CPR & Cardiopulmonary Practice Dummy.

WARNING: After every use, always wipe the mouth area of the CPR dummy with some alcohol to "prevent" the spread of germs from one-person-to-another.

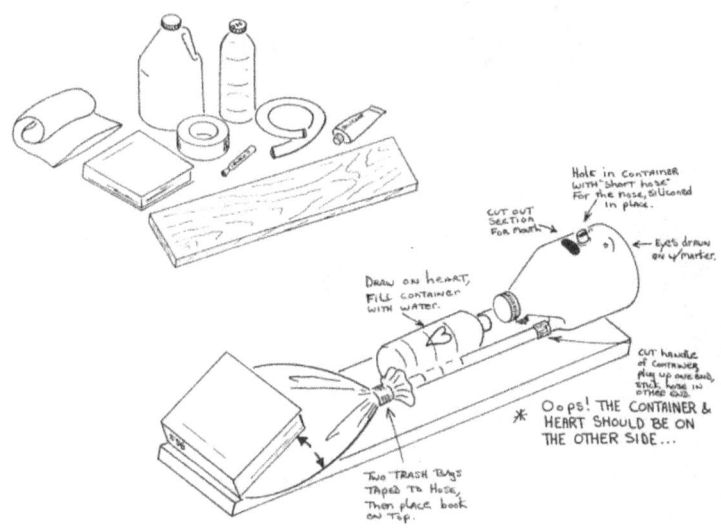

## A DESERT SHELTER "SHADE" TIP

Hey, wanna know how to keep a bit more cooler or less hotter out in the desert? Build a double-decker covered shelter with either two ponchos or two poncho liners.

The top cover protects you from the sun's rays, and the second cover deflects the heat from the first cover away from you.

And if you place the floor of the shelter a 1/2 meter "above" or "below" the desert surface, it'll make it a bit more cooler. Try it!

NOTE: Don't have two poncho's or two poncho liners to use? If ya use only one, the higher it's erected - the less heat you'll feel and deflected off of it. The lower to the ground it's erected -the more heat that'll be deflected and felt. Go figure!

*Rick F. Tscherne*

## GUN MART

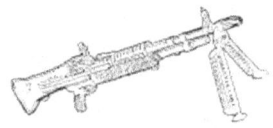

Hey Guys & Gals! I just discovered a neat new magazine, it's called "GUN MART." And although it's published in England, you can still subscribe and get it delivered to your home.

What makes this British magazine different from other (American) gun magazines? Well, for starters it contains over 250+ pages of advertisements and interesting articles, and a lot of it military related. From firearms, equipment, books, videos, clubs, to the latest products available on the market worldwide.

It even has a section on military airguns, replicas, and deactivated (non-firing) machineguns that you can purchase without a license. No BS! Ya gotta see the BB guns that look like real M16s, M60s, MP5 H&K, etc. Neat!

Now don't confuse this British "GUNMART" magazine with the American "SHOTGUN NEWS" newspaper, (see *Ranger Digest III*), these are two entirely different publications. But if you like Shotgun News, you'll love Gunmart. It's terrific! Their website: http://www.gunmart.net/

Because payment must be made in British currency, the pound, you'll need to get an "international money order" from a bank or pay for it with a Master, Visa, or American Express credit card. (Note: $1 USA = 1.70 British Pound, L32.20 = $19.00). Send To:

```
         GUN  MART
    Attn: Subscription Dept.
        Bradley Pavillion
    Bradley Stoke North, Bristol
    BS120BQ         United Kindom
```

## FIELD SPOT LIGHTS

When I was a squad leader and platoon sergeant, I often took my portable car spotlight with me to the field. Really! Where did I plug it in? Into a PRC-77 radio battery. No BS!

After cutting off the cigarette lighter adapter and removing some of the rubber coating from the end of the wires, I plugged it into the battery's positive (+) and negative (-) terminals. Or I'd remove the battery's cardboard box/cover and attach it directly to the red (+) and black (-) "internal" wires.

Today, because there's several models to choose from, you don't have to go through this hassle. You can buy'em from 12 volt plug-ins to battery operated and from 50,000 to 1,000,000 candle power, etc, etc.

What can you use 'em in the field for? A lot of things. But we used my spotlight for squad and platoon ambushes, raids, and defensive operations. The powerful light will not only blind and ruin the enemy's night vision (human & mechanical) capabilities, but expose and illuminate their positions too. Which is impossible to do with parachute flares and star clusters without exposing, illuminating, and jeopardizing your own night vision capabilities and positions. Makes sense don't it? Try it and find out.

NOTE: When using handheld spotlights in the tactical mode, position them (and or the user) a safe distance away so your unit remains in total darkness and will not be exposed or draw unnecessary enemy fire. Get my drift?

# SOME SMART ADVICE FROM A (RANGER) MEDICAL OFFICER

*Submitted By: 1LT Jimmy Deak*

Dear Ranger Rick,

1) Every soldier should carry two field dressings in his or her first aid pouch along with a partially completed (name, ss#, unit, etc) DD Form 1380 Field Medical Card. This will not only help medics to save on using their own field dressings to treat the wounded, but to speed up aid and evacuation procedures too.

2) Every soldier should carry his or her own IV kit, complete with needle & tubing in case they themselves become wounded. A medic can only carry so many of these kits in his aid bag before running out and needing to acquire some more from the battalion aid station.

3) Every soldier should carry either a "blue" or "red" chem-light so they can identify themselves at night when they need immediate medical attention. As it's almost impossible for a medic to determine (at night) from a distance who is and who isn't wounded. A soldier can either pop the chem-light himself (if he's conscious) or another individual can do it for him.

4) Every 1SG should carry in his track or vehicle an emergency backup "Class VIII Medical Resupply Chest" just for his company's use. Items to be carried should include bandages, i.v. fluids, tubing, needles, water

purification tabs, hypochloride, ointment, chloriquine, medicine sunscreen, foot powder, sunscreen, syringes, catheters, and more.

5) Every unit medic or tasked litter team should carry a SKEDCO LITTER (http://www.skedco.com/Military/sked-rescue-stretchers) rather than a POLELESS LITTER. Unlike the poleless litter, it only takes a 2-3 man litter team to comfortably carry and transport a patient.

With the poleless, it's not only uncomfortable for the patient, but for the litter team too. As the straps and carrying handles keep cutting into the hands of the carriers thus forcing them to stop, switch, rotate, and take turns more often.

For more information on how your platoon or company can improve their emergency medical readiness, contact your battalion aid station.

# MEDICAL AID BAG TIPS

*Submitted By Michael Chase*

Dear Ranger Rick,
Enclose you will find some ideas that my team mates and I have both found to be very useful, and I'm sure your readers will too, especially if they're medics. I wish I could remember the name of the para-rescue operator who provided us these useful tips, but I don't remember.

BATTLE PACK:
To take care of multiple patients with gunshot and or major bleeding wounds can be difficult, especially when more than one medic or assistant is rummaging through your aid bag looking for a particular first aid item.

To remedy this problem, always carry several pre-packaged individual "battle packs" with the necessary items needed to treat bleeding and sucking chest wounds. This way, you can quickly toss them to your on-the-scene helpers so you'll be able to keep your aid bag closely nearby.

Below is a list of the minimum medical items that should be pre-packaged in a battle pack;

2 X zip-lock bags for double bagging & water proofing and as an outer dressing for burns, sucking chest wounds, etc.
2 X first aid field dressing, as in most gunshot wounds, one for the entry wound and the other for the exit wound.
1 X petrolatum gauze bandage, for sucking chest, burns, and penetration wounds.
1 X kerli roller gauze, for dressing and as a packing bandage.
1 X 5x9 surgi-pad for pressure dressings.

2 X 4x4 cover sponges, 12ply, more dressing material 1 X triangle bandage - for use as a tourniquet or constriction of limb amputations, uncontrolled bleeds, etc.

M-5 MEDIC BAG MODIFICATIONS TIPS:
If you remove the external straps and add a D-ring to the Alice pack shoulder straps and to the side mounted set of D-rings on the shoulder straps, it'll make it easier to carry on long distance road marches. Add a couple of M-60 MG cordura ammo cases to the outside portion of the aid bag and you'll be able to carry a few more IV fluid kits.

# THE MANY USES OF MAALOX

*Submitted By: Dr. David A. Williams*

One of the safest and best all around medicines to carry to the field is liquid Maalox, either regular or extra strength. And a few of the many illnesses and discomforts it can relieve....

1. Poison Ivy - rub a small amount of maalox on the affected areas and it will neutralize the poison, dry it, and stop the itching within minutes.

2. Jocky Rash or "Great Balls of Fire" - rub some maalox on the affected area and you'll soon be relieved of the burning, itching, and discomfort.

3. Athlete's Foot or "Swamp Foot" - rub a small amount of maalox on the foot, including in between the toes, and the itching, irritation, and odor will soon be neutralized.

4. Blisters, Scalds, Abrasions, & Cuts - will heal quicker if treated daily with a little bit of maalox, just rub on a small amount and let it dry.

5. Sunburns & Windburns - rub maalox on like a lotion and it will quickly relieve the pain, discomfort, and irritation.

6. Hemorrhoids & Rectal Irritation (Due to Diarrhea) - pour a small amount of maalox on a piece of toilet paper, wipe or rub on the affected area and the bleeding, itching, and irritation should subside within minutes.

7. Feeling Constipated? - take several (3-5) teaspoons of maalox and your stomach should settle down in no time, also works as a laxative too.

8. Got a Cut That Just Won't Stop Bleeding? - put a few drops of maalox on it and cover with a bandaid or bandage.

Ranger Rick's Comments: Attention Readers - Use At Your Own Risk.

# EMERGENCY PROTECTION

*Submitted By: Dr. David A. Williams*

Always carry in your rucksack a small bottle of "olive oil," the type that comes with an eye or noise dropper. Because it can always be used....

a) To prevent foreign matter from entering the eyes and ears should you need to swim or dive in salty or filthy water and you don't have a mask or a pair of ear plugs for protection.

Just place a few drops in the eyes and ears before entering the water and the salt and foreign matter won't stick to the surface of the eyes nor enter the sensitive areas of the inner ear.

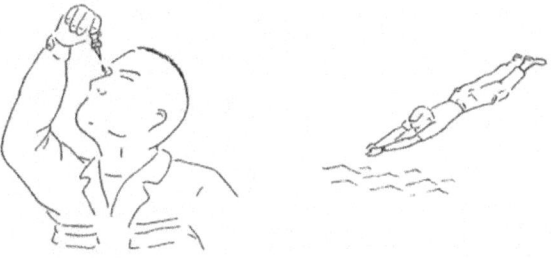

b) Works great for relieving earaches. Simply warm up the bottle of olive oil with a match or lighter, just slightly above body temperature, and place a drop or two in the affected ear. To keep it in place, use a piece of cotton, tissue, and or even a gun cleaning patch. Should it become infected and very painful, just add 1 x drop of iodine to 5 x drops of olive oil and you have an excellent field remedy.

*Rick F. Tscherne*

## LOCK UP YOUR WEAPON

When you're out in the field and NOT on tactical/alert status, does your chain of command always make ya carry your bullet launcher wherever you go? Like to the shitter, chow line, wash area, etc?

Yea? No kidding, so did mine. But what I use to do so I wouldn't have to drag my weapon with me everywhere I went, is secure it. No, NOT at the arms room and NOT with one of my buddies neither, but with a "bicycle cable lock."

Lock the bolt of your M16/M203 to the rear and run the cable through the magazine well and upper receiver dust cover so they can't be separated from one another. And then wrap the weapon and cable around a fixed, non-movable and or not-so-easy-to-move stationery object such as to.....

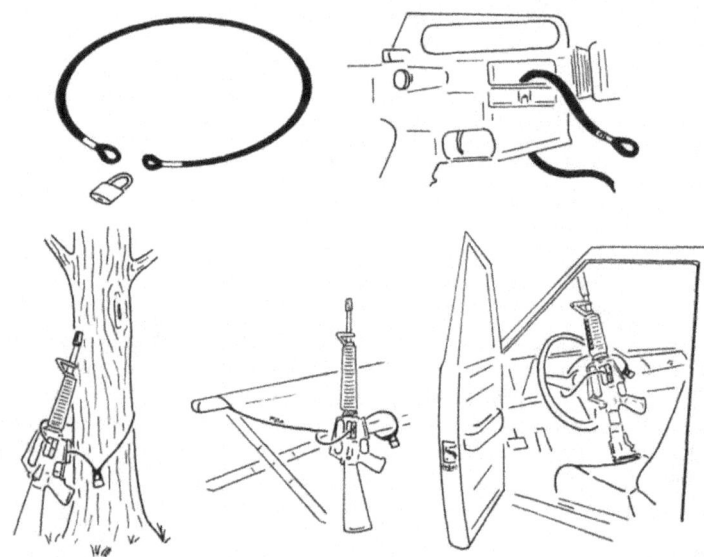

## DO I HAVE A SENSE OF HUMOR OR WHAT?

At my last duty assignment in Vicenza - Italy as the US Army SETAF G-3 War Plans Division NCOIC and Operation Sergeant. I got tired of seeing all these damn classified documents come across my desk each day. So I decided to come up with my own official stamp in how they should be classified and filed. Well, this is what I came up with, (see below).Everyone got a big kick out of it (the EMs, NCOs, Lts, & captains)"except" my boss (an 0-5/Ltc) and a couple of field grade officers who worked in my office. I wonder why?)

Circle your Recommendation:
1. KEEP ON FILE    2. SHRED IT
3. SELL TO KGB    4. RTN.TO SENDER
5. DONT GIVE A SHIT


Well, what da ya know about that, another dissatisfied Ranger Digest reader. The second complaint and bad report card rating I have ever received since starting this book series back in the 1980's. Boo-hoo, boo-hoo, boo-hoo. Whaa!

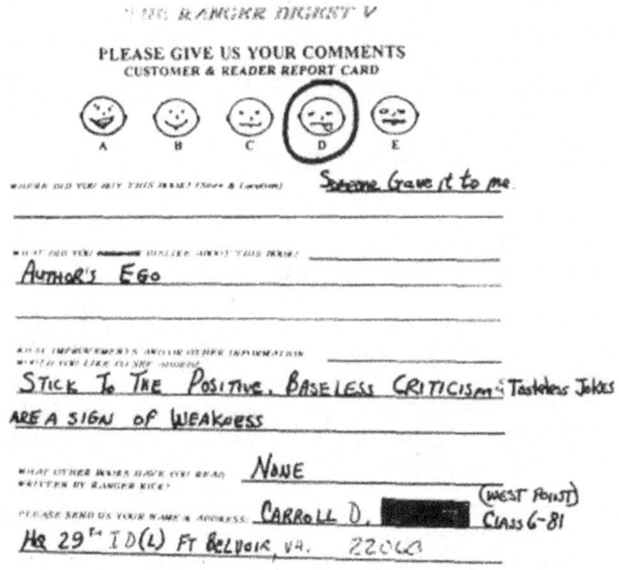

THE RANGER DIGEST V
PLEASE GIVE US YOUR COMMENTS
CUSTOMER & READER REPORT CARD

Where did you buy this book? — Someone gave it to me.

What did you like best about this book?
AUTHOR'S EGO

Any improvements and/or other information would you like to see added:
STICK TO THE POSITIVE. BASELESS CRITICISM & Tasteless Jokes ARE A SIGN OF WEAKNESS

What other books have you read written by Ranger Rick: NONE

Please send us your name & address: CARROLL D. ▇▇▇▇ (WEST POINT) Class 6-81
HQ 29th ID(L) Ft Belvoir, VA. 22060

Rick F. Tscherne

# WHAT THE %^#$%&??

### Air Force nominee under fire
Clinton supportive despite claims of errors, deceit

By TOM RAUM
The Associated Press

WASHINGTON – The Clinton administration is sticking with its nomination of a Florida state senator and former fighter pilot to be Air Force secretary as Republicans question his uneven flight history and business dealings.

The Senate Armed Services Committee planned to meet again early next week on the nomination of Daryl ████ after a contentious 10-hour hearing Thursday.

Defense Secretary William Cohen reiterated his strong support. "He is a man of proven ability who understands the challenges of public service," Cohen told the committee in a letter. And at a photo session Friday, Cohen told reporters, "We're still supporting Daryl ████.

At a Pentagon briefing, spokesman Kenneth Bacon shrugged off charges made against ████ at the hearing. "He has the full confidence of the secretary," Bacon said.

President Clinton nominated Jones last October to replace Sheila Widnall, who resigned.

Witnesses said Thursday that ████, as an Air Force Reserve pilot, almost ran out of fuel on one occasion, flew the wrong way on another and damaged his aircraft by scraping the plane's tail on as many as four occasions – twice on the same day.

"I was scared. I realized I could no longer fly with him," said retired Maj. Allan F. Estis, a flight instructor who accompanied Jones at Homestead Air Force Base near Miami. Estis testified he quit the Air Force Reserve rather than continue working with ████.

Jones did not dispute his bad flying record but blamed it on his devotion to public service. He insisted it would not interfere with his ability to oversee the Air Force.

Republican sources, speaking on the condition of anonymity, said Thursday night that ████ was likely to win majority support of the panel and the Senate despite the controversy. They said a key holdout, however, was Sen. John Warner, R-Va., a senior member of the panel.

Warner, who has been ████ most persistent critic, could tie up the nomination. He did not tell reporters whether he would seek to block confirmation.

████ told the committee Thursday that he voluntarily stopped flying F-16 fighters in 1991 because of his crowded business life. He disputed testimony by his former...

Is this %#@& up or what? Can you believe the ignorance of the idiots who tried to get this guy nominated for Secretary of the Air Force????

Whoever suggested or recommended Mr. Daryl ___, should be lined up in front of the Viet Nam Wall Memorial in Washington D.C. and shot at sunrise. What the %#@& were they smoking and drinking the day his name popped up, COW MANURE and RUBBING ALCOHOL?

It's like...

Nominating an old, cranky, tightwad, and very disliked anti-military geezer to Secretary of Defense. (Oops, we did one time.)

Nominating a mother who believes in teaching masturbation to young school children to be the Surgeon General of the United States. (Oops, we did one time.)

Nominating a dope smoking, lying, adulterer, who has never served in the military to be the Commander & Chief of the Armed Forces. (Oops, we did one time. OUCH - we did twice!)

You'd think by now these damn politicians would learn from their previous mistakes, wouldn't you? I guess not.

And as I was reading through a Soldier of Fortune magazine, I came across an interesting article written by Col.(Ret) David Hackworth Check out what he' s got to say.. (next page)

"TELL'EM HACK"...

## *Bad Leaders Equal Bad Units*

Excerpt:

Hooray and hallelujah! Daryl Jones, Bill Clinton's nominee for Secretary of the Air Force, got shot down in flames by a brave Senate committee.

John Warner (R.-Va.) spearheaded this maneuver, ably backed up by Charles Robb (D.-Va.) and Robert Byrd (D.-W.Va.). They and other panel members put America ahead of partisan politics and stopped Jones from further zapping an Air Force already under siege.

Senate panel members — where a majority put conscience over cronyism — will catch a lot of flack from politically correct hand-wringers who think affirmative action and quotas are more important than putting the right person in the right job when it comes to defending America.

Jones' nomination was by far the worst of a long string of bad Clinton Pentagon appointments: Les Aspin, William Cohen, John Dalton, Sheila Widnall and Togo West — none of whom could lead a frog to a puddle on a hot day.

Jones would have clobbered an Air Force already on the mat because of bad senior leadership, an Air Force trying to do too much with too little and weakened by an exodus of good people walking out the front gate.

Jones' former commander, Colonel Thomas A. Dyches, told the panel, "I believe there are serious questions about Daryl Jones' personal integrity." Trust is something Jones doesn't exactly inspire. People who have served with him openly call him a liar.

To pump fresh air into a deeply troubled Air Force it needs a strong leader who can take charge, kick butt and clean house. The extraordinary fine young folks who serve in our Air Force deserve no less. Warner says Jones was unable to even "inspire those who serve under him."

My computer almost burned up with Jones' honor stories:

• An Air Force colonel: "We're in a steep decline both in operational capability and morale. Putting that self-serving nitwit in charge would have put us into the proverbial 'smoking hole.'"

- An Air Force major: "How did he draw four years' flight pay when he wasn't on flight status? If a grunt tried this stunt he'd be making license plates at Leavenworth prison."

- A senior Air Force sergeant: "The USAF is in far worse shape than even you write about. Jones would've made it worse. I served with him. His judgment stinks. He'd probably put Amway products in the BX [Base Exchange] and march us there to buy them."

It's pathetic that a person with so flawed a character could even be considered for such an important national security assignment. But then, as one colonel said, "Look who nominated him."

COL. DAVID HACKWORTH
PASSED AWAY IN 2005

---

You can read more about him here:

http://hackworth.com/

AND

The organization he started:

*Soldiers For The Truth*

Now called:

Stand for the Troops

SFTT.org

*A few of his books...*

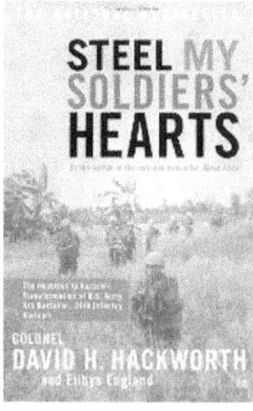 

## DOG TAG SILENCERS

*Submitted By: Sp4 Gregory A. Barker*

First, find yourself some "heat shrinking tubing" about 1/2 - 3/4 inch in diameter. Cut off two pieces about the same length as a dog tag and then flatten and stretch it on over them. Then hold it over a small flame (match, candle, lighter, etc) until they shrink and become tight around the dog tags. Then grab a small nail and poke a hole through the tubing where the holes of the metal dog tags are located so you can slide through the chain. This method will not only prevent the dog tags from making "clanging sounds," but make them subdued in appearance too.

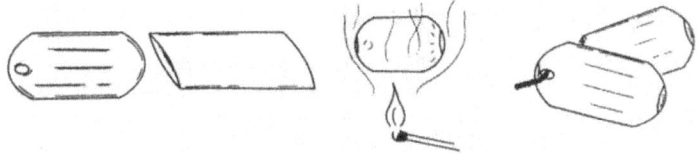

## SURGICAL GLOVE USES

*Submitted By: Sgt.Frank D. Gilliland*

Hey Ranger Rick,
Here's a tip I thought you could use for your *Ranger Digest VIII* that I use to teach to all my survival school students.

Always carry to the field a pair or two of disposable surgical gloves. You can get them from your unit aid station or from almost any convenient store, they come in vacuum sealed packages about one set per package.

They can then be used not only to prevent you from catching a bacteria, disease, or virus from an ill or sick patient/casualty. But as an emergency water storage bag, as a tourniquet, or for water proofing small items. And if you cut off one (or more) of the glove's fingers, you can use them as an improvised "dust & dirt cover" for the barrel of your weapon.

Rick F. Tscherne

# TIPS FROM AN LT.

*Submitted By: 1LT Karl L. Mims*

Dear Ranger Rick,
I thoroughly enjoy all your Ranger Digest handbooks. They not only teach soldiers a lot of useful field tips, tricks, and ideas, but they also allow them to share some of their own techniques and experiences too. So here's a few of my own...

1. Before going on a long deployment purchase a package of pre-stamped "blank" postcards from your local post office. They'll not only be easier to pack and carry in your rucksack and or deployment bag, but they're much easier to write home on and mail from the field.

2. When possible, try to acquire as many mini bars of soap and shampoo bottles as possible. You know the type, the kind found in almost all the hotel and motel bathrooms? Then store and carry them in your personal field hygiene kit that's kept in your rucksack. They'll be lighter to carry then a 3 x month bar of soap and or a bottle of shampoo.

Ranger Rick's Comments: Another way you can lighten the load is by cutting the bar of soap in half or less and carrying your liquid shampoo in one or more empty 35mm film containers.

And if you're one of those that like to shave in the field with disposable razors. Try cutting the plastic handles down to size so you can fit several inside your plastic soap dish container along with the cut-down bar of soap and 35mm shampoo container. Talk about compact and light, this'11 do the job.

*The Complete Ranger Digest: Vol. VIII*

# AN ANTENNA SET-UP TIP

*Submitted By: PFC Christopher Watson, USAR*

Dear Ranger Rick,
Something I picked up from one of my instructors at 38A AIT, was how to quickly and correctly set up an antenna. It's not as easy as you may think even for a highly trained individual, as it must be set up in a certain sequence. But thanks to the Army's infinite wisdom, there's no easy way to determine which piece or section goes where, especially when some of them look the same.

So the next time you go to the field and set up one of those OE292, OE-254, etc, antennas. Before pulling it apart and packing it away, wrap some 100 mph tape around all the ends of the pieces and number them in the sequence they need to be assembled with a permanent marker. Start at the bottom, (the base of the antenna), and work your way up.

So the next time you or someone else has to set up the antenna, all you have to do is follow the numbers in sequence. It will not only be a whole lot easier, but a lot faster to set up too.

Also, when ever you take down an antenna, be sure to carefully wrap up the tie-down and wires before packing it away. I can't tell you how many times some of us have been pissed off because some lazy s.o.b. before us didn't take the time to untie all the knots and carefully wrap the wires before packing it.

Ranger Rick's Comments: Another technique and trick that I've heard works pretty well is to color code the ends of the pieces.

The Army has about 6 different types of colored marking tape; red, yellow, green, blue, white, and black. You can either mark the ends of the pieces, red-to-red, blue-to-blue, etc. Or mark a particular section of the antenna all the same color and then number them; blue-1, blue-2, blue-3, etc.

NOTE: Never use black tape because it's difficult to see what color it is at night with a red filter flashlight. And use only a medium or fine point alcohol pen rather than a thick magic marker, it'll be a lot easier to see and read the numbers on the tape.

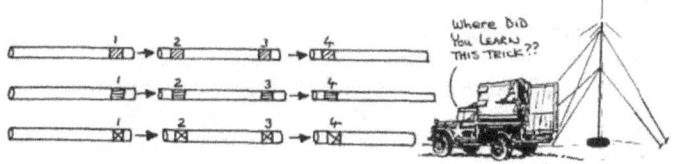

Rick F. Tscherne

# CANTEEN WATER HEATER

Yours truly, "ol' McGyver" has done it again. I came up with another handy-dandy idea, I call it a CANTEEN WATER HEATER.

Now I wonder who's gonna steal this idea and send it to the Department of the Army or to an Army publication and claim it was their idea. Or market and make a profit from it. To name only a few of my ideas that have been stolen from my Ranger Digests...

> A. My poncho liner with a zipper - a company that makes and sells'em calls it a "D——— Liner".
> B. My disposable MRE shit box - the Army Times published it in one of their editions back in 1995 and someone else took credit for it. Even though it was published in my *Ranger Digest V* "years before" they printed it.
> C. My disposable MRE water container - as you can see from the previous page, some captain wrote an article to a military magazine claiming it was his idea. Again, even though it was published in my *Ranger Digest V* years before he got wind of it.
> D. My BDU patrol hat with a hidden interior pocket - a company that makes and sells 'em calls it a "M-- P——— BDU Cap."

These were only a few of my ideas that were stolen and taken out of my *Ranger Digests*. Should I be mad and sad because someone stole'em? Well, I don't mind anyone using my ideas, nor profiting from them. Provided; (a) they ask my permission to use them, or (b) they mention the source as to where they got the idea(s). And that's all I ask, "give credit where credit is due."

So now, before publishing any of my tips, tricks, & ideas in future *Ranger Digests*, I forward the best ones to BRIGADE QUARTERMASTER to review. And if they like any of them, I give them permission to use'em. But as to this day, I have never, ever, profited from any tips, tricks, and or ideas that my readers have sent to me, NEVER! Nor will I ever take credit or money for an idea that wasn't mine.

Fear not my fellow readers, honesty and integrity are my strongest trademarks, and I challenge anyone, ANYONE to prove differently.

OK, enough BSing around, now about my canteen water heater. What you'll need is an "aluminum cigar container," which is what special and expensive cigars come in. A tube of "clear silicone" and preferably an "NBC" canteen cap. But if an NBC cap ain't available, you can get away with using a regular ol' canteen cap instead. Though an NBC canteen cap would be much better to use because it's more durable, etc, here's what ya gotta do to it....

(a) After removing the "inner guts" of an NEC cap, get yourself a drill bit about the same size as the cigar container and drill a hole right through the center of the cap. If you can't find one, no problem, just grab a "round" file and file it out until you can slide inside the cigar container.

(b) Grab the tube of silicone and seal the cigar container in place inside the cap. BUT! Only squeeze out enough to hold it securely in place. Not too little and definitely not too much, or you won't be able to screw the cap back onto the canteen.

Now all ya gotta do is fill the canteen with water, screw on this water heater, place inside some fuel, and wait for it to heat up.

If you notice, I didn't say anything about "igniting the fuel." Because though I recommend that you only use the flameless fuel from the MRE "food heating packets." As an alternative, you can also use some "crushed" heat tabs or b-b-q charcoal fuel too.

If you do use one of these fuels, to avoid burning and damaging the plastic, the canteen should be completely full of water up to the rim "after inserting in" this water heater. And then place and burn only a "small amount" of fuel at a time.(Ya gotta experiment)

In a military tactical environment where fires are not permitted, the fuel from an MRE food heating packet is the best to use. Why? Because it doesn't give off any smoke or flame. Which means you can use this handy-dandy canteen water heater at a clandestine (ambush, recon, etc) site to make coffee, soup, or hot chocolate without compromising your position. Is this neat or what? Huh?

*Rick F. Tscherne*

## DON'T BLAME ME IF YOU PANICKED & SOLD OUT

OK, OK, OK, I know what I said in my last *Ranger Digest* (VII). But after receiving some nasty letters on how much $$$$$ some of you lost during Wall Street's September/October 1998 stock market correction. Well...

As I stated in my previous Ranger Digests, I'm NOT a financial advisor. And past performances of any stock or mutual fund does not guarantee the same future results year-after-year.

But if you're in it for the long haul, I guarantee you're gonna make a lot more money investing in stocks and mutual funds than in cds, saving bonds, and treasury bills. But if you're in it for the short haul and you're trying to make a fast killing ($$$$$). Well, I suggest you try playing the slot machines instead. Or...

If you don't like my investment tips, then do your own research:

### INVESTMENT MAGAZINES

MONEY
WORTH
KIPLINGER

### INDIVIDUAL INVESTOR INVESTMENT SITES

WWW.CNNFN.COM
WWW.BRILL.COM
WWW.HOOVERS.COM
WWW.FUNDALARM.COM

I hope most of you didn't panic and cash out of the stock market back in September and October 98. Because if you did, you probably did lose a whole lot of $$$$$, right? I'll betcha did.

But if you held on and didn't cash out like everyone else, by December '98 you should have fully recovered all your losses, plus a little bit more. What drove the stock market down? Besides some negative news about both, the US and global economies, PANIC & GREED!

Am I invested in the same mutual funds that I recommend to all my readers? You betcha! Did I lose any money when the stock market took a dive in September/October 98? You betcha! Did I remain in the stock market and recover all my losses? You betcha! Am I still fully invested in either stocks or mutual funds? You betcha!

*The Complete Ranger Digest: Vol. VIII*

LESSONS LEARNED? If you own a good mutual fund that's had five years or more of excellent performance. Relax, sit back, and just ride out these dips & corrections. Because when the market does drop, you panic, and pull out....you're only gonna miss the train ride back to "profit land," and I guarantee you will too.

Here's an updated list of the best mutual funds to be invested in:

8 YEAR RETURNS (1990-98) As of Dec. 31, 98

| FUND (Symbol) | CATEGORY | % ANNUALIZED GAIN since Jan. 1, 1990 | VALUE of $1,000 Investment |
|---|---|---|---|
| 1. Fidelity Select Home Finance (FSVLX) | SF | 30.1% | $9,372 |
| 2. Fidelity Select Electronics (FSELX) | ST | 27.5 | 7,896 |
| 3. Fidelity Select Computers (FDCPX) | ST | 26.7 | 7,490 |
| 4. Fidelity Select Regional Banks (FSRBX) | SF | 26.0 | 7,108 |
| 5. Hancock Regional Bank B (FRBFX)¹ | SF | 25.7 | 6,984 |
| 6. Invesco Strat. Financial Svc. (FSFSX) | SF | 25.7 | 6,971 |
| 7. Fidelity Select Software & Comp. (FSCSX) | ST | 25.6 | 6,875 |
| 8. Fidelity Select Brokerage & Inv. (FSLBX) | SF | 25.3 | 6,797 |
| 9. PaineWebber Fin. Svc. Gro. A (PREAX) | SF | 24.7 | 6,537 |
| 10. Fidelity Select Financial Svc. (FIDSX) | SF | 24.6 | 6,466 |
| 11. Seligman Comm. & Info. A (SLMCX) | ST | 24.5 | 6,456 |
| 12. Fidelity Select Health Care (FSPHX) | SH | 24.3 | 6,427 |
| 13. Spectra (SPECX)² | LG | 24.2 | 6,325 |
| 14. Alliance Technology A (ALTFX) | ST | 24.2 | 6,299 |
| 15. Invesco Strat. Technology (FTCHX) | ST | 23.9 | 6,179 |
| 16. Fidelity Select Technology (FSPTX) | ST | 23.8 | 6,148 |
| 17. Oppenheimer Main St. Inc. & Gro. A (MSIGX) | LB | 23.5 | 6,024 |
| 18. MFS Emerging Growth B (MEGBX) | MG | 23.3 | 5,926 |
| 19. Amer. Cent.-20th Cent. Ultra (TWCUX) | LG | 23.0 | 5,801 |
| 20. T. Rowe Price Science & Tech. (PRSCX) | ST | 22.7 | 5,675 |
| 21. AIM Aggressive Growth (AAGFX)¹ | SG | 22.6 | 5,670 |
| 22. Vanguard Spec. Health Care (VGHCX) | SH | 22.6 | 5,669 |
| 23. CGM Capital Development (LOMCX)¹ | MB | 22.5 | 5,662 |
| 24. FPA Capital (FPPTX)¹ | MV | 22.3 | 5,553 |

16 YEAR RETURNS (1982-98) As of Dec. 31, 98

| FUND (Symbol) | CATEGORY | % ANNUALIZED GAIN since Aug. 1, 1982 | VALUE of $1,000 Investment |
|---|---|---|---|
| 1. Fidelity Select Health Care (FSPHX) | SH | 23.6% | $29,975 |
| 2. Spectra (SPECX)² | LG | 22.7 | 25,957 |
| 3. Fidelity Select Financial Svc. (FIDSX) | SF | 22.6 | 25,514 |
| 4. Fidelity Magellan (FMAGX)² | LB | 22.4 | 25,045 |
| 5. CGM Capital Development (LOMCX)² | MB | 22.4 | 24,983 |
| 6. Fidelity Destiny I (FDESX) | LV | 22.3 | 24,748 |
| 7. Davis NY Venture A (NYVTX) | LV | 21.8 | 23,085 |
| 8. Alliance Technology A (ALTFX) | ST | 21.6 | 22,640 |
| 9. Amer. Cent.-20th Cent. Ultra (TWCUX) | LG | 21.5 | 22,181 |
| 10. Sequoia (SEQUX)² | LV | 21.4 | 21,890 |
| 11. Guardian Park Avenue A (GPAFX) | LB | 21.2 | 21,304 |
| 12. Federated Capital App. A (FEDEX) | LB | 21.1 | 20,997 |
| 13. AIM Constellation A (CSTGX) | MG | 20.9 | 20,656 |
| 14. Fidelity Contrafund (FCNTX)² | LB | 20.7 | 19,965 |
| 15. Putnam Voyager A (PVOYX) | MG | 20.6 | 19,731 |
| 16. IDS New Dimensions A (INNDX) | LG | 20.5 | 19,499 |
| 17. AIM Weingarten A (WEINX) | LB | 20.1 | 18,394 |
| 18. United Income A (UNCMX) | LB | 20.1 | 18,375 |
| 19. SIFE Trust A-I (SIFEX) | SF | 20.0 | 18,246 |
| 20. Enterprise Growth A (ENGRX) | LG | 20.0 | 18,139 |
| 21. FPA Capital (FPPTX)² | MV | 19.8 | 17,729 |
| 22. Century Shares (CENSX) | SF | 19.7 | 17,429 |
| 23. Washington Mutual Investors (AWSHX)² | LV | 19.5 | 17,184 |
| 24. Selected American (SLASX) | LV | 19.5 | 17,020 |

Rick F. Tscherne

# STRUGGLING???

## Troops Struggling financially

Survey: More than half have money troubles

The survey looked at three basic issues — how often young servicemembers have financial problems; various factors that raise or lower the risk of experiencing those problems; and what kind of help is available and who is actually getting it.

Among the findings:

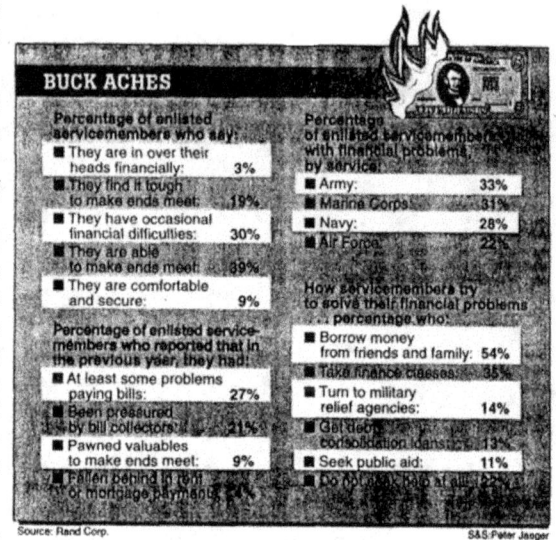

• About 27 percent of enlisted troops have trouble paying their bills. A similar survey of civilians found that only 6 percent reported such problems, although analysts noted that the civilian data, from a survey done in 1991, might be outdated.

• Twenty-one percent had been pressured by bill collectors in the past year, while 9 percent had pawned or sold valuables to help make ends meet.

• Soldiers and Marines are more likely to have financial problems than their counterparts in the Navy and Air Force. Fully one-third of Army troops and 31 percent of Marines reported such problems, compared to 28 percent in the Navy and 22 percent in the Air Force.

• First-term troops with children seem to run a much higher risk of financial problems than those without kids - 36 percent vs. 27 percent. Troops with children are also 2'/2 times more likely to have gotten an emergency assistance loan or a loan from a military relief agency.

- More than half of enlisted troops deal with their financial problems simply by borrowing money from friends and family. Thirty-five percent sign up for financial

Ranger Rick's Comments: What the survey failed to reveal was how young service members get themselves into financial trouble. How? Paying off expensive monthly...

*Car Loans   Credit Card Bills   Charge Accounts   Car Insurance*

Or they spend too much money on...

*Impulse Buying   Eating out   Boozing   Gambling   Partying*

Whose fault is it? Well, if you can't control your own $$$$$ and spend it wisely on only what ya need, then there's no one else to blame but yourself. And the sooner you take control of your $$$$$, the sooner you're gonna be (and stay) out of financial trouble.

## PRODUCING WATER FROM EMPTY TRASH BAGS

As I mentioned in previous Ranger Digests, you should always pack and carry a few trash bags in your rucksack and or on your lbe, as they will always come in handy for something out in the field. And one such use, is for producing water in an emergency "life & death" survival situation.

These techniques are nothing new, in fact, almost every military and civilian survival manual shows you how to make 'em. But what they fail to tell ya, is how much water they can produce in a day. Which depends entirely on the type of terrain, climate, temperature, and the sun's ultraviolet rays.

Now if you only have "one" trash bag, you can only use it for "one" water producing method. But if you got "two" trash bags, you can use 'em for "two" water producing methods. "Three" trash bags, "three" water producing methods, etc.

The bottom line.... the more trash bags you use, the more drinking water you can produce. Duh! Makes sense, don't ya think? If so, then how come all these high-speed survival handbooks don't teach ya this? Huh?

NOTE: An excellent place to store trash bags, is inside your canteen covers.

Rick F. Tscherne

# SOME FIELD TOYS I KNOW YOU'LL LIKE

As you know from my previous Ranger Digests, I like to go through mail order catalogs searching for unusual products that I think will help soldiers in the field. Well, this time I found me a whole bunch of neat little goodies. Check out this stuff:

FOLDING GRAPPLING HOOK

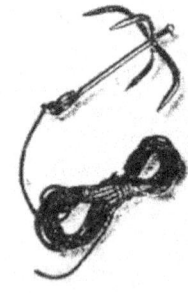

**Specially Designed Folding Grappling Hook**
**800 lbs. Rated**
Specially designed grappling hook folds down for compact and easy carrying. With a few twists, the 4 carbon steel talons can be locked into position assuring a safe assent. Best yet, it is rated to over 800 pounds and comes with 33 ft. of tough braided nylon rope.

Now when I was back in the Army, the only grappling hooks that I ever saw, was during MOUT training. And the only units that had grappling hooks, were engineer units. So if we ever needed one, we had to beg or borrow from the engineers. Now, if they would have had these folding grappling on the market when I was in the Army, I would have bought one or two of them for my platoon. It's better to be prepared than unprepared.

$21.97   Black Widow Folding grappling hook
$24.95   Folding grappling hook

UPDATE: This item is probably made overseas and is not especially strong nor safe for life support use. (The one I saw was so loose, flimsy and cheaply made.) For serious tactical scaling uses you may want to consider one of these better made models below:

- CMI grappling hook (CMI-GEAR.com)
- Capewell Retractable Grappling Hook, https://www.botach.com/capewell-retractable-grappling-hook/ (Army Natick designed/US Made)
- MR Grappling hook  https://www.botach.com/capewell-retractable-grappling-hook/ (USA MADE)

SODA CAN LAUNCHER

When I saw this little device, I said to myself. "Heeey, this would make a nice field toy." An M16/AR15 soda can launcher that propels a full can of soda over a hundred yards away with just a single blank cartridge. Just think of all the fun you could have with this baby on your next FTX. Instead of launch-launching soda cans, try beer Cans. **WARNING - CAUTION - DANGER** Soda & beer cans when used in this device can be just as harmful and deadly as a bullet. Use extreme care and caution when using this device.

UPDATE: Still sold by: AmericanSpecialtyAmmo.com
Also, Golf ball launcher attachment: BloomAutomatic.com

BRIGHT EYES REFLECTIVE TACKS

Here's a nifty little item that will help make your tactical night operations run a bit smoother, a bag of glowing tacks. Whether you gotta lead a column of troops through the dark woods, or you just want to mark off where your unit boundaries or fighting positions are located. Just take a handful of these glow tacks and stick'em in the trees, bushes, or on the ground. You can even stick'em in the heels of your boots, a lot easier to see at night than cat-eyes. No one in your unit will have any problems seeing them, can be seen as far as 100 meters away. $4.00 at Lewishunting.com

BANG MECH TRICK SET

Stay Alert - Stay Alive. Want a way to train your troopies in teaching them to be careful in what they touch and or pick up out in the field? Better to learn by the sound of bang than the feel of lead, you know what I mean? Just load the bang mech with a loaded cap, set the spring, place a little weight on top of it and wait for a victim. $0.96 each Order from: mjmmagic.com

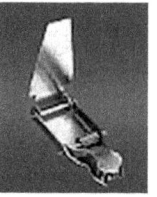

SOLAR FLASHLIGHTS

Never again carry extra batts to the field, a solar power flashlight. A variety of cheap and moderately priced ones can be found here: http://www.dx.com/s/solar+flashlight

10 X 20 CAMO NET

A huge camo net with a reversible brown and green pattern. What makes this camo net so special? The price, dummy, it's only $29. Get a few of your buddies or squad members together, split the cost, cut it up and use 'em as mini cammie nets. Or cut a hole in the center and wear'em as ghillie suits Order from: Tapco, PO Box 2408 DEPT C83 Kennesaw GA 30144 (check availability before ordering)

### DELUXE SOLAR BATT CHARGER

**DELUXE SOLAR BATTERY CHARGER WITH METER**
This is the best solar battery charger we've ever offered! Our new deluxe model charges 11 popular battery sizes including the new, flat prismatic batteries. You are able to charge your NiCad AAA, AA, C and D-Cells with free solar energy. This deluxe weather-resistant unit is now equipped with a charge meter which indicates: how much solar energy is available, if you have a short and the approximate charge time for each different battery size. A built-in blocking diode is included to prevent battery discharge at night. Easy to maintain and use. Great for everyday use, camping or emergencies when batteries are unavailable. Measures 7" x 3" x 2¼". Harness the power of the sun! (Batteries not included).

When you light infantry grunts (Ranger, Airborne, SF, etc) can't carry enough batts to the field to keep your life support system (am/ fm, cd, cassette, radio) going, it might be wise to invest in one of these. Though I talked about 'em in a previous RD handbook, this solar batt charger is a lot better. It's got a meter so you don't have to guess when they're fully charged. And it also takes several size batts too. Only $17.95
UPDATE: No longer avail. from listed supplier. Instead try these from: Sundancesolar.com.

### Ml6 COMBO - 4x20 SCOPE

Something every soldier in the US Army should be issued, an M16 rifle scope and mount. Better to "reach out and touch someone" at long distances than to engage them at close range, ya know what I mean? Installs quickly & easily onto an M16 rifle and has it's own integral "see-thru" mount. This high quality scope is also fully adjustable for windage/elevation. And it's only $67.95. Order from: SARCO (http://e-sarcoinc.com/arm16scope4x20milldot.aspx )

ARIS/M16 MILITARY TYPE 4x20 COMPACT SCOPE
Installs easily using its own integral see-thru handle mount. High quality optics feature precisely applied multi-layer coatings, nitrogen filled tube and bullet drop calculator. Fully adjustable for windage and elevation. Similar to the ones used by the military. (OPT0900) $39.99 each, or get 3 for only $35.99 ea.

### REUSABLE HEAT PACK

**Instant, Portable Re-usable Heat Pack** at the push of a button

Have ya seen these yet? They work great! All ya gotta, do is push a button on this hand warmer and it heats up to an amazing 130° for about two hrs And without any pre-heating, batteries, and or external power. No BS! And unlike other hand and feet warmers, it's reuseable and never needs to be thrown away. How much is it? $14.50, Order from: Quantumheatpacks.com - a Veteran owned company.

*Rick F. Tscherne*

## SWISS ARMY SNIPER VEIL

If you haven't used one of these yet, you don't know what you're missing. I first encountered them when I was attending Belgian Commando School, and I brought back a bunch'em for my Recon Pit buddies. Not only do they make great cammie veils, but you can stretch'em over your fighting positions to cammou-flage'em. They also make excellent scarves and drive-on rags too, the holes in'em allow the body heat to gradually seep out). Only $12.00 Order from: Swisslink.com

## BACK-UP EMERGENCY MAGS

("Oh Shit Mags")

I was looking through a military supply catalog the other day and saw something called a M16 Magazine Stock Pouch. (See below.)

But it's really nothing new. In fact, during the Korean War many of the troops who were issued Ml carbines use to carry extra magazine clips in a pouch worn around the wooden buttstock of the weapon.

Now according to some old timers, these magazines were not suppose to be used unless they were almost out of ammo. And when they got that low and had to use their reserve magazines, you can bet they said, "OH SHIT!"

Now this M16 magazine stock pouch they're selling on the market, I wouldn't even think of buying and attaching one to my weapon. Why? I'd be afraid of either getting it caught on the weapon's charging handle or on some tree branches, bushes, or wait-a-minute vines.

But instead of buying one of these high-speed magazine stock pouches, why don't ya just attach the magazine itself to the V-JS. stock. How? With a couple of thick rubber bands made from a cut-up "mountain bike" inner tube. (NOTE: Mountain bike tire inner tube, NOT a regular bicycle tire inner tube, you're gonna need something a bit more thicker and stretchable for this tip.)

These rubber bands will not only hold the magazine more securely to the weapon, but reduce the sound the magazine will make should it accidentally come in contact with something solid, like a tree,

IMPORTANT: When mounting the spare magazine to your weapon, mount it on the side of the butt stock nearest to your stomach. Then stretch the bands over the entire "top and bottom" portion of the magazine so they'll keep out foreign matter and reduce the chances of getting it caught and snagged on things.

UPDATE: A buttstock mag pouch for $9.95 is now available from Specopsbrand.com in Tan or O.D. (http://www.specopsbrand.com/tactical-gear/clearance-blow-outs/ready-fire-mode-buttstock-mag-pouch-m4-olive-drab.html ) Made in USA

## RANGER BAND USES

Hey guys and gals, check out the many uses of "Ranger Bands." Did my imagination go wild or what? And if you got a better idea in what you can use 'em for then come on and send 'em on into me and I'll publish it in my next Ranger Digest handbook. Hooaah!

You can use Ranger Bands for:

1) Attaching extra magazines to the butt stock of a weapon.

2) As shock absorbers for rifle scopes, (on the ends).

3) Better grip and hold of the pistol grip.

4) Attaching magazines together side-by-side.

5) Grasping & holding firmly a weapon forearm stock

6) Grasping & holding firmly a weapon forearm pistol grip
7) Attaching mag-lites to the forearm of a rifle.
8) Waterproofing a radio handset "mouth & ear" piece
9) Silencing the moveable parts of a canteen cup.
10) Added protection for binos, making 'em more shock resistant.
11) Attaching "survival containers" to knife sheaths.
12) Added protection for cellular phones, cassettes, handheld GPSs, and two-way radios to prevent them from getting damaged.
13) For grasping pistols more firmly, (especially when their wet)
14) As shock absorbers for compasses, (wrapped around the sides).
15) As mag-lite 'mouth-holders,"(wrapped around the bottom).
16) As dog tag silencers.
17) For grasping and holding firmly e-tools, tools, & knives.

## YEP, ANOTHER RANGER BAND USE

Oh mama, mama, mama. Am I on the roll or what when it comes to Ranger Band uses? Check this @&%%-out....

I discovered this technique by accident when I went to a ski resort here in Italy and slipped and fell on some ice while wearing my ol' military combat boots. So I had a couple of these Ranger Bands in my pocket (because I use 'em to attach my skis together) and slid them over the bottom of my boots. And guess what? Yep, they worked, I didn't slip or fall on my ass anymore.

But, but, but, be advised this technique is good only for short walking distances on "solid ice" and or "snow & ice." Because if you wear 'em over your boots on a hardball or dirt road where there's some "exposed" rocks or small pebbles. They'll create holes in the (rubber) Ranger Bands and cause 'em to split & rip apart. So if you use this technique, use 'em only for short walking distances, and when you get off the ice - take'em off. Hooah!

# THE TOP TEN...

### ...THINGS THAT SOUND DIRTY IN GOLF, BUT AREN'T.

10 - "Nuts...my shaft is bent.
9 - "After 18 holes, I can barely walk."
8 - "Wow, you really wacked the hell out of that sucker."
7 - "Mmmm-mmmm, look at the size of his putter."
6 - "Just keep your head down and spread your legs."
5 - "Mind if I join you in a threesome?"
4 - "Stand with your back turned up and drop it."
3 - "My hands are so sweaty I can't get a good grip of it.
2 - "Nice stroke, you must have been practicing."
1 - "Hold up...I need to wash my balls first."

### ...THINGS THAT SOUND DIRTY IN LAW, BUT AREN'T

10 - "Have you looked through her briefs yet?"

9 - "He's one hard judge."

8 - "Counslor, let's do it in the chambers."

7 - "Her attorney withdrew at the last minute."

6 - "Is it a penal offense?"

5 - "Better leave the handcuffs on."

4 - "For $200 an hour, she had better be good!"

3 - "Can you get him to drop his suit?"

2 - "The judge gave her the stiffest one he could.'

1 - "Think you can get me off?"

*Rick F. Tscherne*

...THINGS THAT SOUND DIRTY AT THE OFFICE, BUT AREN'T

10 - "I need to whip it out by 5."

9 - "Mind if I use your laptop?"

8 - "Just stick it in my box."

7 - "If I have to lick one more...I'll gag!"

6 - "I want it on my desk...NOW!"

5 - "Mmmmmmmmm...I think it's out of fluid!"

4 - "My equipment is so old, it takes forever to finish."

3 - "It's an entry-level position."

2 - "When do you think you'll be getting off today?"

1 - "It's not fair - I do all the work while he justs sits there."

## FIGHTING POSITION OVERHEAD COVER

When I first saw one of those new Fighting Position Overhead Covers (FPOC) in Soldier Magazine. The first thing I said to myself was, "Hey, great, more f——— shit for the light infantry to carry, like they ain't got enough stuff already."

OK, yea, big f——— deal. So what if this FPOC reduces or eliminates a soldier's dependency on indigenous material such as logs, plywood, sheets of plastic, etc. And who cares if it can withstand air bursts from a 155mm artillery round and even a direct hit from an 81mm mortar. Do you really believe it's worth carrying to the field and into combat?

I can see it all now....

Soldiers in formation at attention with their rucks on their backs, a weapon in their right hand and their FPOC in the other.

Or a platoon of soldiers moving tactically through the woods with their weapons in one hand and their FPOC in the other. When all of a sudden a burst of enemy fire opens up. "Quick, what are ya gonna do now soldier? Drop it and run, or get behind it?"

Knowing some, they'll probably try to assault through the damn enemy position carrying their FPOC in front of them. Hey, why not? If it can withstand a direct hit from an 81mm mortar, I guess it can withstand a direct hit from a 7.65mm and even an RPG, Right?

I wonder what's going to happen to a soldier that loses one.

Private: "Sergeant, I can't find my FPOC. I could have sworn I laid it down over there, but now it's gone and I've looked everywhere for it. What am I gonna do, sergeant?"

Sergeant: "Now calm down, soldier, relax, take it easy, I don't think we're gonna get hit with any artillery or mortars today. Maybe tomorrow, but not today. In the mean time, we'll get you a brand new one just as soon as you sign this here statement of charges."

I don't know, maybe it's just me. But I believe if you can't carry it in your rucksack or attached to your lbe. Well, it's just another worthless piece of shit that you don't need. I don't think FPOC stands for Fighting Position Overhead Cover, but rather another "F———— Piece Of Crap" that the Army dumped on the light infantryman. Ya know what I mean?

When deployed *(above)* the FPOC can withstand a direct hit from an 81mm mortar round. Folded *(right)*, the device is about the size of a large suitcase.

*Rick F. Tscherne*

# BUNGY QUICK RELEASE

*Submitted By Chris Ayers*

Some bungy cord attached to the outside portion of a rucksack can come in pretty handy for attaching all sorts of things, like jackets, poncho liners, aiming stakes, foam mattresses, etc.

Instead of stopping to open up a rucksack to place something inside, you only have to lift up on the bungy cord and slide it underneath to hold it in place. Hell, even your buddy can attach something to your rucksack while you're still on the move.

There's several ways you can attach a bungy cord to a rucksack. Either by wrapping it around the aluminum part of the frame, or by cutting the bungy cord in half and permanently tying it to the top or bottom portion of the frame.

But before cutting a bungy cord in half, wrap some 100 mph tape around the cord where you intend to cut it. This will prevent the outside material (that's wrapped around the rubber cord) from unraveling. Then burn and melt both the ends where the tape is wrapped to prevent it from further unraveling.

NOTE: Never remove the 100 mph tape from the bungy cord, or the material that's wrapped around the rubber cord could unexpectedly unravel, leave it in place.

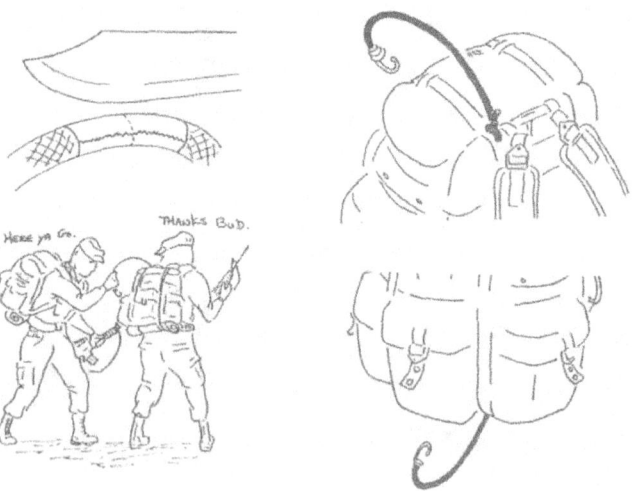

*The Complete Ranger Digest: Vol. VIII*

# FIRST AID POUCH TIP OR TRICK

*Submitted By: PFC Ben Donaldson*

Don't have enough pockets to store all your nice-to-have little goodies? Then buy yourself a couple of extra first aid/compass pouches and sew 'em to the side of your canteen covers.

What can ya store in 'em? Extra s—— like first aid dressings, water purification tabs, camouflage sticks, ear plugs, compact emergency space blanket, and whatever else you can think of to carry in 'em.

NOTE: When attaching the first aid/compass pouch to the side of a canteen cover, DO NOT sew it in place along the edges. This will cause the pouch to become too flat, tight, and flush against the canteen cover and you'll lose a lot of the storage space. Instead, sew it in place along the back side where the belt clip is located

Rick F. Tscherne

# A URINAL DIRECTOR FOR FEMALE SOLDIERS

Submitted By: Friedrich W. Eickelen

When one of my readers sent me this tip, I was laughing my ass off and said to myself, "Get the f—— outta here." But after I stopped laughing and read his letter a second time, I said, "Yea, well, ok, maybe he's got a point there."

Points to ponder? How many times have you seen a female soldier squatting in the woods to take a piss? Your answer had better be none, unless of course, you're a female soldier yourself. Unlike us male soldiers, they need a bit more privacy to urinate, you know?

Female soldiers can't just go around a tree or behind a vehicle to take a leak, not without exposing their buns to everyone within eyeball range. So they have no other choice but to go for a long walk away from the rest of us.

If you're assigned to an integrated/co-ed unit, how many times has your CO or 1SG stopped the convey so everyone could get out to take a leak? And who were the last ones to come running out of the woods to get back on the vehicles? I'll bet it wasn't a male soldier.

Unless, of course, your 1SG erected a special latrine just for the females. You know the type I'm talking about, the kind you set up between two trucks or vehicles with a couple of ponchos.

Well, there's a device on the market today called a "Freshette,"or what's also known as a "Female Urinary Director" that makes pissing in the woods for female soldiers a bit more comfortable.

How does it work? A female soldier only has to open her BDU trousers just wide enough to squeeze inside the "trough" and she relieves herself through an "extension tube" that hangs outside of her pants like a penis. Convenient for everyone? Well, almost...

If more female soldiers start to use these freshettes out in the field, I'm sure the guys will instead start to go for a long walk in the woods just to take a leak. After all, we'll have a lot more to show and expose than the females will. (Our ding-a-lings!)

This urinary director will not only protect a female's genital areas against unsanitary conditions, such as dirty restrooms and latrines. But it will make it easier for them to urinate in bad weather (rain, wind, & snow) and unsuitable "squatting" terrain.

The cost? Only $20. Available from: Campmor Inc

UPDATE: No longer avail., but try these similar and less expensive products:
Lady J: https://www.campmor.com/c/lady-j-80976
GoGirl: https://www.campmor.com/c/gogirl-female-urination-device

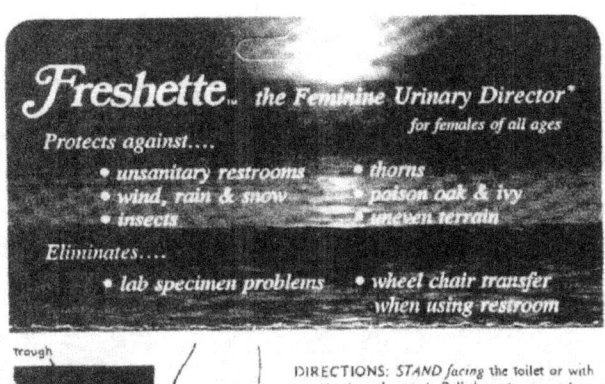

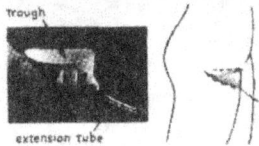

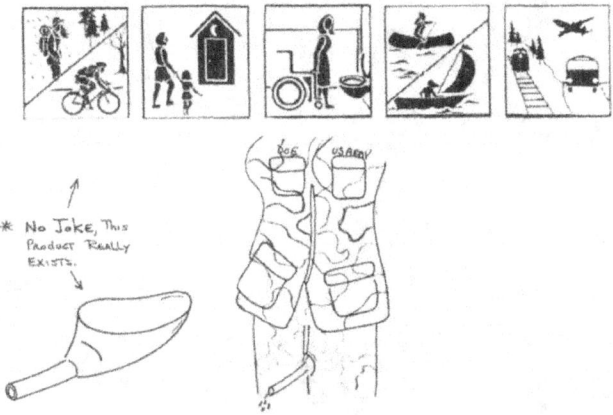

## WHAT TO DO WITH THOSE OLD BDUs

When your old BDUs are no longer serviceable due to wear, tear, and discoloring. Instead of trashing them, cut 'em up into long narrow strips and attach 'em to your lbe, rucksack, and kevlar helmet. They're just as good for camouflaging as cut up burlap bags and real foliage (leaves, weeds, etc). Check it out.

Rick F. Tscherne

## Bandits burned when bush said 'freeze'

Detective Earl Feugill shows off the disguise he wore when he foiled an armed robbery attempt Tuesday in Pembroke Pines, Fla.

Four would-be robbers were surprised when a dried-up bush outside a fast-food restaurant turned out to be a shotgun-wielding former Marine decked out like Rambo.

Police Det. Earl Feugill, who had been blending into the shrubbery for 90 minutes, was staking out a Checkers drive-through window Tuesday night in Pembroke Pines, Fla., after a series of restaurant robberies in southern Broward County.

When the four — who were armed and wore masks — approached the back door, Feugill sprang up, raised his shotgun and ordered them to freeze. "They were quite surprised," Feugill said.

The disguise was made of burlap attached with silicone to a camouflage outfit. He used it in his Marine days and still hunts with it.

"Scottish gamekeepers used these suits to cull their game herds, Feugill said.

James ___, 19, and three juveniles were charged with armed robbery and resisting arrest. An employee also was charged with armed robbery because she was an accomplice and opened the door for the bandits, police said.

# TWO QUART SIDE-PACKS

*Submitted By: Ssg. Steve Brittian*

For those of you who are assigned to a mechanized infantry or wheeled vehicle unit, I'm sure you'll appreciate this next tip.

Are ya tired of getting in & out and sitting uncomfortably in your track or HMMVEE because of that damn f—— butt pack keeps poking ya in your back?

Well, Ssg. Steve Brittian got fed up with it too, so he doesn't wear one anymore. Instead, he bought himself a two quart "canteen cover" and attached it to the side of his pistol belt and carries about the same amount of stuff he would normally carry in his butt pack.

For those of you who like to carry a bit more stuff, buy 2 x two quart canteen covers and attach them to the left and right side of your pistol belt. Not only will it be a lot more comfortable getting in & out and sitting in your track or vehicle, but carrying a rucksack on your back too. Smart idea, uh?

*Rick F. Tscherne*

# A FEW MINUTES WITH RICK

Ranger Rick's Commentary Today's topic guys & gals, is on 2nd lieutenants. *Hooah!*

Like many platoon sergeants, I've always hated to get a brand new "butter bar" straight outta school. Why? Well, not because they're young and inexperienced, but because we (PSGs) don't have time to baby sit and train 'em the way they should be trained.

Being a platoon sergeant ain't an easy job, ya know? In fact, it's more like being a father with forty-something kids, which is why your troopies call ya the "platoon daddy." And like a real father, you're responsible for their health, welfare, discipline, and training twenty-four hours a day, seven days a week. (Waah!)

To become a platoon sergeant you gotta work your way up through the ranks from Pvt/E-1 to Sfc/E-7. Officers on the other hand, become (make-believe) "instant" platoon leaders once they've completed their officer training program and receive their commission. And although they may have completed the required education and training to become . an officer, they are NOT yet (platoon) leaders.

Personally, I believe every officer should serve time as an enlisted before becoming an officer. And only after they have earned the rank of Sgt/E-5 and have proven they can lead soldiers, then (if they want) let 'em become officers.

By starting at the bottom they will not only learn to respect and care more for their men, but they'll have done everything that an enlisted soldier has done. From picking up cig butts, cleaning latrines, pulling guard duty, to learning how to become a leader.

Unless you've started at the bottom, you'll never know what it's like to be an enlisted, what it takes to be an NCO, and how to lead "real soldiers." Which is entirely different from leading a group of "classmates and buddies" in an officer training program.

Now let me tell ya about the best G— damn platoon leader that I've ever had. His name was 2LT GREGORY HIEBERT and he came to my platoon ("C" Company, 1st Battalion 509th Airborne Battalion Combat Team) straight out of West Point in 1982.

Now to be honest with ya, I thought he was gonna be your typical clumsy, inexperienced, think-he-knows-all butter bar. But boy, was I ever wrong about him. He was an exceptional individual, a quick learner, and a fine leader. He wasn't assigned to the platoon a month before picking up the ball and running with it and doing everything an "experienced" platoon leader would know how to do.

Unfortunately, my battalion commander (Ltc."Nuke'em Needham") saw in him what I did. And about a year or so later, he snatched his ass from my platoon and assigned him to another company and platoon that had leadership problems.

I knew of another butter bar who's career ended after he failed to take control of a "leadership confrontation" between himself and his Platoon Sergeant. His unit was out in the field for about a month, hadn't slept in 36 hours, and his PSG disagreed with a tactical decision he made. And after unsuccessfully trying to persuade him to change his plans, the PSG lost his temper and screamed...

"G— damn it, sir, get your f———— head out of your ass and listen to me! I know what the f—— I'm talking about, you don't! You're just some f——— dumbass cherry 2d lieutenant who thinks he..."

Well, this poor lieutenant didn't know what to do after his PSG degraded, embarrassed, and ridiculed him in front of the entire platoon. So he gave in and went with the platoon sergeant's plan.

When they returned to garrison, the lieutenant was too ashamed to report the incident to his company commander. And it wasn't long before gossip and rumors started to spread as to what his PSG did to him out in the field. Finally, his commander called him into his office to find out if all the gossip and rumors were true.

Embarrassed, he confessed the rumors were true. Unfortunately, because his name and reputation was already smeared and ruined in the unit, the commander had no choice but to reassign him, to the mess hall as the unit dining facility officer. And a few months later, he chose to resign his commission and get out of the Army.

What should this 2d LT had done? What an "experienced" leader (NCO or officer) would have done when challenged by a lower subordinate

a) Order the individual to "AT EASE," and if he failed to obey the order, stand firmly and wait until he has depleted his anger and..

b) Order the individual to "STAND AT ATTENTION" and verbally reprimand him in front of the same group of soldiers that he has challenged you in front of. And if he (again) fails to obey your order...

c) Immediately relieve him of his position, explain why he is being relieved, order him to the rear of the formation and report the incident to the 1SG and company commander as soon as possible.

The moral of the story? It takes not only experience, but balls to be a good platoon leader. And if you ain't got a pair of balls, then you had better find some, because you're gonna need 'em in place of "leadership experience."

Til next time kiddies…

*Sgt. Rick F. Tscherne*

U.S. Army, Retired (E-7 1/2)

# BOOK NINE

 *FOREWORD*

**Hooah!** Another Ranger Digest, more tips, tricks, and ideas than ever before. And because I've been getting a lot of inquiries and orders from hunters, campers, hikers, and survivalist, I've had to expand and touch on other subjects too, in particular - outdoor survival techniques. But don't worry this stuff is still useful and valuable whether you're in the military or not.

As you read over this latest edition you'll notice my spelling and grammar has improved quite a bit because I bought me a new computer. Not just because my old one was an antique, but because the Y2K bug got to it too. Boo-hoo-hoo... So long old friend.

Well besides some new tips, tricks, and ideas, I've got some other things to tell ya about. A year or so ago I received a letter from a disturbed and concerned *Ranger Digest* reader saying he saw my books being sold at several gun shows and fairs. And the copies that were being sold were obviously bogus reproductions and not the originals. And after he forward me a few of them to see for myself, sure enough they were fake, bogus, bootleg copies.

To make a long story short, I tried to get in contact with this copyright infringement culprit, but he wouldn't respond back to none of my certified letters. So I sent him one last letter, and just when my lawyer was about to slap him with a copyright infringement lawsuit, he replied back and agreed in writing to cease reproducing and infringing on my copyrights. But this wasn't the only scumbag to get caught, another concerned and caring reader helped me catch yet another one.

Readers Beware: If you encounter someone selling bogus, reproduced copies of my *Ranger Digests*, I'd appreciate you letting me know. And if you help me to file a copyright infringement lawsuit against the culprit(s), I'll share some of the lawsuit money with you. How can you help? I need copies of the books, the vendors calling card, a receipt for the books listing the titles purchased, and a copy of the cashed "personal check " that you made out to the vendor. And if you can get all this and more, like his vendor stand on film showing my books, we've got a solid case for a lawsuit and we can both make some money. So keep your eyes out for these scumbags.

Here's something else you're not going to believe. I received some reports that a few senior (active duty & retired) NCOs and officers have been trying to get my books removed from the store shelves of some Army installations because they consider 'em "inappropriate reading material for soldiers." Can you believe this?

Well ain't it nice to know there are some communists within the ranks who are trying to violate my Constitutional rights under the First Amendment -

Freedom of Speech & Freedom of the Press. And as I always tell people, "Hey, if you don't like my books, hell, then just don't buy 'em."

Anyway, just thought you'd like to know this bit of info. Well I hope ya all like this latest edition, and if you got some free time on your hands, then why don't you drop me a line and let me know what you think of it, because your tips, tricks, letters and comments DO COUNT! Til next time .

PS: Now don't forget, if someone asks you where you learned these tips & tricks from, ya tell 'em, "from my buddy Ranger Rick."

AUTHOR'S DISCLAIMER: The author and his contributors cannot be held liable nor responsible for any injuries or deaths caused by the use of any of these tips, tricks, and ideas. You are advised to use them at your own risk.

## SPECIAL THANKS

As always I dedicate a special page to all those who took the time to share with me and my readers their favorite field tip, trick or ideas. And if it wasn't for these contributors, there wouldn't be a *Ranger Digest IX* today.

David A. Williams M.D   Cpt Sheran L. Benerth   Sfc Mark E. Porrett

Ssg James D. Hunt   Ssg R.D. Cowgill   Sgt William Johnvin

1 Lt John Davis   Joseph Ricker   David White

Lt Oliver Fladrich

AND AN EXTRA SPECIAL THANKS TO A FEW OF MY RANGER DIGEST BUDDIES

Sgt Frank D. Gilliland - For Drawing Most Of The Sketches/Artwork

Sfc Mark Baker -  For Allowing Me To Use Some Of His Pvt Murphy Cartoons

Friedrich W. Eickelen - For Keeping Me Informed & Updated On New Products/Ideas

Joseph A.Laydon Jr. - For Sharing With Me Some Of His KISAP Survival Secrets

COMING SOON...A "RANGER RICK WEBSITE"
Yep, that's right, and it's well overdue too. And I'm working on it right now to get it up and running so I can share with you some of my latest field tips, tricks, and ideas and other interesting (and sometimes controversial) information.

*Rick F. Tscherne*

**WANTED - TIPS, TRICKS & IDEAS**
That's right, I still need your help and assistance in keeping the Ranger Digest series alive. So if you've got an outdoor field tip, trick, or idea that you haven't seen in any of my Ranger Digest handbooks, by all means send them into me via e-mail or snail mail. And if you would like to receive one of my calling cards (see below), just send me a self addressed stamped envelope and I'll send you one. I'll be looking forward to hearing from you soon. Hooah! *(Tips no longer accepted)*

## RANGER DIGEST UPDATE

And as the song goes, "and another one bites the dust..." And it gives me great pleasure to report the Army is finally cracking down and punishing senior NCOs and Officers who fail to live up to the military standards and regulations that they expect everyone else to follow except themselves. And as I stated in my previous *Ranger Digest*;

"It seems we always hear about the little fish (E-l - E-6) getting caught but rarely about the bigger fish (E-8, E-9, O-6 & above) getting caught, charged, and prosecuted far doing something wrong or inappropriate."

Why? Well most NCOs and officers probably won't agree with me, but it's because senior leaders are known to look out after their own. And it's a known fact among top ranking leaders that it's better to allow a fellow senior NCO or officer to retire (if they're eligible) than to prosecute them so they can retire with full military honor, benefits, and to save them (and the Army too) some embarrassment.

But as you can see from this recent newspaper headline, senior NCOs and officers are now getting what they deserve, and I'm glad they are too, Hooah! Because as 1 stated in my last book, I'm a strong believer in "the more rank you have, the more severe you should be punished because you should know better." But again, I'm sure most leaders will disagree with my philosophy on this, especially if they're a senior NCO or officer.

### In the military
# Army sgt. major under scrutiny

Investigation under way in sexual misconduct case

By MARION CALLAHAN
Heidelberg Bureau

**Attention Readers:** Csm Riley Miller use to be the USASETAF Command Sergeant Major here in Vicenza, Italy. And personally, I believe any senior leader who gets caught doing something wrong or inappropriate has probably done it a few times in the past before finally getting caught.

And as of December 24th, 1999, Csm Miller was found Not Guilty of all but one charge -Fraternization! Do I think it was a fair trial? Well, I think Csm Miller and the victim both should have been given a lie detector test and the results made available to the public. And regardless of the court's verdict, like in the OJ Simpson trial, whoever failed the test would always be looked upon as the guilty culprit & liar for the rest their life. It's called "**justice in the form of public shame and condemnation.**"

# RD VIII UPDATE:
# B-B-Q AMMO CAN

If you're in the military I'm sure you won't have any problems finding a 5.56mm or 50 cal. ammo can laying around somewhere. And if you're not in the military, then you'll probably have to buy 'em from a military/outdoor supply store or mail order catalog. And if you do, don't worry they're not very expensive, they cost only about $5.00. Which is pretty darn cheap considering all the many things you can use 'em for. (See *Ranger Digest VIII* page 14-19.)

As I mentioned in my previous Ranger Digest (VIII), one of my favorite uses is to use 'em as a portable barbecue. But a few troops deployed "down range" to Kosovo wrote;

Dear Ranger Rick,
Your b-b-q ammo can tip sounds like a great idea, unfortunately were not allowed to make any holes in our ammo cans because our unit recycles them, is there another way?"

Yep, there sure is. All ya need is some chicken wire for a grill, some aluminum foil for the inside portion of the ammo can so the charcoal, wood, etc won't burn or discolor the metal, and of course, some food and you're ready to start barbecuing.

And you know what I really like about this barbecue ammo can, you can keep everything you need right inside of it. Not only the charcoal/wood, aluminum foil, and chicken wire/grill, but even the food. And because it's small and compact, it can be easily stored in your car, boat, camper, or plane. Pretty cool, huh?

Now if you want to make it a bit more fancier, you can also attach a one-legged stand to it so you can barbecue "standing up" instead of sitting down. All ya need for this is a 3 x 2 x 1.5 inch piece of wood, a 1.5 x 1 inch piece of rubber from an old bicycle inner tube tire, a couple of GI wooden tent poles and study the following photos & drawings below and on the next page. Enjoy!

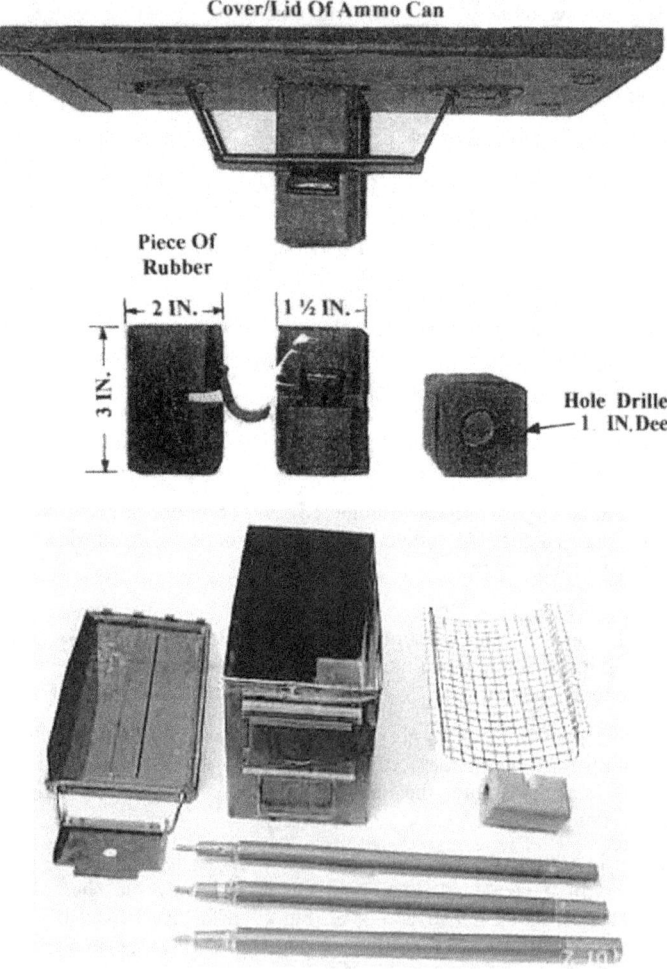

What your BBQ Set Should Contain

How To Use It

Place Aluminum Foil In Ammo Can

"Cheers Ya All!"

# HAVE YA EVER HEARD OF A DAKOTA FIRE HOLE?

One day while reading the American Survival Guide magazine I came across an interesting article about a field expedient fire pit called a "Dakota Fire Hole." Now I've heard of them and I've also used them a few times myself too, but I didn't know it was an old Army technique in making small clandestine out-of-sight fires for cooking, drying clothes and staying warm during military tactical operations.

Nope, I didn't believe it so I looked it up in an old US Army Field Manual (FM 21-76) and sure enough there it was in "black and white." And as the magazine article points out, it says in the manual to "build it at the base of a tree so the branches and leaves above It will disperse the smoke."

And as the magazine article also pointed out, this is kinda dumb to do because you're obviously going to waste a lot of time cutting, digging, and pulling up a lot of tree roots. And when you finally do have your Dakota Fire Hole built and going, chances are some underground tree roots will probably catch fire, burn, and smolder for many hours well after you have finished using it or moved out? Well, as the saying goes, "if there's an easier way and a hard way in doing something - the Army always wants you to do it the hard way. Ain't this the truth?

Now listen up boys & girls... it'll be lot easier and faster to dig a Dakota Fire Hole if you dig'em away from trees and bushes. And if you do need to disperse the smoke so the enemy won't see it, hell just break off some big tree branches and stick'em in the ground "above the hole" so they'll disperse the smoke closer to the ground rather than way above it. Make sense, don't ya think?

A Dakota Fire Hole should be dug about 12 x 12 inches wide and about 16 inches deep and then another hole dug about 4x4 inches wide at a "slight angle" all the way to the bottom of the first hole on the "upwind side." This smaller hole is called an "airshaft" and it will ensure air flows continuously to the bottom of the main hole so that the fire will burn steadily. Then after you have dug these two holes, place some very fine shredded tinder at the bottom of the main hole and then stack some small dry sticks on top of it "teepee style." Then all ya gotta do is light the tinder and when the teepee starts to burn, add more wood to it.

You can also build a Dakota Fire Hole in the snow, but you gotta build a base at the bottom of it so the fire won't keep sinking further and further down into the snow as it burns. To do this, you need to use "green logs" as a base because they won't burn as easily as dry logs. And if you do use green logs, then your fire hole should remain at the same depth that you built it. To extinguish the fire all ya gotta do is fill in both holes simultaneously and the fire should go out.

If you want to carry something lightweight that'll help get a fire going in no time, find an old bicycle "inner tube" tire and cut it into 1x1 inch pieces and put as many as you can inside an empty 35mm film container. Then when you need to start a fire, pull one piece out, light it, place it on top of the tinder and it'll burn for approximately 3 minutes, which is more than enough time needed to get a fire going. Try it, it works great!

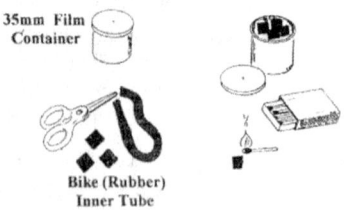

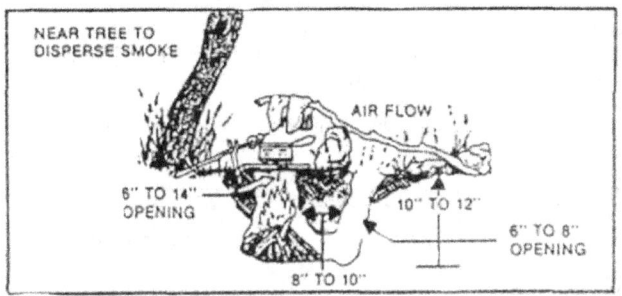

# HOW TO MAKE A SURVIVAL VEST

In almost every outdoor survival handbook there's a chapter on "what you should pack and carry in a survival kit" But what most of these survival handbooks fail to explain is the difference between the basic "must have" necessities and the "nice-to-have" items that can make your survival situation less life threatening, more comfortable, and yes... even enjoyable too.

And whether you're in the military or not, a hunter, camper, hiker, survivalist, or some other type of outdoor enthusiast, or you frequently travel to remote and off the road places, you need to carry a survival kit And as Joseph A. Laydon Jr. of Intensive Research Information Services and Products (IRISAP) points out in his books; (http://survivalexpertbooks.com/)

A survival kit MUST BE constructed to satisfy the eight elements of survival, which are Fire, Water, Shelter, First Aid, Signal, Food, Weapons, and Navigation. And If a survival kit does not contain the proper items to satisfy ALL (and not part) of these eight elements of survival, then it's NOT a true survival kit. UPDATE: See more general info on survival kits at Ranger Rick's website: http://www.survivaloutdoorskills.com/survival_kits_tips.htm

Well, after personally reviewing about dozen or more survival handbooks, these are the basic "must have" necessities they all recommend you should carry in a survival kit.

| *ITEMS* | *QTY* | *WHAT IT'S USED FOR* |
|---|---|---|
| Candle | 1 | Starting fires and as a light source |
| Compass | 1 | Determining directions |
| Flashlight | 1 | Light source and for signaling |
| Wire Saw | 1 | Cutting wood for fires and shelters |
| Safety Pins | 3 | Securing clothing & used as fishing hooks |
| Pocketknife | 1 | Preparing game and field craft needs |

| Signal Mirror | 1 | Signaling rescue parties during daylight |
| Wire (10 yards) | 1 | Making animal traps and snares |
| Fishing Line & Assorted Hooks | 6 | Catching fish and birds |
| Fire Starter (Matches, lighter, or flint) | 1 | Starting fires |

And these are the "nice-to-have" items they all recommend you should carry in a survival kit:

| ITEMS | QTY | WHAT IT'S USED FOR |
| --- | --- | --- |
| Whistle | 1 | Signaling Rescue Party |
| Sewing Kit | 1 | Repairing clothes |
| First Aid Kit | 1 | Treating cuts and wounds |
| Zip-Lock Bags | 2 | Gathering water |
| Nylon String (50 yds) | 1 | Building shelters & other field craft needs |
| Large Plastic Trash Bags | 2 | Making solar stills |
| Water Purification Tabs (Bottle) | 1 | Purifying water for drinking |

What should you carry these items in? Well if you're in the military you should carry them in an extra ammo pouch on your belt or inside your butt pack or rucksack. And if you're not in the military, then you'll probably want to carry them in a small nylon or canvas pouch or rucksack. Unless you want to carry'em me same way military aviators carry theirs, in a Survival Vest

Now there are several types of vests on the market, from military flight and assault vests to safari and traveler vests. But what I don't like about these vests is the price - they're too damn expensive. Except one type, a fisherman's vest Which I much prefer because they're less expensive, lightweight, and they have an assortment of small, medium and large pockets for all your survival goodies.

And when you need something, you can quickly get to it without having to dump everything out. And if you do forget which pocket you put it in, you just have to feel for it. And best of all, this vest can be easily stored under the seat of a car, boat, plane, or snowmobile. Not bad, huh?

And in addition to these "must have" and "nice-to-have" survival items, I personally carry in my survival vest the following "luxury items."

| ITEMS | QTY | WHAT IT'S USED FOR |
| --- | --- | --- |
| Sponge | 1 | Gathering water from wet /moist leaves, |
| 5 50 Paracord | 1 | Making shelters and other field craft needs |
| Space Blanket | 1 | Maintaining body warmth |
| Assorted Nails | 1 | Making spears and arrows |
| Snake Bite Kit | 1 | Treating snake bites |
| Small Bar of Soap | 1 | Maintaining personal hygiene |
| Pocket Rain Jacket | 1 | Protection from weather |

| | | |
|---|---|---|
| Sheet of 4 X 6 Plastic | 1 | Building shelters, gathering rain water |
| Water Bottles (1 x Liter) | 2 | For carrying drinking water |
| Emergency Strobe Light | 1 | Signaling rescue parties during darkness |
| Sling Shot/Elastic Band | 1 | Acquiring game |
| Binoculars (Compact Type) | 1 | Look for help, routes, game |
| Meal-Ready-To-Eat(MRE) | 1 | Emergency Meal |
| Drinking Cup (Folding Type) | 1 | For solar still and gathering water |
| Solar/Dynamo AM/FM Radio | 1 | Reducing boredom and maintaining sanity |
| Hand Sanitizer (Ethyl Alcohol) | 1 | For personal hygiene and starting fires. |

Although all this may seem like an awful lot of extra stuff to carry in a survival vest, no matter what type of terrain or weather environment I may find myself in, I guarantee these additional survival goodies will no doubt enhance my survivability TREMENDOUSLY. Think about it.

### FISHING VEST

### FRONT

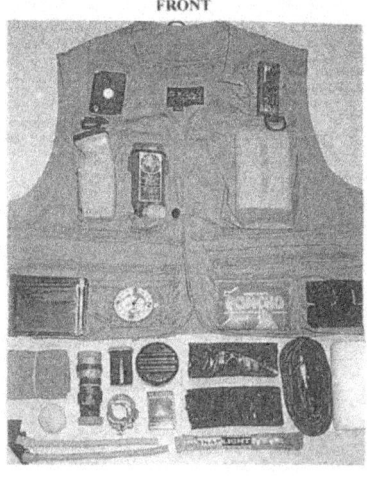

### BACK

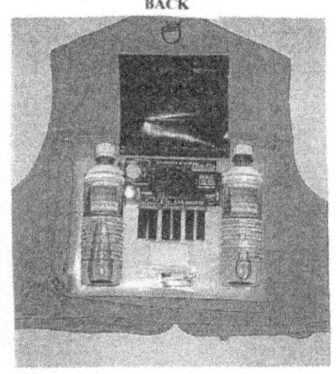

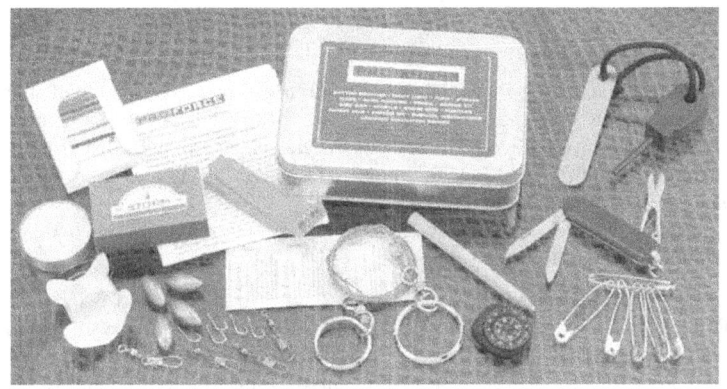

**SAS Combat Survival Tin**
This compact and durable tin features a compass, whistle, candle, brass wire snare, a fishing set, instructions, matches, a pencil, sewing kit, wire saw, water tablets, safety pins, a fire lighter, folding scissors and a striker blade. The tin can also be used for cooking purposes! United Kingdom. Measures 4⅞ x 3⅝ x 1½". Wt. 8 ozs.

# THE MANY USES OF FIBERGLASS TENT POLES

Many years ago while on a camping trip I was a bit B-O-R-E-D. So bored that I removed the fiberglass poles from my tent just to see what else I could use 'em for. Now unlike the military issued wooden tent poles, these commercial "fiberglass" poles are a lot more durable, flexible, and lighter in weight too. And because they're flexible, have a hole drilled down the center and a metal adapter on the ends for connecting several poles together, they can be used for other things too. For example:

AS A WALKING STICK - Just connect 2-3 poles together, wrap some 550 parachute cord around one end for a handle grip, jam a nail in the other end to keep die hole free of dirt and then wrap some wire or string around it to prevent it from splitting should you poke something a bit too hard with it, like a rock.

AS A SNAKE CATCHER - Just connect 2-3 poles together, run some nylon string or wire down the center, make an "adjustable loop" on one end and tie a large knot on the other to prevent the string/wire from coming out. To capture a snake with it, just slip the adjustable loop over it's head, pull tight on the other end and lift up. To release it, just loosen up your grip and the snake (if not yet choked to death) will (angrily) wiggle out of the loop by itself, then watch out!

AS A FISHING SPEAR - Connect 2-3 poles together, find a large nail about the same diameter as the hole in the poles, cut off the head of the nail, bend it "slightly" in the middle and then "jam it" in one of the ends. Then take a large fishing hook, bend it in the middle (as shown in the drawings/photos), position it over the nail and then tie it securely in place with some wire/string. Note: When a fish is speared, the fishing hook helps keep the fish from coming off the nail.

AS A HUNTING SPEAR - Connect 2-3 poles together, find a large nail about the same diameter as the hole in the pole, cut off the head of the nail, bend it "slightly" in the middle and "jam it" in one of the ends and then wrap some wire or string around it to prevent it from splitting should you hit something a bit too hard with it, like a rock or tree.

AS A HUNTING BOW - Connect 3 poles together, run some "nylon string" down the center, (preferably 4-5 nylon inner strands from some 550 paracord), bend the pole into a quarter of a circle/moon and then tie the two ends of the string together. Then take some 100 mph tape or 550 parachute cord and wrap it tightly around the center portion of the bow so you'll be able to grasp and hold it more securely in your hand.

AS A FISHING POLE - Connect 2-3 poles together, run some fishing line or string down the center, attach a fishing hook, grab the pole in one hand, wrap the excess fishing line around the other and cast out.

AS A GRAPPLING HOOK - Connect 2-3 poles together, find a large fishing hook, cut or break off the "eyelet," bend the hook in the middle and "jam it" securely in one end of the pole.

IMPORTANT: Should the poles not fit together very well, just take a pair of pliers and slightly bend the metal adapters on the ends until they do fit more securely and tightly together.

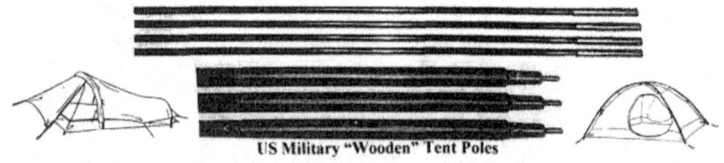

US Military "Wooden" Tent Poles

*The Complete Ranger Digest: Vol. IX*

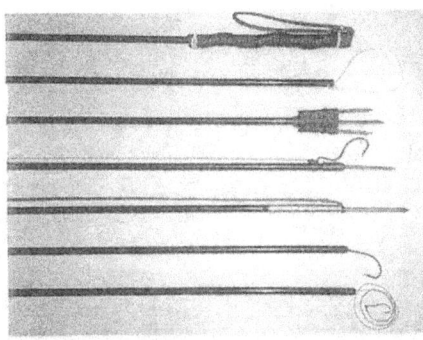

- Walking Stick
- Snake Catcher
- 3 Prong Fish Spear
- Fishing Spear (w/hook)
- Hunting Spear
- Grappling Hook
- Fishing Pole

Aim/Throw Slightly Below Fish
(To offset the reflection of the fish in the water)

Retrieve Spear

Big deal, anybody can do that, Ranger Rick

And Presto....A Fish Dinner
(Note: Not as easy as it looks.)

Rick F. Tscherne

Fiberglass tent poles available from CAMPMOR, 800-525-4784 or www.campmor.com (Listed as tenet "replacement poles", make sure they are fiberglass, not metal.)

## WALKING STICKS

Check out this ad, is it me or what? But would you pay $225 for a walking stick? I don't give a flying crap if you can use it as survival staff, baton, blow gun, lance, penis or whatever else it can be used for. And if you're willing to pay this much money for a walking stick, well you're either filthy rich or dumb. Or BOTH! Com'on, give me a break, $225 for a f—-walking stick? It's ridiculous!

Well, I'm no expert on walking sticks, but I've made a few and know how to use one too. Like who doesn't? Duh! And one particular walking stick that I've had for many, many years and still use is from an old broken movie camera tripod. I found it in a trash dumpster one day, the mount was broken but the three telescopic legs were still OK. And so I removed the aluminum legs, bought me a set of rubber bicycle handle grips and slid one over a leg. And yep, it works pretty good as a collapsible folding walking stick.

I took it to Bosnia with me too, never left my quarters without it, never gave a class without it, and used it as a walking stick, pointer and a baton too. And today I still have it and always keep it in my camper.

But before I had this one I had a wooden walking stick that I use to keep downstairs in my garage and take with me when ever I went for walks in the woods. It was a five foot long wooden stick with all the bark removed from it and I had my name engraved on it too. And what I did was make it into a survival stick. How? By simply attaching a few survival items onto it so I didn't have to carry them in a pouch or a rucksack. Check it out and see how I made it.

**THE SURVIVAL STAFF**
By Pat and Wes Crawford
Handmade for 15 years

Hiking Staff
Walking Stick
Baton
Blow Gun
Lance

All in one package
Made from Hard Aircraft Aluminum
$224.95 - Ready for delivery

**CRAWFORD KNIVES**
205 N. Center Drive
West Memphis, AR 72301
(870) 735-4632
www.crawfordknives.com

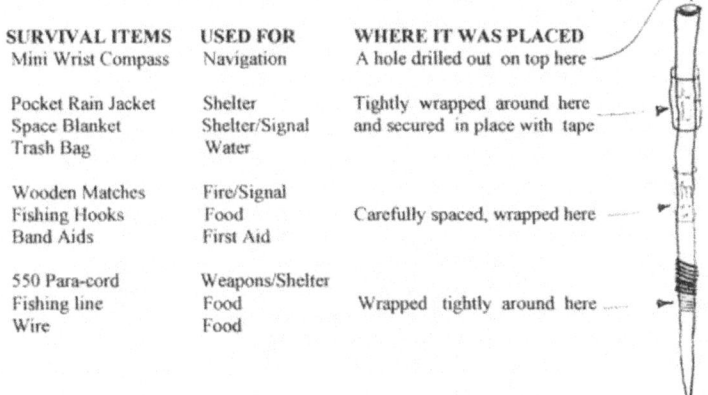

| SURVIVAL ITEMS | USED FOR | WHERE IT WAS PLACED |
|---|---|---|
| Mini Wrist Compass | Navigation | A hole drilled out on top here |
| | | |
| Pocket Rain Jacket | Shelter | Tightly wrapped around here |
| Space Blanket | Shelter/Signal | and secured in place with tape |
| Trash Bag | Water | |
| | | |
| Wooden Matches | Fire/Signal | |
| Fishing Hooks | Food | Carefully spaced, wrapped here |
| Band Aids | First Aid | |
| | | |
| 550 Para-cord | Weapons/Shelter | |
| Fishing line | Food | Wrapped tightly around here |
| Wire | Food | |

Rick F. Tscherne

# NEW SHELTERS?

Have you heard what the Army's been working on or searching for lately? You haven't? Well check out this article from the *Army Times*:

# New shelter may debut by 2000

## Four prototypes in the running

**By Matthew Cox**
Times staff writer

As early as next year, soldiers could be trading in their shelter halves and ponchos for a new level of hooch technology.

For years, soldiers of all ranks have fashioned field-expedient shelters using ponchos, bungee cords and tent stakes to avoid carrying the Army's outdated canvas pup tent.

Officials are looking to change this by searching for an off-the-shelf system that will become the Improved Combat Shelter.

"The current shelter half system is rarely deployed by the soldier simply because of its weight. It's difficult to set up [and] it still has a number of shortcomings," said Jimmy Hodges, project director for Product Manager, Enhanced Soldier Systems.

Fueled by a barrage of complaints from soldiers and commanders, officials began searching last year for something already made that soldiers could use as a rain shelter as well as a ground cloth, equipment tarp and makeshift litter.

In addition to the shelter half, officials say the new system could replace the poncho, since the Army already issues a two-piece rain suit. "We believe for the most part ponchos are used for ground cloths, to cover equipment and to line the inside of tents," Hodges said.

Four prototypes are being tested at the Army's Joint Readiness Training Center at Fort Polk, La. The test shelters range from free-standing designs to versions that require flexible poles and stakes for support.

Two of the test models have separate mosquito netting and all but one have built-in floors.

While the shelter would be intended for one-man use, officials said they want the technology to let soldiers connect two or three together based on mission requirements.

Officials hope to wrap up testing late this year and field the new shelter by fall of 2000.

One of the top requirements is to select a design that weighs about a pound and half less than the 5.4 pound shelter half.

But in the end, Hodges said, the weight factor may be a trade off if the soldiers favor one shelter over the others.

"It has to come back to the soldier's desire to use the item," he said. "Obviously lighter is better, but if they are carrying what they have today because they don't like to use it, it may be worth a few ounces increase in weight."

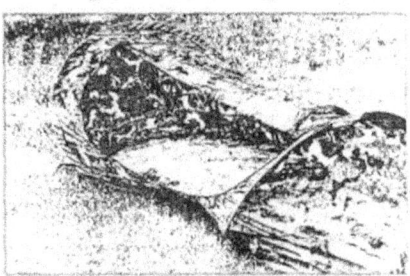

Officials at the Army's Joint Readiness Training Center at Fort Polk, La., are testing four prototype Improved Combat Shelters to replace the outdated and heavy standard-issue canvas pup tent. The prototype designs range from free-standing shelters to those requiring poles and stakes.

*Rick F. Tscherne*

Well it's about f—— time the Army started developing or searching for another shelter. These old Gl "canvas" pup tent were not only too heavy & bulky to carry, but when they got wet they took a loooong time to dry out too. And in my 20+ years in the military the only time I've ever used one was during basic training, and that was the last time I've ever used one. And if you can't wait for the Army to get their heads out of their butts in deciding which shelter they want, then buy yourself some fiberglass tent poles and use 'em to make one these several types of poncho hooches. Hooah!

**Important**: Before you can use the fiberglass tent poles with your poncho, you'll need four (4) nails and four (4) bolts. The nails must fit snugly & tightly inside the fiberglass end of the poles so they'll keep the dirt out, and the four (4) bolts must fit snugly & tightly inside the "metal adapters" at the other end so you can attach & connect the poncho to the poles. But if you can't find any bolts that will fit snug & tight inside the metal adapters, just wrap some "black electrical tape" around the bolts until they do fit. (See drawings & photos)

Construct One Of These Poncho Shelters Based On The Terrain & Weather Conditions (Note: For more details see *Ranger Digest I* page 35.)

## CATCHING FISH & GAME WITH AN IMPS-NET

If you haven't bought one of these Individual Multi-Purpose Survival (IMPS) Nets from Brigade Quartermaster, you don't know what you're missing. Check out this advertisement.

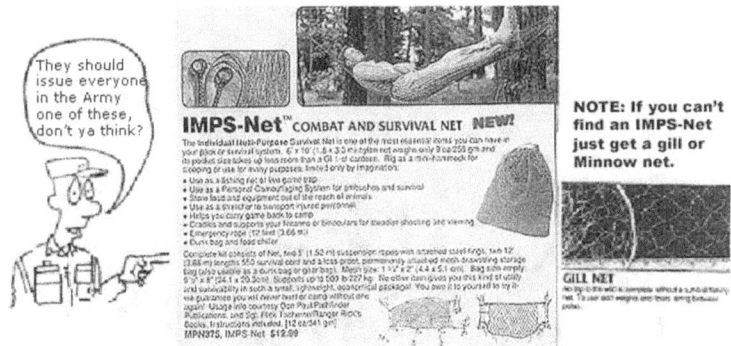

UPDATE: Not available at Brigade Quartermaster anymore. Try this alternative: https://colemans.com/shop/sporting-goods/u-s-g-i-survival-net/

I bought me one of these years ago and I gotta tell ya, "they're pretty damn handy." And whether you're in the military or not, I guarantee you' ll find a lot of uses for it. To name just a few...

CATCHING FISH: When used as a fishing net, wait until the fish are directly over the center portion of the net.

Then quickly raise it up. Note: I used my fiberglass tent poles and some 550 parachute cord to make this particular net.

*Rick F. Tscherne*

**CATCHING SMALL GAME:** Make a circular or square wooden frame for your net, prop up one side with a stick, attach a long string to it, find a good hiding place where you can observe it and when an animal is underneath it.....

...pull the string and quickly run up and kill the animal before it has time to escape or chew a hole through the net. Note: Again, I used my fiberglass tent poles and some 550 parachute cord to make this small game trap.

*The Complete Ranger Digest: Vol. IX*

# RANGER RICK'S
# LIGHTWEIGHT INDIVIDUAL CAMOUFLAGE KIT

You know what's wrong with the Army's camouflage net system? Besides needing a vehicle to transport the damn thing around, the "color & pattern" doesn't always match the surrounding terrain you're operating in. No BS. And if you ask any soldier who has ever set up one of these he'd tell you *"They're af-— pain in the ass. "*

You'd think by now the Army would've come up with a smaller, lighter, and easier-to-transport camouflage net system, wouldn't you? Sure, but they haven't. And until they do come up with a better one, you just might want to try my **Ranger Rick Lightweight Individual Camouflage Kit** (L.I.C.K.). And here's what you need:

1 x **IMPS-NET** - Order from **Brigade Quartermaster**
4 x fiberglass tent poles - Order from any camping supply catalog.
2 x net spreaders - **Make-it-yourself** out of wood, circular or X type.
5 x wire tent stakes - Or you can use wooden sticks.
2 x short, thick bolts - Complete with nuts & washers

Note: If you can't find an IMPS-NET, you can use a Gill or Minnow Net.

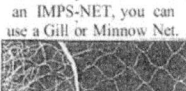

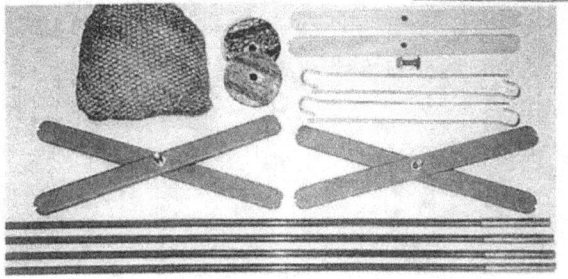

To make a couple of spreaders for your net, find some wood and cut out either two round circular spreaders or four flat narrow spreaders as shown in the drawings & photos. Then find yourself an electric drill, a drill bit about the same diameter as the bolts and drill a hole in the center of the spreaders and then slip in the bolts and secure 'em in place with the nuts & washers.

The bolts must fit snugly inside the "metal adapters" at the end of the fiberglass tent poles. And if they don't, men take some "black electrical tape" and wrap it several times around the bolts until they do fit snug as a bug.

You know what I really like about this L.I.C.K? You can quietly set it up and take it down "without making any noise." Just lay out the net, secure the corners, connect the poles, attach the spreaders, lift up the net, slide the poles & spreaders underneath the net and camouflage it.

I have no doubt all you light Infantry Grunts, Airborne, Ranger, Special Forces, etc will love my' Ranger Rick Lightweight Individual Camouflage Kit And if you do, write and let me know. And if you don't, write and let me know why ya don't, OK?

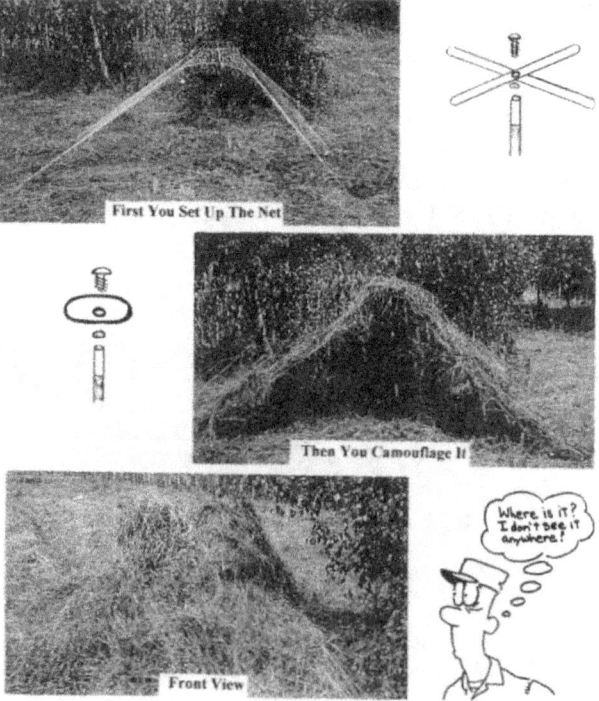

# PACK A FEW NAILS IN YOUR SURVIVAL KIT

While playing around with a home-made bow and arrow one day, I tried to figure out how I could make some simple arrow heads for my arrows. You know, other than the primitive and complicated ones they show you how to make in some survival manuals.

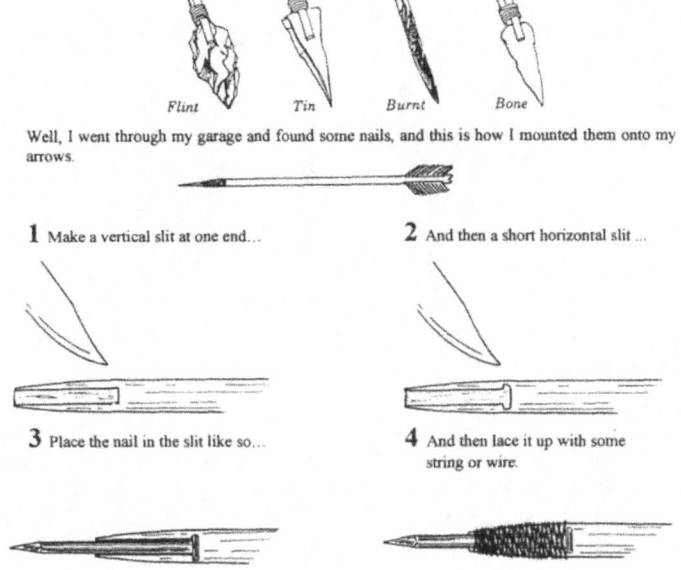

Well, I went through my garage and found some nails, and this is how I mounted them onto my arrows.

1 Make a vertical slit at one end...

2 And then a short horizontal slit...

3 Place the nail in the slit like so...

4 And then lace it up with some string or wire.

Not a bad idea, huh? And since this worked out pretty good, I decided to see if I could use some large nails as spear heads. And yep, this worked out real good too. So boys & girls, here's a few more things I recommend you pack in your survival kit, five 3-inch nails and three 5-inch nails. But sharpen them a little bit more with a metal file or grinder before packing 'em away in your survival kit.

Rick F. Tscherne

# CAMELBAKS

*Submitted By: Lt. Oliver Fladrich (Douglasville Police Dept)*
By now I think everyone has either heard of them or bought one, a CAMELBAK. It's a small backpack that holds approximately 2 quarts of water (or more) and you drink from it through a flexible tube that has a "bite valve" mouthpiece on the end of it.

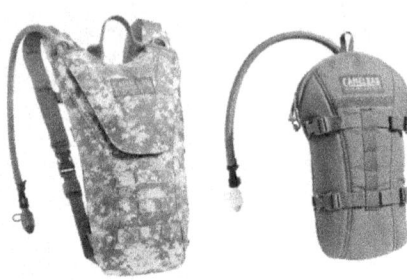

Well, I received an interesting letter from a Police Officer named Lt. Oliver Fladrich and he wrote,

*Hello Ranger Rick,*
*I am a Georgia Police Swat Team member and have been putting a lot of "Ranger Rick Tips & Tricks" to good use as we utilize a lot of US military gear. And it seems like a lot of outdoor enthusiasts, police snipers and soldiers are now using the "Camelbak" Hydration System or the "drinking tube " attachment for their military one or two quart canteens.*

*Now the only problem with this, is the "exposed mouthpiece " can get pretty damn dirty. But an easy way to get around this is to simply take a plastic 35mm film container, punch a 5/16 inch hole into the lid, remove the mouthpiece, slide the drinking tube through this hole and slip back on the mouthpiece. Then take the film container, punch a small hole through the bottom of it, take a single strand of nylon string (from the inner strands of some 550 para-cord) and slide it through this hole. Then tie a knot on the end of it, attach the other end to the drinking tube (so you can't lose it), connect the film container back onto the lid and you now have a quiet, cheap, tactical cover for your Camelbak mouthpiece.*

Ranger Rick's Comments: What an excellent idea, no doubt all you "desert bound" troopies will appreciate this tip. And I'm sure someone will probably steal this tip from my book, send it to the Army's Suggestion Program, take credit for it and get an "Attaboy Award." But we'll all know who's idea it was, right? Police Officer Lt Oliver Fladrich

*The Complete Ranger Digest: Vol. IX*

 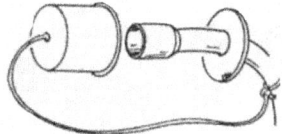

## RANGER RICK'S MULTI-PURPOSE HANDSAW

Here's a handy-dandy tool you'll never want to leave home without, I call it my Ranger Rick Multi-Purpose Handsaw. What you'll need is:

(A)   1 x Saw Bade - the type used for cutting small trees & branches.

(B)   1 x Piece of 550 Parachute Cord - or other type of nylon cord.

(C)   l x Fiberglass Tent Pole-must have metal adapter on one end.

(D)   1 x Narrow Bolt - complete w/ wing nut & washer.

(E)   4 x Circular D-Rings-the type used for attaching & carrying keys.

(F)   2 x Pieces of Wood - 6 x 1 x 3A Jnches in length/thickness.

Note: Most of these items can be purchased from almost any hardware store

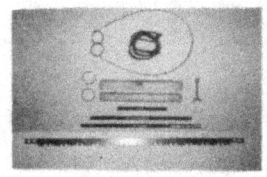

When you have acquired these items, just study the photos to figure how to make one – it's easy!

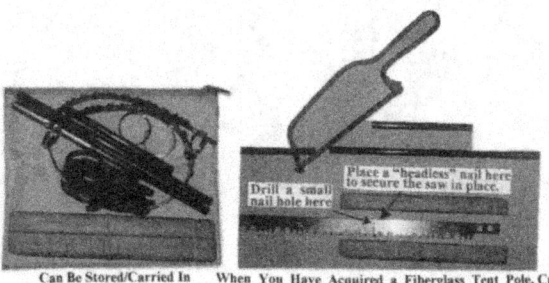

Can Be Stored/Carried In A Cheap Carrying Case

When You Have Acquired a Fiberglass Tent Pole, Cut It Down Into Three Pieces (Note: Study photos carefully)

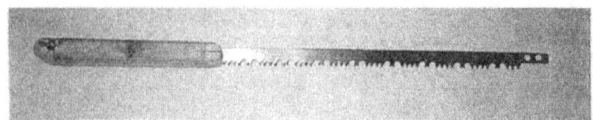

Light-Duty Handsaw

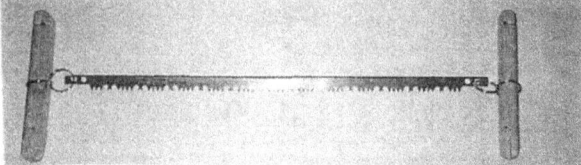

Two Person Handsaw

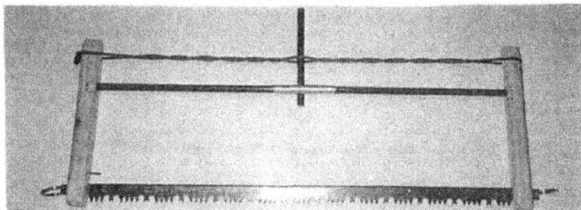

Heavy Duty Handsaw

*The Complete Ranger Digest: Vol. IX*

# WHAT YOU CAN DO WITH AN OLD PARACHUTE

Submitted By: Ssg. James D. Hunt

Unless you're a Paratrooper, Ranger, or Special Forces, you're probably not aware of the many things you can use an old parachute for. Well Ssg. Hunt wrote:

Hey Ranger Rick,
I had problems using my "fish net type " hammock for my hooch because the buttons on my BDUs kept constantly getting caught and entangled in the net and pissing me off. So everyone in my squad chipped in and bought an old parachute from Fort Bragg's DRMO for about $20 and we cut it up on the outside of the "gore seams, "sewed" two panel together, " one on top of the other and now we all have nylon hammocks. We didn't do the sewing ourselves, one of the guy's wife did it, but any sewing shop can do it. The nylon material is strong, lightweight, packs down pretty tight, and it's warm in the winter and cool in the summer.

Ranger Rick's Comments: If you don't weigh more than 200 lbs, you can get away with using a single panel instead of a double panel for a hammock, but no doubt two panels sewn together are much better than one.

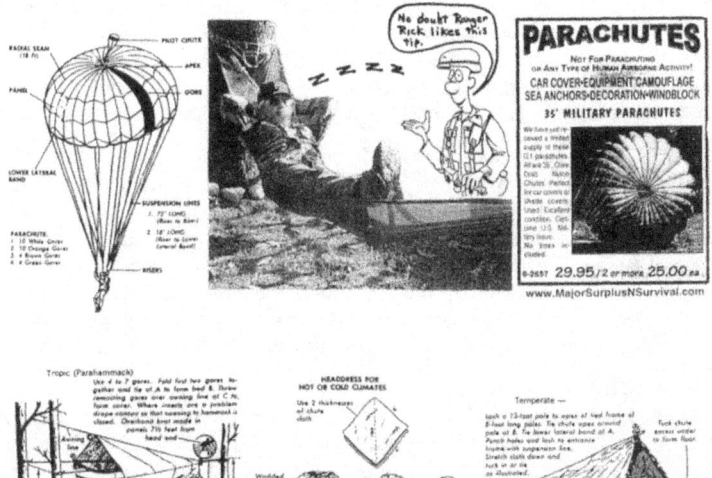

Rick F. Tscherne

# SOME SURVIVAL HANDBOOKS WORTH BUYING

As you know from my previous *Ranger Digests*, I like going through mail order catalogs searching for unusual products & books that I think will help soldiers out in the field. And after personally reading these books, I strongly recommend you acquire them for your personal library.

THE SAS SURVIVAL GUIDE - The author, John Wiseman, a former SAS instructor covers everything you need to know on how to survive in the wilds "anywhere in the world." It's the smallest (3 1/4 x 4 1/2) and most comprehensive (385 pages) survival handbook I have ever read. And unlike other survival handbooks, you can carry it comfortably in your shirt/pants pocket. Contains lots of pictures, drawings, & sketches, particularly on the types of plants and animals you can safely eat worldwide. Revised Edition
http://www.amazon.com/SAS-Survival-Handbook-Revised-Situation/dp/0061733199

SURVIVAL - The authors, Chris & Gretchen Janowsky teach you the art of how to survive in the wilderness of Alaska, Canada, and the lower 48 states. They explain their outdoor survival techniques in simple understandable terms that you feel like you're reading a novel and not a survival manual. Most of the information is based on the authors personal experiences in the Alaskan wilderness, who by the way operate a wilderness survival school there too. Contains lots of sketches & interesting stories.
http://www.amazon.com/Survival-Manual-That-Could-Skills/dp/0873645065

THE EMERGENCY-DISASTER SURVIVAL GUIDEBOOK - The author, Doug King a retired US Army Special Forces Survival Instructor. Explains what to do in the event of an earthquake, flood, blackout, hurricane, riot, tornado, winter storm, and other situations, but mostly in urban environments. Also features how to make a 72 hour survival kit, prepare food, and lots more. If you're on the road a lot or live in a small/large city or town, this book could save your life in an emergency.
http://www.amazon.com/Emergency-Disaster-Survival-Guidebook-Doug/dp/1883736102

# WRAP A RANGER BAND AROUND YOUR MAGLITE

Now even though I've been retired from the Army since 1993, I still wear my ol' Maglite Utility Holster on my belt. And it contains not only my Mini Maglite, but my Swiss Army Knife, a pen, comb, and cigarette lighter. And rarely, and I mean rarely do I ever go anywhere without it. Not to church, a friend's house, a restaurant, not even if my wife begs me to leave it home. "Come on Rick, please one time leave it home, we're going to a wedding! " Sorry honey, been wearing it too many years, feel naked without it and I just might need it.

And believe it or not, there's been many times that I was lucky and glad I had it with me, because I needed it, and urgently too. The only time I never wear it is when I forget to put it on, which is rare. One time I did forgot to put it on my belt and got about a half mile away from my home when I realized I wasn't wearing it. I turned the car around (which pissed off my wife) and went back to get it.

Hell, people are so use to seeing me with it on, if I wasn't wearing it, it would be the second noticeable thing missing from my body. The first would be my "black baseball cap" that I always wear with my mini Ranger tab and Airborne wings pinned on the front of it.

Anyway, a couple times I went to the store and I was stopped by strangers and told, "Hey buddy, your flashlight's on!" Again! Now I know why my batteries keep dying out on me so soon. Because you have to rotate the body of a Maglite to turn it on, it seems like every time it's carried in the holster and gets bumped a few times it somehow gets accidentally "rotated on." I tried placing a little piece of tape around it so wouldn't turn on so easy, but it just doesn't seem right that you gotta remove the tape every time you want to turn it on. You know what I mean?

So what I did to solve this problem was cut out a rubber "ranger band" from an old bicycle inner tube tire and wrapped it twice around the two parts of the Maglite where it rotates. And yep, this seemed to solve the problem, now it doesn't accidentally turn on anymore and it still fits snug as a bug inside my holster.

You got this problem too, bubba? Oh yea? Well place a "ranger band" around it and it won't accidentally switch on anymore, try it out.

*Rick F. Tscherne*

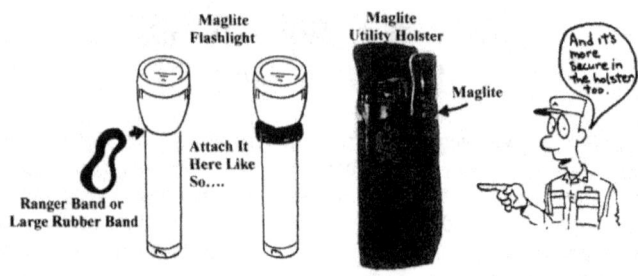

# MINI-MAG LITE SURVIVAL KIT

A lot of people seem to like the US military (Fulton MX-991/U1) angle flashlight just because it has a compartment on the bottom that can be used for storing a few survival items like matches, fishing hooks, sewing kit, etc. But 1 wouldn't buy one just for the storage compartment. In fact, I've never liked mis type of flashlight at all. Why? Because it's too big, bulky and the D-cell batteries weigh too much, especially if you gotta carry a few extra sets, ya know what I mean?

And sooner or later the VIPs at the Pentagon are gonna get their heads-out-of-their-butts and spend some of our tax money on a flashlight that the troops are already buying out of their own pockets - a Mini-Mag Lite. (Note: Uses 2 x AA batteries)

Now just because the Mini-Mag Lite doesn't come with a storage compartment doesn't mean you can't carry a few survival items attached to it. And a way to do this is to buy yourself a Mini-Mag accessory kit and then find an old bicycle inner tube tire and an empty 35mm film container.

Then take the rubber "lens holder" from the Mini-Mag accessory kit and place it over the flashlight. Then take the plastic 35mm film container and hold the "opening part" over a small flame (like a match or candle) for only about 10-15 seconds and then immediately (while it's still hot) slide it over the rubber "lens holder." When it cools down, remove it and you'll see it has molded to the shape of the rubber lens holder. Important: Do not attempt to force the plastic film container over the lens holder without first heating it up or it will split open right away or unexpectedly later on.

Now take the bicycle inner tube and cut off about a 3 inch long (rubber band) piece and slide it over both, the film container and the top portion of the flashlight. This will secure the film container to the flashlight so it won't come off when you fill it with your survival goodies. Then when you need to use your Mag Lite or some survival items, just pull off the film container along with the rubber band.

Here's what your Mini-Mag Lite Survival Kit should look like and contain:

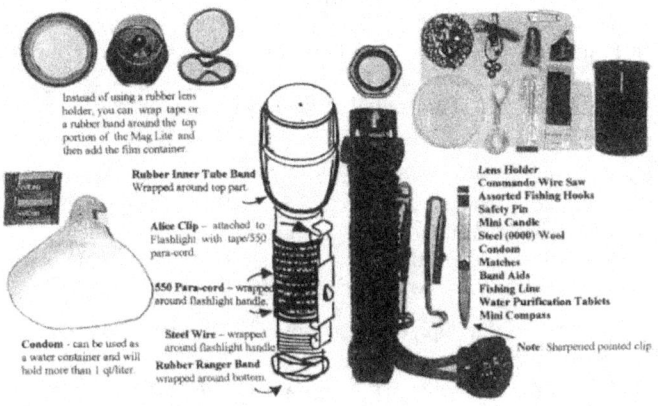

## A FLASHLIGHT COULD SAVE YOUR LIFE

This next tip was forward to me by a retired U.S. Army Special Forces instructor by the name of Joseph A, Laydon Jr. And like myself, he also writes, publishes, and markets his own line of handbooks, but under the name of Intensive Research Information Services and Products. (See end of this book for more information.) He wrote:

*Dear Ranger Rick,*
*Did you know a flashlight could save your life? In particular, a military issued Fulton MX-991/U flashlight. It's not only durable, waterproof, and versatile, but it floats too, (when empty or loaded with rechargeable batteries). Here's a few of the many things it can be used for in a life or death survival situation.*

*\* Why waste your precious matches or the fuel from your cigarette lighter to ignite your tinder when you can use the reflector from your flashlight and the power of the sun. Just place a cigarette through the rear opening of the reflector, aim it at the sun, and move the cigarette in & out until you find the smallest and sharpest concentration of sunlight and it starts to smoke.*

Ranger Rick's Comments: Though I showed you in my last *Ranger Digest (VIII)* how to do this, be advised this technique works best with "cigarettes." I'm not saying you can't start a fire with other dry stuff, but due to trying to find the "smallest and sharpest concentration of sunlight" (the same way you use a magnifying glass) on whatever you're trying to bum. If you can't get it to ignite, then try some other type of dry stuff, and if you still can't get it to ignite - then more than likely the sun isn't bright enough.

To determine if there's enough sunlight, try this: Place your "pinky finger" in the bottom of the reflector and push it inside the hole as far as it will go. Now face it in the direction of the sun, move your pinky slightly in & out and if you start to feel some heat and then a "burning sensation," men you've got plenty of sunlight and you shouldn't have any doubts this flashlight reflector trick works. Get my drift?

* Now that you know how to start a fire with a flashlight reflector, what if there isn't any sun? Well, if you purchase some Super Fine #0000 Steel Wool and keep some of it in the bottom compartment of your military flashlight, regardless of the weather conditions, rain, sleet, snow, wind, etc, you II be able to start "several fires " with it in just seconds.

How? Just unscrew the top portion of your flashlight, turn it on, take a very small piece of steel wool and "fluff it up. " IMPORTANT: A light, compact wad of steel wool will NOT work, it must be "fluffed Up!" Then position your flashlight downward and directly over the tinder, place the fluffed up steel wool inside the flashlight and simultaneously touch both the positive (+) and negative (-) prongs/terminals of the flashlight until it ignites.

**WARNING/DANGER**: You MUST immediately release it as soon as it ignites or you will be severely burnt, so be careful and watch out what you 're doing at all times. Then let the ignited steel wool fall out onto the tinder and lightly blow on it to get it burning.

Ranger Rich's Comments: This steel wool fire starting technique works great, only wish I had thought of it first before Mr. Laydon.

* Need to signal a ship, plane, or a rescue party and you don't have a mirror? Try using the reflector from your flashlight, just aim it in the desired direction and move it up, down, & sideways to attract attention.

Ranger Rick's Comments: Though it's not as good as a regular mirror, but if you've got nothing else to use - why not?

* Need a drinking cup or something for gathering water? Just take your flashlight apart and remove the batteries, coil spring, reflector & bulb and reassemble it with all the waterproof o-rings and fill it up and drink from it from the bottom.

Ranger Rick's Comments: Before using, carefully inspect the inside portion of the flashlight to insure there are no "battery acid stains," And if there are, you must thoroughly clean it before it can be utilized. **WARNING** - Failure to properly clean it could cause severe illness.

READERS NOTE: If you don't have a military MX-991/U flashlight, not a problem, you can use any type of D, C, or AA battery cell commercial flashlight. And if it doesn't have a storage compartment for the steel wool, seal it in some 100 mph tape, flatten it out, remove the batteries and place it on the bottom of the flashlight "underneath" the coil spring. Warning: Insure the steel wool is securely wrapped up in the tape and there's none sticking out or your entire flashlight could burst into flames "unexpectedly."

Wanna learn some more "unusual" survival tips & tricks? See SF Joe Laydon's advertisement at end of this book.

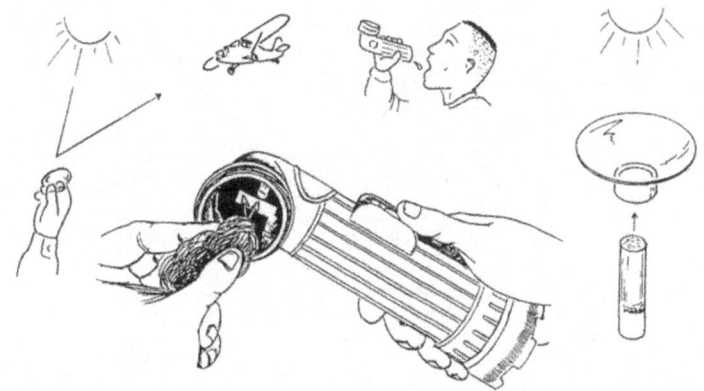

## HOW TO MAKE A FIELD EXPEDIENT COMPASS

Remember when you were a little kid in school and your science teacher showed you how to "magnetize a nail" with a 6 volt battery and some wire? Remember? He or she wrapped some wire around the nail, attached it to the positive (+) and negative (-) terminals of the battery and then showed the class how it was now capable of picking up other nails with it. Remember?

Well, here's a trick that can be done with almost any flashlight, but you'll need some wire, booby trap or some other commercial type of wire, and a piece of paper or a leaf.

Then unscrew the top portion of your flashlight, remove the reflector & light bulb and put it away inside your pocket. Now take the cover & lens and carefully fill it with water and then place it off to the side. Take the wire, cut off about 1 1/4 inch piece wire, bend one end 1/4 inch inward [——] and wrap several coils of wire around it making sure "none of the coils" touch one another.

Then turn on the flashlight, connect the two ends of the wire to the two (+/-) prongs at the opening top of the flashlight and hold 'em in place for about 1 minute. Then remove the piece of wire from the coils, making sure it's straight as a nail, tear off a piece of paper/leaf and place it in the water. Then take the magnetized wire and place it on top of the piece of paper/leaf and the "bent (thick) end" of the wire should point NORTH.

Got a sewing kit with you? Great! A sewing needle is much easier to magnetize than a piece of wire, provided you've got at least 2 volts of electricity and nothing less. And when it's magnetized the "eyelet (thick) end" should point NORTH.

Note: Be aware the piece of wire and sewing needle will not stay magnetized very long, so once you've got it magnetized quickly place it on the piece of paper/leaf before it weakens. Do not use straight pins because for some reason they do not magnetize very well.

Want another tip? Always carry inside your flashlight (taped alongside your batteries) a couple of sewing needles and about 6 inches of wire just in case you have to make another one of these field expedient compasses again. Sound like some good advice? You betcha!

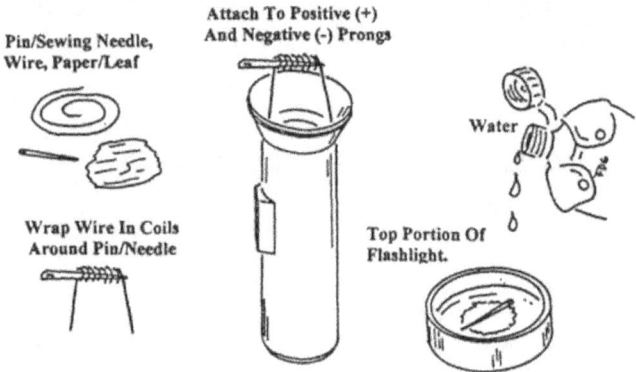

# USING TREES TO DETERMINE GENERAL DIRECTIONS

Did you know you could use trees to determine directions? Well not all trees, it depends on what type they are, where they're located, and how much sunlight they receive. You didn't know this? OK, everyone knows the sun rises in the East and sets in the West, right? So which side of a tree do you think gets the most sunlight? Well, if you live in the Northern Hemisphere -

it's the South side, and if you live in the Southern Hemisphere - it's the North side. You follow me so far?

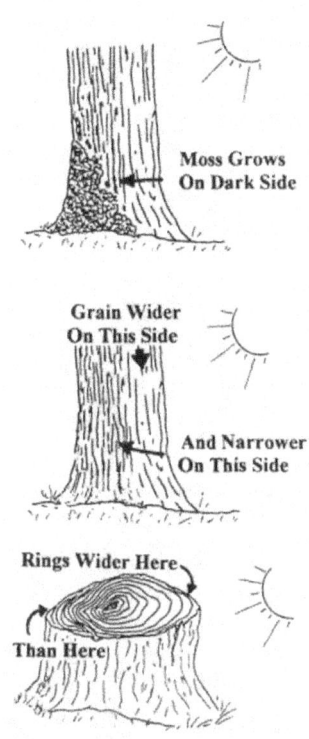

Now if you found moss growing on a tree, which side would you think it would grow best on? According to the Encyclopedia, moss grows in dark and damp places, and which side of a tree would you think that would be on? The side that received the least amount of sunlight. So if the sun raises in the East, goes down in the West and you're located in the Northern Hemisphere, what side of the tree would receive the most sunlight? The South side! So then the darkest and dampest side of the tree must then be the North side and where you would find moss growing on it. But what if moss was growing all around it? Whichever side has the most and or the greenest moss growing on it would be the darkest and dampest side of the tree, which then must be the North side.

OK, now most trees such as pine, walnut, oak have a rough, grainy bark with noticeable "gaps" in between layers. And depending on which side of a tree gets the most sunlight, the grain on that side of the tree will have a "much wider" grain pattern than the other sides. And the side that gets the least amount of sunlight will have a much tighter grain pattern than the other sides. So now that you know how to determine which side of a tree receives the most and least amount of sunlight, if the grain pattern/gaps are much wider on one side of the tree and you're in Europe, what side would that be? The South side! Got it? Are you sure?

OK, one more. If you ever stumble across a good size tree that has fallen down, or better yet one that has been cut down with a saw, look at the stump. You'll notice the tree rings are much wider on one side than the others. Why? Because that side of the tree received more sunlight than the others. Now if you were in Canada what side of the tree would that be? The South side! Got it? Understand?

NOTE: Trees found in low areas such as in valleys or thick forests will not receive as much sunlight as those found on hilltops, ridges, etc, therefore they will be much more difficult to use in determining directions. And if you must use trees to get from point A to point B, remember what side of the trees you must keep your eyes on and follow while on the move, get my drift?

## Pvt. Murphy Joins the Army

A NEW comic strip about the misadventures of a soldier will be made available for print in Army installation newspapers worldwide starting Aug. 1.

"Pvt. Murphy's Law" is a comic strip created for soldiers, about a soldier and by a soldier. SFC Mark Baker, a signal intelligence analyst in the Military District of Washington, created the cartoon in the early 1990s after making his first jump after airborne school at Fort Bragg, N.C.

"I hit the ground — feet, knees, face — but I felt no pain because of all the adrenaline," said Baker. "I woke up hurting all over the next day and said to myself, 'You know, there's got to be a cartoon in here somewhere.'"That was the birth of PVT Murphy.

In 1993 Fort Bragg's newspaper began printing Baker's comics. "Pvt. Murphy's Law" ran weekly at Fort Bragg for two years. "Murphy's whole purpose for existence is to make soldiers laugh," said Baker. "It's not an editorial cartoon—it's just for fun."

Baker is a self-taught artist and began drawing cartoons in high school. His talent was evident to Arizona State University, which offered him a full scholarship in 1986 — he turned it down. "I just wanted to join the service," said Baker. "Besides, I wasn't quite ready for college."

Turning down the scholarship didn't stop him from succeeding, though. He published his first paperback Pvt. Murphy collection in1997, and it's still available in military clothing sales stores, post exchanges or direct from the author.

The Army News Service will provide "Pvt. Murphy's Law" to editors of Army publications monthly.

Did you know "Pvt. Murphy" has his own website? You didn't? Then check it out at:
http://www.pvtmurphy.com/
There are books, t-shirts & prints available for order too !

## ATTENTION "PVT. MURPHY" FANS!

Have you seen Pvt. Murphy in your local base newspaper? No? Not yet? It's a hilarious comic strip about the misadventures of a soldier. And regardless of your MOS, everyone will see a little bit of themselves in Pvt. Murphy. Well, almost everyone... not the officers, of course.

And the cartoonist, Sfc Mark Baker has had to pay his dues in ass chewings to get to where he's at today. All because of some soldiers (mostly officers) found his cartoons to be more "offensive" than funny.

Come 'on people, give me a f—— break and lighten up, will ya? As one Sergeant Major by the name of JOEL PEARSON once told me, "If you can't laugh at the Army once in awhile, then you shouldn't have joined in the first place because sometimes it's just one big f——n joke" And I'm sure most of you know where he's coming from too, right? Especially if you're working under or for an incompetent leader. In other words, you gotta have a sense of humor if you're gonna make it in the Army.

Well here's a few of Sfc Baker's cartoons:

# WEB BELT ALICE CLIPS

*Submitted By: Ssg. R.D. Cowgill*

Dear Ranger Rick,

After reading in your *Ranger Digest VII* (page 18) on how to attach a web belt "alice clip" to a Maglite so you can wear it on your lbe, I decided to try to attach it to something else a can of pepper spray. After being attacked by a very large, vicious dog one time while riding my motorcycle, 1 never leave home without it and wear it where I can easily get to it - on my belt.

Ranger Rick's Comments: A few more useful things that you can attach an "alice clip" to so you can wear it on your belt, lbe or rucksack...

*The Complete Ranger Digest: Vol. IX*

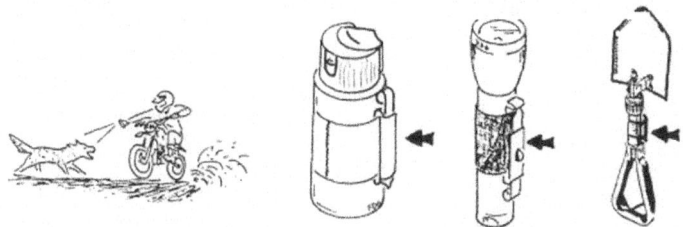

Note: Due to the weight of these items, 100 mph (duck) tape may not be enough to secure the clips, reinforce it with some wire to make sure it doesn't come off unexpectedly.

Wanna know how to turn an alice clip into a weapon? First separate the pieces by sliding the sleeve" forward and vigorously pulling out the sliding portion of the clip. Then take a metal file and sharpen both sides of its lower edges and then sharpen the round tip into a "sharp point" Be careful not to sharpen/remove too much metal from tip or you won't be able to use it as an alice clip for attaching things to your belt. Just sharp enough so you can cut things with it and pointed so you can attach it to a stick and use it as a spear (head).

Ain't got no file? No problem! Just find yourself a smooth flat rock and rub it back & forth until it becomes sharp and pointed. Though this method will take a bit longer, it'll still get the job done.

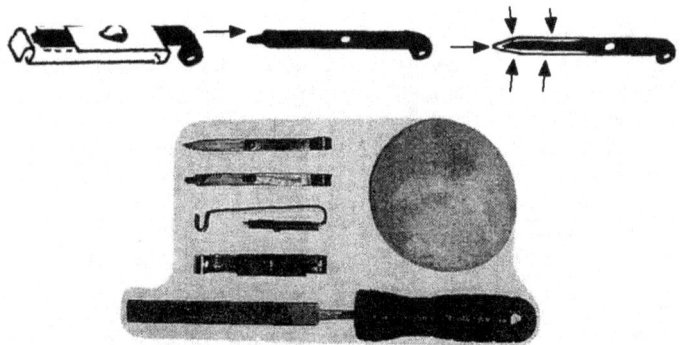

## YOU MIGHT BE A SOLDIER....

When you see a war movie you ruin it for everyone by pointing out the unrealistic military scenes

When camping with your family you check first for "good fields of fire" before setting up the tent.

You've been assigned overseas so many times that your kids speak three different languages.

Your wife responds to "Hooah" and understands what it means "regardless" of how you use it.

You use a poncho liner as a bed spread and have a bottle of hot sauce ready for every meal.

When in a strange place and need to go to the bathroom, you always ask "where are the latrines?"

You convince your wife that all ten of your guns are necessary for home protection.

You keep a case of MREs at home and in the trunk of your car in case of an emergency.

No one understands none of your war stories because of all the military acronyms you keep using.

You've seen "Patton" so many times you can recite his speech forward and backwards.

You'd rather live on post just so you can hear reveille and retreat everyday.

When your kids are getting rowdy and noisy you always yell to them "AT EASE!"

Every time your two year old sees someone in BDUs he says "Daddy!"

When you go to the store you always "back in" your vehicle into parking spaces.

## LITTLE JOHNNY

On Christmas morning a cop on horseback was sitting at a traffic light when Little Johnny, the son of a Command Sergeant Major, rode past him on his brand new bicycle.

The cop yells, "Hey kid, come over here."

Little Johnny turns his bicycle around and pulls up beside the cop.

The cop asks Little Johnny, "Nice bike kid, did Santa Claus bring that to you?"

Little Johnny says, "Yep, he sure did."

The cop says, " Well next year tell Santa Claus to put a tail light on it." And he issues Little Johnny a $20 safety violation ticket.

Little Johnny takes the ticket, but before he rides off he says to the cop,

"By the way, that's a real nice looking horse you got there, did Santa Claus bring him to you?"

Wanting to humor Little Johnny, the cop says, "Yep, he sure did kid"

Little Johnny then says, "Well next year tell Santa Claus to put the "dick" underneath the horse instead of on "top."

### TWO FEMALE 2D LIEUTENANTS

Two female 2d lieutenants were walking down the street when one of them finds a small compact on the sidewalk. She picks it up, opens it, looks into the mirror and says, "Hmmm, this person sure looks familiar."

The other lieutenant says "let me see" and the lieutenant hands it to her. "You dumb shit" the other lieutenant screams, "That's me!"

## PORTABLE SOLAR BATTERY CHARGERS

If you're like most soldiers, campers, hunters, and survivalists, you probably don't leave home without a Life Support System - AM/FM Radio, Cassette, or CD Player. Right? Yep, I know ya don't. And when it comes to batteries, I'll bet ya either pack "too many" or "not enough," right?

Well here's a product that'll take the guess work out of how many batteries you think you'll need to pack for the field, it's called a PORTABLE SOLAR "AA " BATTERY CHARGER.

If your life support system runs on 2 x AA batteries, as a minimum you should pack at least three sets of rechargeable batteries. While you're using one set (2 x batteries), the other two sets (4 x batteries) should be recharging. Makes sense, don't ya think? Sure!

Now the only bad thing about these solar battery chargers, is that they weren't designed to recharge batteries while on the move. In other words, you can only recharge them when in a stationary (non-moving) position. And according to the instructions that come with these solar battery chargers, it takes...

    2-3 hours to recharge 1 x battery
    4-6 hours to recharge 2 x batteries
    7-10 hours to recharge 3 x batteries
    10-14 hours to recharge 4 x batteries

As you can see you're gonna need "every minute of sunlight" to fully charge your batteries. And if you don't charge 'em while on the move, it's gonna take "several days" instead of several hours to fully recharge 'em. Ya know what I mean? What the damn manufacturer should have done is added a belt clip or a carrying strap so it could be worn on the belt, over the shoulder or attached to a rucksack. This way you'd be able to recharge your batteries while on the move and NOT just in a stationary position. Makes sense, don't ya think? Sure it does.

Well if you want your batteries to get maximum charging time from the sun, then here's what you should do;

(a) Add a small screw to the left and right side of the solar battery charger, attach some 550 parachute cord to them and then you'll be able to wear it around your neck, over your shoulder or on your rucksack.

Or... (b) Buy a G.I. nylon first aid or compass pouch, cut out a rectangle hole along the top portion, slide in the battery charger leaving the "solar panel" exposed on the outside of the flap, close the flap, and then you'll be able to attach or wear it on your belt, lbe, or rucksack.

NOTE: If your flashlight operates on 2 x AA batteries, then you'll get DOUBLE USE out of this solar battery charger. But make sure you take along a few more extra batteries, at least another set

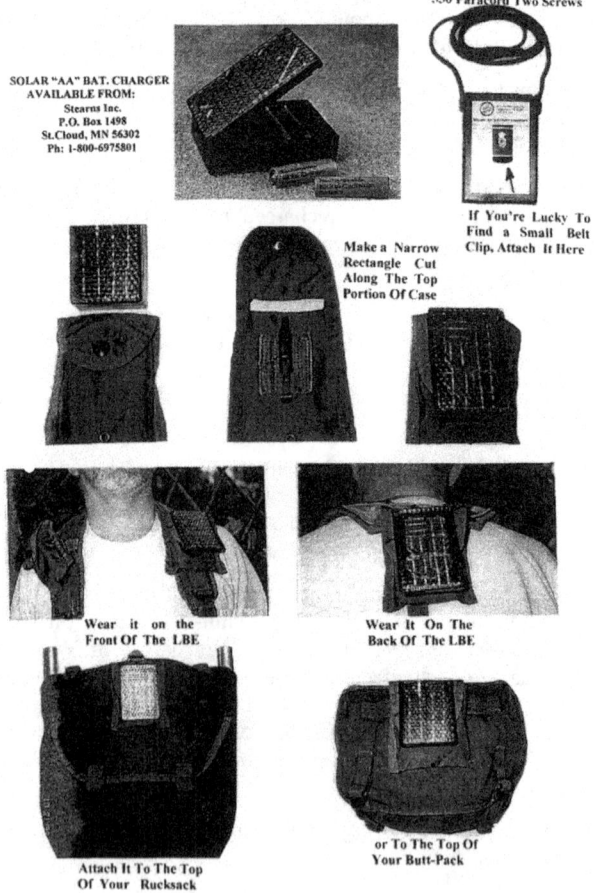

*The Complete Ranger Digest: Vol. IX*

# SAS SURVIVAL TIPS?

*Submitted By: An Anonymous British SAS*

Greeting Mate,

For some time I've been acquiring your Ranger Digest handbooks from an American friend of mine, a Special Forces chap. Love 'em! Enclosed you'll find some survival cards that I hope you can use in your next edition. Keep up the great work, Ranger Rick Cheers!

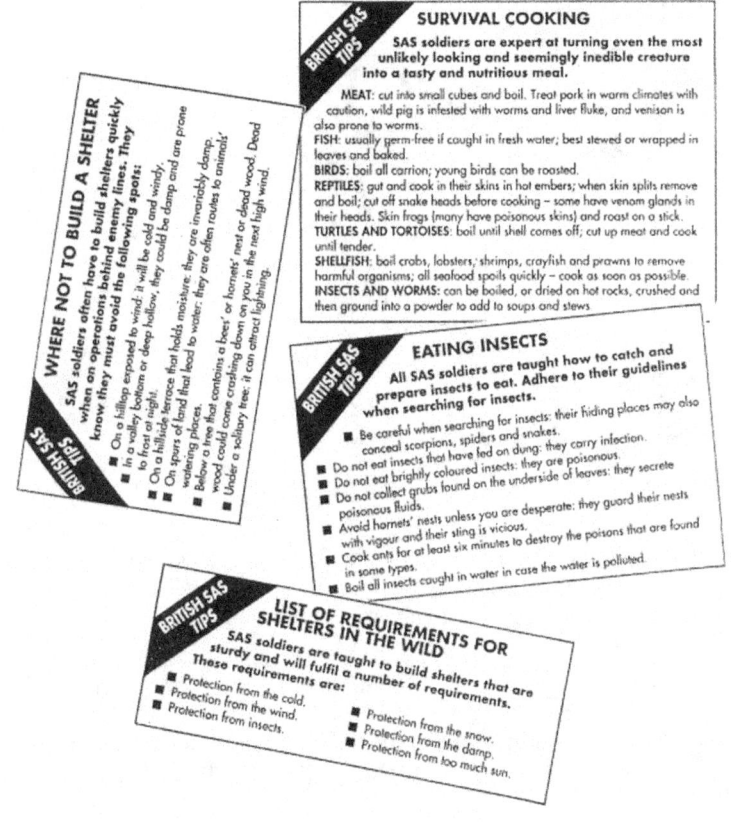

Rick F. Tscherne

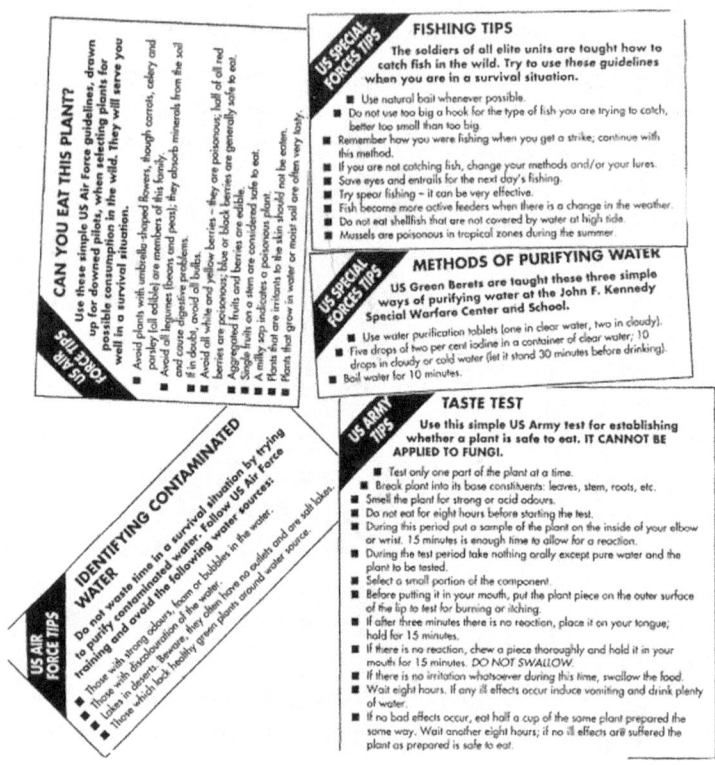

Readers Note: By coincidence, a few days before this book went to press I came across a book called "The Survival Handbook" by Peter Darman. And guess what, I came across these very same survival tips.

Hmrnm, I wonder where this British SAS fella got his tips from? And after reviewing Mr. Peter Darman's book, I recommend you add his book to your library collection too.
**Order From**: Stackpole Books, 5067 Ritter Road, Mechanicsburg, PA. 17055 or  http://www.amazon.com/The-Survival-Handbook-Peter-Darman/dp/155046194X

*Special Thanks: To the copyright holders (©1994), Amber Books Ltd of London, England for allowing me to publish this information.*

*The Complete Ranger Digest: Vol. IX*

# CATCHING MINNOWS FOR FOOD

In a life or death survival situation when you're not having any luck catching fish or small game, minnows can be just as tasty and fulfilling as a trout, rabbit, or pheasant. Provided of course, you can catch enough of them.

Now I've read a lot of survival manuals and I gotta tell ya, most of these damn books make it sound pretty easy in how to catch a fish. All ya gotta do is tie some fishing line to a hook, attach it to a stick, bait it, drop it in the water and presto - you'll soon catch a fish. Buon Appetite!

Yea, right, keep on dreaming. You ain't gonna catch a fish that easy unless you're fishing on a fish farm where they breed 'em, ship 'em and sell 'em to grocery stores, restaurants, and private fishing holes. It ain't gonna be that easy, bubba. And as I stated in my Ranger Digest VI;

*"Though I'm sure you much prefer to catch a trout, catfish, or a salmon, you had better set your menu on something a bit more smaller and realistic, like minnows and sunfish. Chances are they'll be more plentiful and a lot easier to catch than a trout or salmon."*

Makes sense, don't ya think? Now I'm not saying don't waste your time fishing for the big ones when there's plenty of minnows around. I'm just saying don't rule out catching and eating minnows if you're not having any luck catching other fish or game. Ya know what I mean, bubba?

Here in Italy where I live, eating "fried minnows" at wine festivals is about as common as eating french fries at carnivals back in the states. In fact, I don't recall never seeing fried minnows being sold right alongside with french fries. And if you go to a restaurant here and order a "mixed seafood" meal, you'll usual find a handful of minnows mixed in with your meal. It's that common.

Though they taste much better fried, you cook 'em just like french fries, you can also boil or cook 'em right over an open fire. And you don't need to gut 'em if they're less than 3 inches long, only if their fat, chubby, and longer than this.

What I find amusing is how Italians here catch minnows, with a looong telescopic pole and a very tiny little fishing hook. No BS! Really! Now where I come from, Berwick - Pennsylvania, we don't catch minnows with a hook & line, we catch 'em with a minnow net or trap.

Now one time I gave my dumb ass Italian brother-in-law three wire minnow traps and I told him, (a) you gotta place some bait in 'em, (b) you gotta place 'em where you can see minnows swimming, and (c) you gotta leave 'em out overnight.

About a week later he comes back to me and says, "Rick, you Americana minnow trap no workie. " So I asked him, "Did you put bait in 'em before putting 'em in the water ?" He said "I try, but the worms, bugs keep fall thru wire. " I then asked, "Did you place 'em where you could see minnows?" He says, "No, I not see, but no mean no fish there. " And then I finally asked him, "Did you leave them in the water overnight?" He says, "No, I scare maybe someone find and steal, I leave in water only one hour. " Is my Italian brother-in-law a dumbass or what?

Anyway, if you can find a couple empty plastic transparent water bottles, here's how you can make 'em into a field expedient minnow trap.

Cut off the top part of the plastic water bottle...

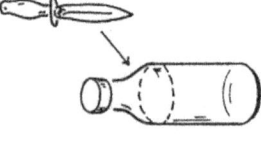

Place it inside the bottle **upside-down**...

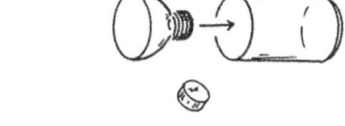

Punch a few small holes all along the sides...

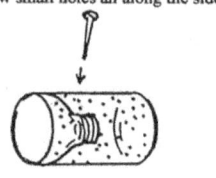

Place some rocks inside of the bottle...

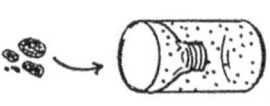

And place it in the water only where you can see minnows swimming.
(Note: It doesn't do any good to put it where there aren't any minnows, ya know?)

And here's how you can make a field expedient "minnow scooper" out of a sock and stick.

# SOLAR STILL WATER BOTTLES

Here's a technique you won't find in any military or outdoor survival manual, I call it a **Solar Still Water Bottle**.

One day while sitting in my back yard I decided to try an experiment on some empty plastic and glass bottles. So I grabbed a bunch of them out of my neighborhood dumpster, placed some green vegetation inside of em, sealed the top of em closed and then I lined 'em up in my backyard where the sun could beat down on them all day long. And yep, my experiment worked, I was able to produce safe drinking water inside the bottles.

I know what you're probably thinking, "Wow, big f——n deal. Ranger Rick!" Right? I know you are. Well, it is a big f——n deal because if you're ever in a life or death survival situation where you can't find "any water" and you don't have any plastic trash bags to make a solar still. But ya were able to find a few empty glass or plastic water bottles laying around somewhere, you can use 'em to produce safe drinking water. And the more bottles you can find, the more water you can produce, provided (of course) you're in a warm weather environment and can find some green vegetation to place inside of em.

Oh, so now you get it, huh? But you're still wondering, "So where do the bottles come from?" Well, no matter where you go in the world, you'll more than likely come across some empty bottles laying around somewhere. Trust me, you'd be surprised where you'll find 'em.

The next time you're watching CNN and they show television pictures of some people starving and dying of thirst in some third world country like

Africa, Somalia, Kosovo, etc. Look closely at your television screen to see if you can see any trash laying around on the ground, in particular, empty discarded bottles. And men just imagine yourself being there without any water and now ya know how to produce safe drinking water with empty bottles. Get the picture? Bravo! I knew you would eventually get it..

When using empty bottles to produce safe drinking water...

a) Use only bottles that you're sure once contained a "safe drinkable substance" such as water, soda, juice, etc. And if you're not sure what it once contained, then DON'T USE IT! Because if it did contain something harmful or dangerous, you just might be putting your life in "Harm's Way."
b) Place inside the bottle only green vegetation that have long stems and leaves and try not to have it touch the bottom of the bottle. Because if it does, as the water begins to form at the bottom of the bottle the vegetation will "absorb and drink it" before YOU DO.
c) Make sure the bottle is completely "airtight sealed" OR no moisture or water will form inside of it. To seal it, just plug the hole with a piece of wood or something.
d) Always place the bottles in a location where the sun will beat down on them all day long with no interference and then empty 'em into a single container at the end of the day and drink wisely.

UPDATE: Clear plastic water or soda bottles (PET type) can also be effectively used for solar water disinfection, using a method called **SODIS**. (wikipedia article: http://en.wikipedia.org/wiki/Solar_water_disinfection )
The basic technique is to just place the bottle in the sun and expose it to natural UV sunlight for 6 hours (or longer if sky is cloudy). Numerous test have shown this to be very effective treatment for bacterially contaminated water.

Note that if the water is cloudy or turbid it should be filtered first as cloudy water prevents the even penetration of UV rays throughout the bottle. Also, do realize that this disinfection method does not remove toxic chemicals that may be present in the water.

And remember: the standard **SODIS** method is designed for tropical and subtropical countries but is not efficient enough in more northern or southern latitudes during winter months.

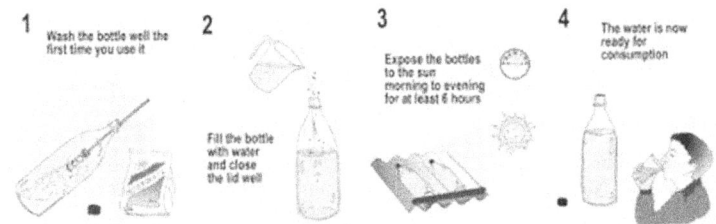

## SODIS FAQ

**What water can I use for the SODIS method?**
The SODIS method needs relatively clear water. You can find out with a simple test whether the water is too turbid. The source of the water (well, surface water) does not matter.

Water that has been polluted with chemicals (poisons, fertilisers, etc.) must not be used. The SODIS method only kills germs. The chemical composition of the water remains unchanged.

**Does the SODIS method change the taste of the water?**
The SODIS method improves the quality of the water without changing its taste. The bottles are closed while they are being exposed to the sunlight, so the oxygen dissolved in the water cannot escape. The water still tastes fresh. On the other hand, if we boil water, the oxygen dissolved in the water escapes. This gives the boiled water an unpleasant aftertaste. The use of chlorine also makes the water taste less good.

**Where should I place the bottles?**
Lay the bottles on a clean surface in the sun, where no shadows will fall for the whole treatment time.

If possible, lay the bottles on a reflective surface, like a sheet of corrugated iron, and protect it from cooling by the wind. The reflection and higher temperature will speed up the disinfection process. However, this is not essential for the application. The bottles can be set down on any surface (wood, concrete, clay brick, etc.).

**How long can I store the water that I have treated with the SODIS method?**
If the bottle is kept unopened after treatment and stored in a cool, dark place, it can be stored for as long as you wish. The dead bacteria cannot multiply again. The only things that may grow are algae. However, these do not represent a health hazard.

**How long can I use the same PET bottle for?**
UV-A radiation must penetrate the bottle in order to kill the germs. Clear, unused bottles normally allow more than 60% of the UV-A light over 340 nm to pass through. Experiments have shown that older, used bottles allow less UV-A light to pass through them. Besides the ageing process of the bottle material, scratches on its surface will also mean that it allows less UV-A light through. So it is very important to handle the

bottles carefully. We recommend replacing old bottles and bottles that are no longer transparent after about 6 to 12 months of daily use.

#### How can I tell the difference between a PET bottle and a PVC bottle?
Only PET bottles should be used for the SODIS method because PVC can be harmful to your health. PET and PVC bottles are normally marked accordingly. The labels can vary from country to country, though. If the bottles are not marked, you can only tell the difference between the bottles by setting fire to them.

PET burns quickly and easily when it is held in a flame. When it is taken out of the flame, the fire goes out slowly, or it may keep burning. The smoke smells sweet.

PVC does not burn easily. The material does not burn at all when not in the flame. PVC smoke smells acrid.

#### Is it dangerous to your health to use PET bottles?
Scientific studies have confirmed repeatedly that when the SODIS method is applied correctly there is no danger to health.

#### How can I improve the efficiency of the method?
If the temperature is increased, the efficiency of the SODIS method can be improved. The germs are killed more quickly. In addition, amoebas are also rendered harmless above a temperature of 50°C. To increase the temperature, lay the bottles on a reflective surface, like a sheet of corrugated iron, and shield them from cooling by the wind.

Old, scratched bottles reduce the effectiveness of the method. Therefore, the bottles should be replaced regularly. But with careful handling, they should be usable without problems for 6 to 12 months.

Turbidity in the water also reduces the efficiency of the method. This problem can be corrected easily by filtering the water.

#### Does the SODIS method kill all bacteria?
The SODIS method is used to kill germs in the water. While the bottle is being exposed to the sunlight, other, harmless bacteria and organisms that occur naturally in the environment can grow, for example algae or naturally occurring coliform bacteria. However, these organisms do not represent a threat to human health.

#### What mistakes do new users make most often?
Using green or brown bottles for the SODIS method;
▶ these bottles absorb UV-A light. This is why only colourless, transparent bottles must be used for the SODIS application.
The bottles used are too big;
▶ the bottles must not be able to hold more than 3 litres.
Bottles are placed upright;
▶ the bottles must be laid horizontally in the sun. This increases the area exposed to the sunlight and reduces the depth of the water the light must penetrate. (With turbidity of 26 NTU, only half of the UV-A radiation penetrates farther than 10cm)
After the SODIS method has been applied, the treated water is poured into dirty containers, so the water is immediately contaminated again;
=> The treated water should be kept in the bottle and drunk directly from the bottle, or poured into a cup or glass immediately before it is drunk.

Additional information is available at: http://www.sodis.ch/methode/index_EN

# NOW HERE'S A COUPLE OF USEFUL TIPS

*Submitted By: Sfc Mark E. Porrett*

Dear Ranger Rick,
After buying and reading the first seven *Ranger Digest Handbooks*, I decided to send in a couple of my own ideas that I've been using for the past 14 years. I've used them during deployments to Honduras, Panama, Turkey, Saudi Arabia, the Combat Maneuver Training Center (CMTC), the Joint Readiness Center (JRC), the National Training Center (NTC) and now here in the Sinai where I'm currently deployed. I hope you can use these ideas for your next Ranger Digest, keep up the good work, Hooah!

#1 When an inspection catches you in a pinch, touch up LBE snaps with some flat black spray paint. Just spray the paint into a paper cup and apply it on with a Q-tip. Every company supply room seems to have flat black spray paint and the battalion aid station/medics have plenty of Q-tips.

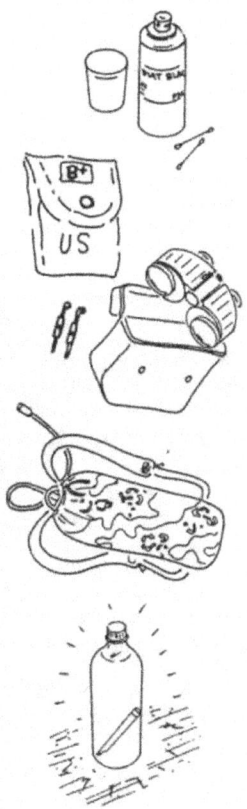

#2 Prior to any field training exercise or real world deployment, I have my soldiers write their blood type on then-first aid pouch or on some tape applied to the pouch. This will save time if someone urgently needs some "juice." [UPDATE: Blood-type patches are available here: http://milspecmonkey.com/store/patches/29-bloodtypes.html ]

#3 One of the many pilferable and easy to lose pieces of equipment is military binoculars. And a great way to protect and have them ready and within easy reach is to carry them inside an empty M-249 ammo pouch attached to your lbe/rucksack with a pair of alice clips.

#4 It seems the "Camelbaks" are the latest thing for the field, and everyone in our task force must have went out and bought one just before they deployed to the Sinai. Now the only problem is the damn brass here don't like seeing so many soldiers out of uniform. So a few of us took some old desert BDU trouser pants, cut off the legs and made desert camouflage pouches for our Camelbaks. [Flash forward a few years; now Camelbaks are standard issue equipment.]

#5 While deployed to Saudi Arabia we had to mark the breached areas so vehicles wouldn't drive off into the minefields, and we had to mark them at night too. And after looking at the mines on the left and right side of the routes, the best method was to drop activated chem-lights inside full bottles of water and then place them alongside the routes. Also works great as a "field expedient lantern".

## A "10 SECOND" KNIFE SHARPENER

Now here's a handy device, it's called a JIFF V SHARP 10 SECOND KNIFE SHARPENER.

And here's how you use it:

1. Always keep your fingers inside the hand guard and your thumb securely placed on the thumb rest at the top of the sharpener, otherwise possible injuries may occur.

2. To sharpen a knife, place the knife on a solid flat surface with the cutting edge facing up. Securely hold the knife in place with downward pressure on the handle. For best results allow about ¾ to 1 inch of the blade to extend past the edge of your flat surface.

3. Starting near the handle of the knife, place the "V" in the head of the JIFF "V SHARP over the cutting edge of the knife. Position the head of the sharpener at a 90 degree angle to the knife's cutting edge. While holding the knife with one hand and the JIFF "V" SHARP in the other, draw the sharpener across the blade from the heel to the point while lightly applying downward pressure. At the end of each stroke continue an outward motion in order to avoid any contact with the blade. You will feel the tungsten carbide cutters removing metal from the knife. Repeat this process 3 or 4 times or until your knife is sharp.

4. Remember, light pressure is sufficient when using the JIFF "V" SHARPENER.. Heavy pressure might cause you to lose control of the knife and could cause possible injury. Also, after finishing the sharpening process always clean your blade thoroughly.

I know a lot of you die-hard Rambo Cowboys aren't gonna like this knife sharpener because it's too easy and it "removes" metal from the blade in a different way than a sharpening stone does. Well, call me lazy if you want but I'd rather use this high-speed fast knife sharpener than a stone.

And while working in Bosnia with MPRI under the US State Department approved "Train & Equip Program," I knew a retired SF Rambo Cowboy by the name of "Buck." And as you can probably guess by his name, yep, he liked knives. And it seems like every time he had nothing else to do he'd pull out his big ol' knife and just sit there and keep on sharpening it over and over and over again.

Well I don't know about you, but once a knife is as sharp as a "surgical blade," why keep on sharpening it? Well my ex-Ranger Buddy, Steve Akana summed it up. He said;

"The only people who carry big knives and keep on sharpening them all the time are those who lack self confidence, self esteem, and think they're some kind of a big badass. But they're not!"

Well you know what? I kinda agree with my ex-Ranger Buddy's statement. And to all you Rambo Cowboys, big badasses, and knife sharpening freaks... ..GET A LIFE!

About $7.00 at Smith's website
http://www.smithsproducts.com/products/product.asp?id=42&cid=10

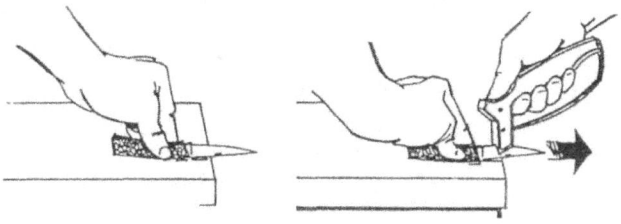

Rick F. Tscherne

# A FEW MEDICAL TIPS & TRICKS

### 35mm Film Container Uses
*Submitted By: David White*

I picked up this next tip while at an Emergency Medical Conference. A safe way medics can carry their Ammonia Inhalant Capsules in their aid bags without worrying about breaking them is to keep them inside a plastic 35mm film container.

### Closing Scalp Wounds
*Submitted By: Cpt. Sheran L Benerth*

In an emergency situation when sutures and stitches are not available, you can close a scalp wound by simply tying the hairs on both sides of the cut "together." And the best knot to use is a "surgeon's knot," it's a variation of a square knot, right-over-left (twice) and then leftover-right (twice).

### Field Expedient Stitching
*Submitted By: Yours Truly - Ranger Rick*

In an emergency life or death survival situation, should you need to stitch up a very bad wound and you don't have a first aid kit, you can always use a fishing hook. Take the smallest fishing hook you have and cut or break off the "eyelet." (Note: If it's a real small fishing hook, then leave the eyelet attached.) Then take your knife and make a few deep nicks and dents at the end of the hook where the eyelet use to be. Then search your clothing for some loose thread hanging off somewhere and carefully remove as much as you can without breaking it, the longer it is the better. Then wrap it a few times around the end of the hook where you made the nicks and cuts (which will help hold the thread onto the hook as you weave it through the skin/wound) and you're ready to start stitching. **WARNING**: Use these medical tips at your own risk.

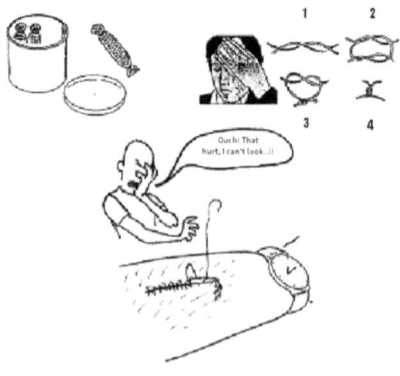

## MAP PROTRACTORS

I know I talked about this in my last Ranger Digest, twit I received some letters from a couple of soldiers asking me if the Army uses only one type of map protractor. Well, the only "military issued" map protractor that I know of is the Department of the Army's GTA Coordinate Scale & Protractor. I have never seen nor used any other type, except a civilian map protractor. And if you don't like the military nor the civilian type, you can always make your own. How?

It's easy, just make a paper photocopy of a military and a civilian map protractor and cut out only the parts you want to use. Then glue 'em to a piece of paper and photocopy it onto a clear sheet of plastic "Xerox Transparency" and then cut it down to size and laminate it. Just like I described how to do it in my Ranger Digest VII (page 43).

Here's a few of my favorite "homemade" map protractors that you can photocopy right out of my book. But before you make a plastic transparency copy, first make a paper copy. Then hold the book page up to the light and align the paper copy directly over the page to make sure the numbers and lines match up perfectly. And if they don't match up perfectly, then you'll need to reduce or enlarge it 1% at a time until they do match or your map protractor won't be accurate. And when you're sure it's a perfect match, then you can make a plastic transparency copy.

Check out some of these map protractors and choose the one that meets your needs or make your own.

*Rick F. Tscherne*

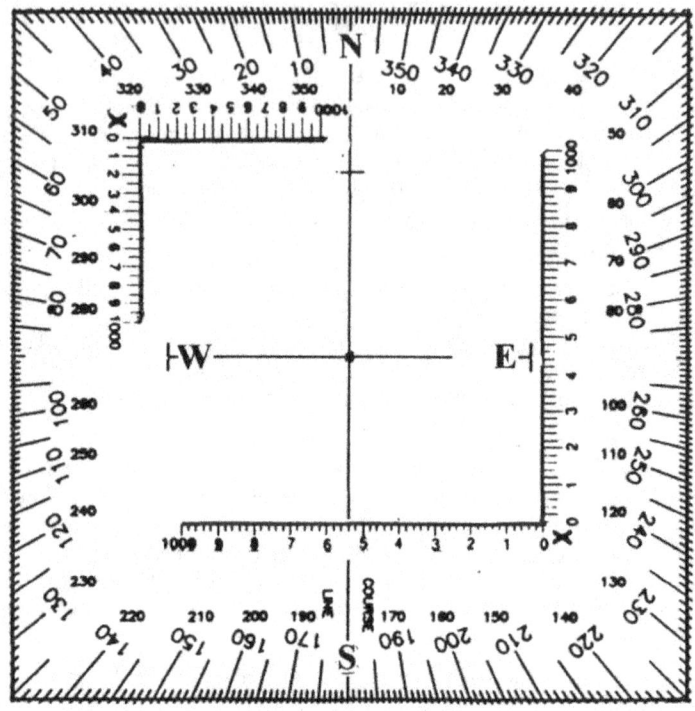

**Use Xerox Transparency**

*The Complete Ranger Digest: Vol. IX*

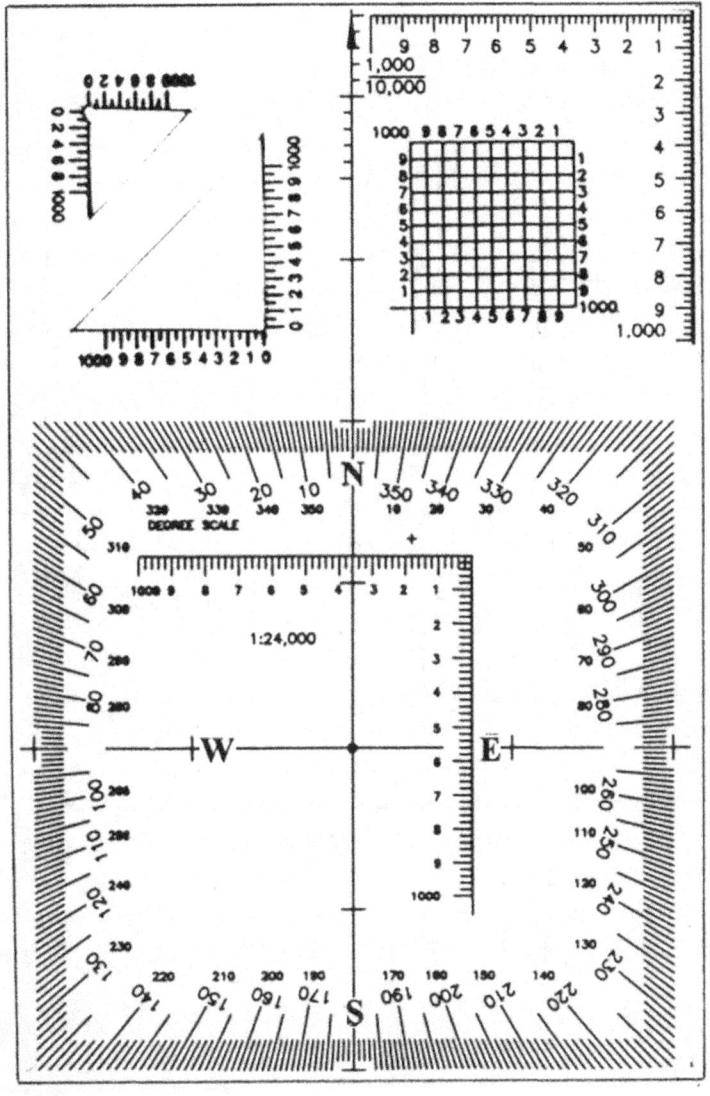

*Rick F. Tscherne*

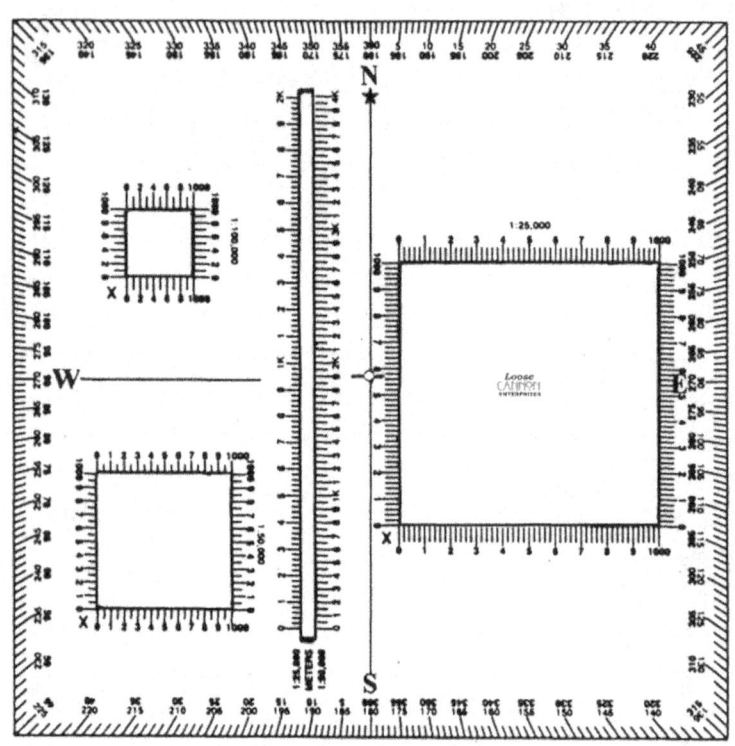

**Use Xerox Transparency**

*The Complete Ranger Digest: Vol. IX*

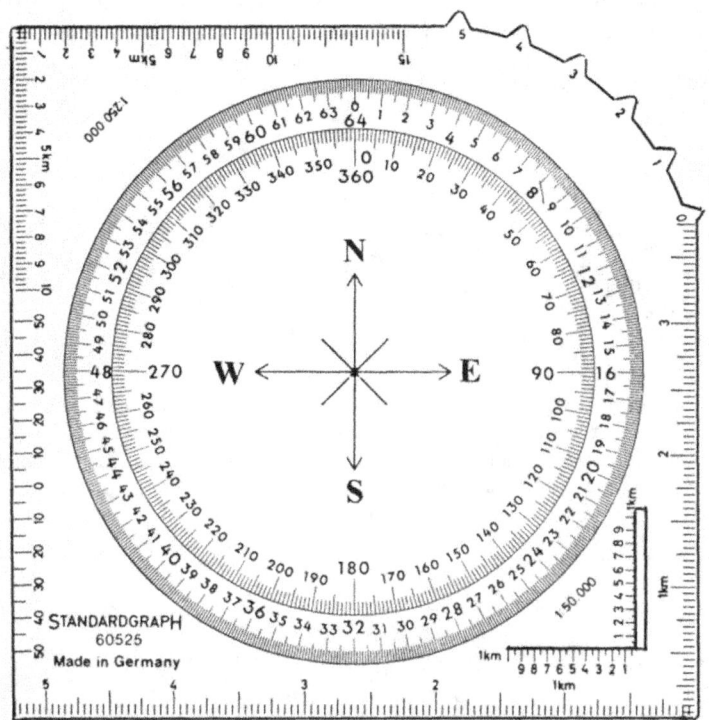

**Use Xerox Transparency**

Rick F. Tscherne

# WHICH ONE DO YOU THINK IS BETTER?

*Submitted By: Friedrich W. Eickelen*

A good friend of mine sent me a German Army First Aid Field Dressing, and after carefully examining it I could not believe why the US military hasn't adopted it yet. It's a field dressing that has 2 x sliding adjustable bandages so it can be used for either one large or two entirely separate wounds such as a bullet wound to an arm or leg. (See below)

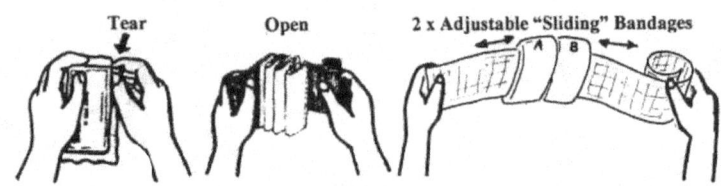

These German Army First Aid Field Dressings are no doubt much better than ours, and they also come packaged in a reinforced sturdy clothe and rubber coated camouflage wrapper that have a much longer shelf life man ours too. And if you think these field dressings are better than ours and you would like to see them in our military inventory, make a photocopy of this page and send it to:

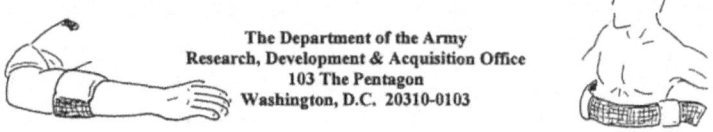

The Department of the Army
Research, Development & Acquisition Office
103 The Pentagon
Washington, D.C. 20310-0103

And maybe, just maybe some brass Pentagon genius will say to themselves, "Hmmm, yea, why don't we have these types of first aid field dressings our inventory? " And maybe, just maybe he or she will get off their fat ass and do something about it. It doesn't hurt to try, ya know?

These German Army First Aid Field Dressings are produced by: PAUL HATMANN AG, C/o LOHMANN GmbH & Co.KG, Paul-Hartmann-Str. 12 , D-89522 Heidenheim, Germany. The NSN# is 6510-12-226-0005

# A SECRET VACCINE?

*Submitted By: David A. Williams M.D.*

Dear Ranger Rick,

Thank you for publishing my article in your last *Ranger Digest* (VII). Enclosed you'll find some more medical research information about a so call "secret vaccine " which the Swiss government administers to their own people for free, it's called BCG-Berna. Among some (but not all) of the things it has been known to do;

1. Reduce and or cure tooth decay and gum disease
2. Help prevent cancer
3. Protect against AIDs, Herpes, and Hepatitis C viruses
4. Control blood pressure
5. Lower cholesterol
6. Prevention of Tuberculosis
7. Restores strength and endurance
8. Prevention and treatment of diabetes

It has been used in over 100 countries since 1921, and not only do all the NATO countries use this vaccine, but also the Russians, Serbs, and the "Vatican" use it too. And many medical experts, both here in the US and overseas strongly believe a terrorist group will someday try to use a deadly chemical or biological agent in Europe or the United States. And one of the best protections you could have is to be vaccinated with this BCG because it boosts the immune system to enormous levels.

Since 1995 our clinic has examined over 3,000 foreign people from 80 different countries and 94% of them were known to have had the BCG vaccination, and all of them were free of AIDS, cancer, and other diseases normally found in most third world countries.

My point is this, if NATO, our enemies and third world countries are using the BCG-Berna, then it just makes sense that we should be using it too. Unfortunately, it can only be acquired "over the counter" in pharmacies in Europe. It not only comes in a "home vaccination kit, " which any nurse or doctor can administer, but in pill form too.

READERS NOTE: Dr. David A. Williams is a (M.D.) Family Physician, Federal Aviation and Immigration Medical Examiner and a USAR (ret.) Bn Surgeon, Anesthesiologist, and a GMO.

And when I received his letter with this information about the BCG-Bema vaccine, I decided to find out more about it. So I contacted a very good friend of mine, an Italian doctor who works at a nearby hospital here in Vicenza (Italy) and I asked him about it. And he told me (quote),

"Doctors who work in hospitals here in Italy are required to be vaccinated with the BCG-Berna because of the many, many patients they come in contact with everyday."

Hmmm, so when I heard this, I decided to see if I could get vaccinated with it myself. And although I couldn't find the injection type, (good, because I hate needles) I was able to find it in pill form and had to take one a day for seven days. And guess what? As far back as I can remember I've always caught a cold, flu, sore throat or some sort of virus every year between the months of November and February. But not this year, this is the first year (1999-2000) that I haven't even coughed, sneezed nor gotten the sniffles. Coincidence? I don't think so.

If you're in the military and stationed here Europe, you're not going to have any problems acquiring this BCG-Berna And if you're not, well then you're gonna have some problems locating it because it's UNAVAILABLE in the USA. But if you can acquire it through "other means," if you get my drift, it's worth it.

Please don't write to me and ask where you can acquire this BCG-Berna, because if you do I won't respond back. Sony, but you're on your own. And if you are able to find it and acquire it somehow, be advised you should consult your doctor first before taking it.

I'm sure if you do some research on the internet for the BCG-Bema, you'll probably find a lot more information about it. Below is what the European "pill form" package looks like. To you guys & gals back in the US, good luck on finding it.

UPDATE: Info per Wikipedia:
Bacillus Calmette-Guérin (or Bacille Calmette-Guérin, BCG) is a vaccine against tuberculosis that is prepared from a strain of the attenuated (weakened) live bovine tuberculosis bacillus, Mycobacterium bovis, that has lost its virulence in humans by being specially subcultured (230 passages) in an artificial medium for 13 years, and also prepared from Mycobacterium tuberculosis.
http://en.wikipedia.org/wiki/Bacillus_Calmette-Gu%C3%A9rin

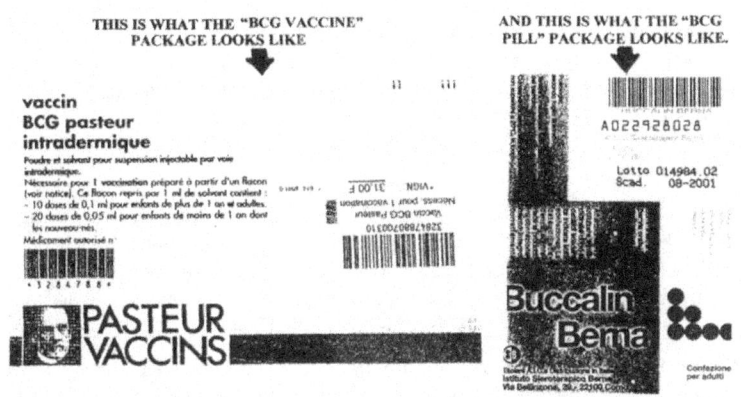

READERS NOTE: Use at your own risk.

# TRY HOOKING YOUR COMPASS TO YOUR LBE

This next tip only works with a military lensatic compass or a civilian "military style" compass.

If you've ever been a compassman, fire team, squad or platoon leader you know how hard it is to hold a weapon with one hand and follow your compass with the other. And some us have even tried to hold a weapon, flashlight, map and a compass all at the same time. But it's not easy, because you either gotta let down your guard to look at/follow your compass or put your compass down to keep up your guard.

Well, I discovered this next technique by accident while "metal detecting" here in Italy up in the Asiago Mountains while searching for some war relics. I decided to take along my old lbe instead of my rucksack because I only had a few things to carry, my metal detector, some tools, a lunch, survival kit, and a map & compass.

And after visiting the Asiago War Museum and carefully studying an old battlefield map they had on display, I choose a site that I wanted to search and so off I went to see what I could find. Well, I got as close to the site as I possibly could with my car, but then I had to hump it the rest of the way on foot. And after following an old foot path for about thirty minutes, it soon disappeared from underneath my feet and I found myself in some thickass vegetation.

Now I was following my map closely and I knew exactly where I was, well, most of the time anyway. But the more I pushed on to the site, the thicker the shit got. And soon I was in some really thick stuff and I couldn't see any prominent terrain features to get my bearings on. So I had to temporarily give up locating the site, study the map, set my compass to a "dead reckoning" setting that would get me out of this shit to a nearby hill top so I could see how far I still had to go

And it wasn't easy neither, the damn 'wait-a-minute' vines and vegetation kept fighting, holding me back and grabbing my compass out of my hand. I needed the use of both my hands so I could push on to "the ranger objective." Hooah! But at the same time I needed to follow my compass very carefully too.

Well, I decided to take a five minute break and study the map. And when I looked down at my lbe to see how I could rig my compass to it so I could free up both my hands, BONG -I got an idea. I opened up my compass, folded the upper part of my lbe strap where the "metal loop" is located and squeezed inside this metal loop the compass "thumb loop". Now I could use both my hands freely and still follow my compass, and fairly accurately too. (See drawings/photos.)

Damn, this worked out so well I wish I had thought of it when I was back on active duty in the Army. I can remember many times stumbling around in the boonies, especially at night trying to hold my weapon, map and compass all at the same time.

Well, to make a long story short. I finally made it out of that thick shit, reached the hill top, plotted a new course to the site, found me a couple nice war souvenirs, got back to my car before dark, drove home, showed my wife what I found, had a nice dinner, got drunk, went to bed and lived happily ever after. *The End*

All Ya Gotta Do Is Look
Down At Your Compass

## POCKET BINOCULARS

Now here's a pair of binoculars I don't mind carrying around with me. They're small (1½ x 2½ x 3½), lightweight (approx. 8 oz), shock resistant (rubber coated), powerful (8 x 21mm), cost about $25 (+/-) and they're available in almost every military, camping and outdoor supply stores too.

### Smith & Wesson Pocket Binoculars

These 8 x 21mm binoculars fold down to just 1¾" x 2½" x 3½" and fit in a handy belt case for go-anywhere action. Binoculars are protected by rubber armor coating and feature flexible eye pieces and neck strap. This is one super deal! 8 X 21mm, Field of View: 5.8".

Smith & Wesson
Pocket Binoculars

And you know what I really like about these pocket binoculars? They fit nicely inside an empty plastic M258 NBC Decon Kit container or a Gl compass/first aid pouch and you can attach'em right onto your lbe. Now that's what I call convenient, check it out.

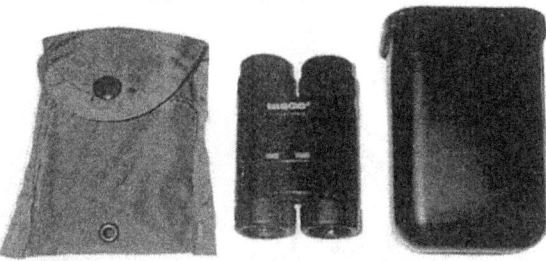

When looking through a set of binoculars, which of these is the correct way to look through them? You'd be surprised at how many people don't know, are you one of them?

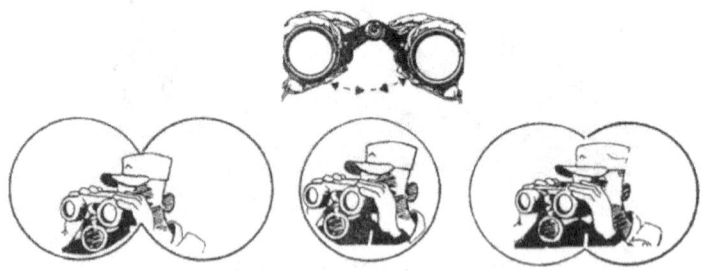

*The Complete Ranger Digest: Vol. IX*

## BULLET LASER POINTERS

Here's something you'll enjoy playing around with, a "bullet laser pointer." The price of these babies have come waaaaaay down that you can find'em selling for less than $10. Really!

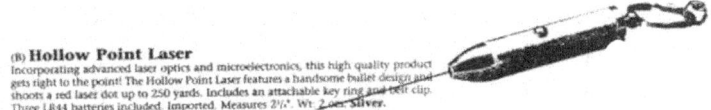

**(B) Hollow Point Laser**
Incorporating advanced laser optics and microelectronics, this high quality product gets right to the point! The Hollow Point Laser features a handsome bullet design and shoots a red laser dot up to 250 yards. Includes an attachable key ring and belt clip. Three LR44 batteries included. Imported. Measures 2½". Wt 2 oz. Sliver.

And if you do buy one, which I'm sure most of you probably will, try mounting it onto a pair of "pocket binoculars." Just open up the binoculars, place the bullet laser pointer next to the focus ring (either on the left or right side) and close that side of the binoculars inward until it rests securely against the laser pointer. Important: Always try to keep constant pressure on that side of the binoculars where the laser pointer is mounted so it won't fall off.

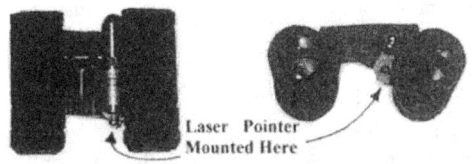

Laser Pointer Mounted Here

Now pick up the binoculars, look through 'em, switch on the laser pointer with your thumb, pick out a target about 10-20 feet away and carefully look for the red laser "dot." If you can't see it, pick out a closer target, and if you still can't see it, then the laser pointer was not properly mounted onto the focus ring. Remove and remount the laser pointer once again onto the binoculars as described above and look for the "red laser dot." And if you still can't see it, play around with it until you do see it. If necessary, depending on the type of binoculars you're using, you might have to wrap a little bit of tape around the laser pointer to keep it straight and level underneath the binoculars. And if you still can't see it, don't give up, just keep playing around with it until you do see it.

Now most laser pointers have a maximum range of about 100-300 yards, but you can't see the red laser dot that far away with the naked eye. Which is how I came up with this idea, I was trying to see how far away I could see the

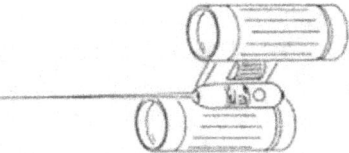

red laser dot during the day and also at night. So I mounted my bullet laser pointer to my binoculars and presto -1 could now see it far away.

You're probably wondering how you can use one of these laser pointers in a tactical situation, right? Well, the only thing I can think of is to use 'em for harassing the enemy. You know, making 'em think you're about to blow 'em away with a laser sight attached to your weapon. If he or she can see the red dot and not you, they just might think a sniper is zeroing in on him.

Another possible use? How about using it for identifying individual sectors of fires and enemy positions at night. Got a couple ideas of your own on what else they can be used for? You do? OK, then write and let me know, and if I publish it in my next Ranger Digest I'll send you a free copy.

## DID WE MISS THE MONEY TRAIN? AGAIN?

Yep, I'll bet ya all ignored my investment advice again, didn't you? I'm talking about the investment advice I gave you in my last *Ranger Digest* (VIII) And the one before that, and the one before that, and the one before that...

Well, if you're gonna keep on snoozing, then you're gonna keep on losing and missing the "money train." And as I've been trying to tell ya all - you don't need to put a lot of money in the stock market to make a lot of money. But if you can put away at least $50 or more a month, men you're on your way in becoming financially secure. The more money you can put away - the more financially secure you'll be for retirement or when you need the cash to buy something really special, like a home, car, or a boat.

And guess what? If you haven't heard already, as I tried to tell ya in a previous Ranger Digest, Uncle Sam has recently raised the cut-off age for drawing social security. That's right, if you were planning on retiring and drawing social security at age 65, forget it, it's now been pushed back to age 67. And ya wanna bet this ain't the last of it? Just wait, by the time you're almost 67 and looking forward to retirement, Uncle Sam's gonna push it back even further, to age 70. That's if you live that long, and if ya don't, Uncle Sam inherits your social security benefits.

Does this get your attention? Now do you see why it's important to start saving, investing, and planning for your retirement now? Because the longer you keep putting it off, the harder it will be to reach your retirement goal, and at an early retirement age too.

Hey listen, I quit school at 17, never attended a day of college in my life, got a GT score of 94, and if I can successfully invest in stocks and mutual funds - ANYONE CAN! And I didn't get started investing in stocks and mutual funds until I was 33 years old, which is when I found out about them. And when I did find out about them, I dumped my US Savings Bonds and Bank CDs and jumped on the "money train."

And because "I got off my ass right away" when I did find out how easy it was to invest in them, and NOT like you guys & gals who keep putting it off, I was able to retire at age 38 when I got out of the Army back in 1993.

And yea, maybe I am bragging a little bit... But I don't have to work another day in my life if I don't want to because I don't owe anyone any money. And I own my own apartment, drive a nice car, and I'm sitting here in my RV right now sipping wine, typing this page and enjoying the scenic surrounding here on Lake Garda (Italy). And the answer is "NOPE," I didn't inherit any money from a rich relative, win a lottery, nor did I rob a bank to get to where I am today -I just invested wisely.

The point I'm trying to get across to you... If a high school drop out and idiot like me can successfully invest his money wisely - ANYONE CAN! But the first step is, YOU GOTTA GET OFF YOUR ASS and START INVESTING NOW. And the sooner you do, the sooner you'll reach your goal in becoming financially secure. Am I getting through to you this time?

Here are the mutual funds that I personally own, look at the "yearly returns" that I've made on my investments and you'll see why I was able to retire. And the earlier you start investing - the more money and earlier you'll be able to retire too, it's that simple.

*NOTE: As of January 1st 2000*

| NAME OF MUTUAL FUND | YEARLY RETURNS | | | RISK LEVEL | MIN. INVEST | LOAD/ CHARGE |
|---|---|---|---|---|---|---|
| | 1 YR. | 3 YR. | 5 YR. | | | |
| FIDELITY SELECT ELECTRONICS | 106% | 52% | 53% | 10 | $2,500 | 3% |
| FIDELITY SELECT TECHNOLOGY | 131% | 64% | 49% | 10 | $2,500 | 3% |
| JANUS TWENTY | 64% | 54% | 45% | 8 | $2,500 | NA |
| JANUS MERCURY | 96% | 51% | 40% | 9 | $2,500 | NA |
| INVESCO TELECOMMUNICATION | 144% | 64% | 46% | 10 | $1,000 | NA |
| AMERINDO TECHNOLOGY D | 249% | 74% | NA | 10 | $2,500 | 3% |
| INTERNET FUND | 216% | 119% | NA | 10 | $1,000 | NA |
| SPECTRA FUND | 71% | 49% | 41% | 9 | $1,000 | NA |
| RYDEX OTC FUND | 100% | 65% | 56% | 10 | $25,000 | NA |

IMPORTANT: Be aware all the mutual funds listed above have a "high risk factor." On a scale from 1-10, a low number means "least volatile & less risk" and a higher number means-more volatile and risk." And as all the mutual funds state in their brochures, "Past performance does not guarantee the same future results."

WHERE TO CALL TO GET INFORMATION ON THE ABOVE FUNDS
FIDELITY FUNDS   800-544-8888        JANUS FUNDS       800-525-8983
INVESCO FUNDS    800-525-8085        INTERNET FUND 888-386-3999
AMERINDO FUNDS 800-832-4386          SPECTRA FUND      800-711-6141
RYDEX FUND       800-820-0888

WHERE TO LOOK ONLINE FOR MUTUAL FUND INFORMATION

Brill's Mutual Funds Interactive:    Morningstar:
www.fundsinteractive.com             www.morningstar.com
Fund Alarm                           MSN Money Central:
www.fundalarm.com                    www.moneycentral.com
Frsthand Funds:                      Mutual Fund Education:
www.firsthandfunds. coin             www. mfea. com
Kiplinger.com:                       The Motley Fool:
www.kiplinger.com                    www.fool.com

# FIELD EXPEDIENT COOKING TIPS

Not long ago I received an old 1987 Virginia Guard post publication titled "*Soldier Craft.*' It's a quarterly printed publication that's distributed to all members of the Virginia Army and Air National Guard. And while reading through it one day I found an interesting article written by General Carroll Childers, so I hope he doesn't mind if I share it with my readers. And although his article is on "How To Heat C-rations Out In The Field," which have been replaced by MREs. You will no doubt find his tips useful in heating and cooking other types of food and beverages.

I found that if troops are shown the correct methods and are given time to prepare their meals, they will usually enjoy them much more than gulping them down cold while on the move. Not to mention, properly prepared meals are also much easier on the digestive system loo. And if you believe the old adage that the army moves on it 's belly, well maybe you will believe that the better you prepare what goes in your belly - the better soldier you will be too.

I do not eat food cold unless the tactical situation precludes a fire. And there are three factors to consider when heating meals: a source of heat, a container in which to heat the ration, and a method of holding the ration or container above the heat source.

There are several ways to heat meals in the field, commercial heat tabs, vehicle engines, gasoline, and of course a good old fashion fire using wood, paper, etc. I like to be able to start a fire, set my food over it and then not have to worry about it until it's hot. So the trick is to have something to support your food or canteen cup above the fire until it's ready to eat.

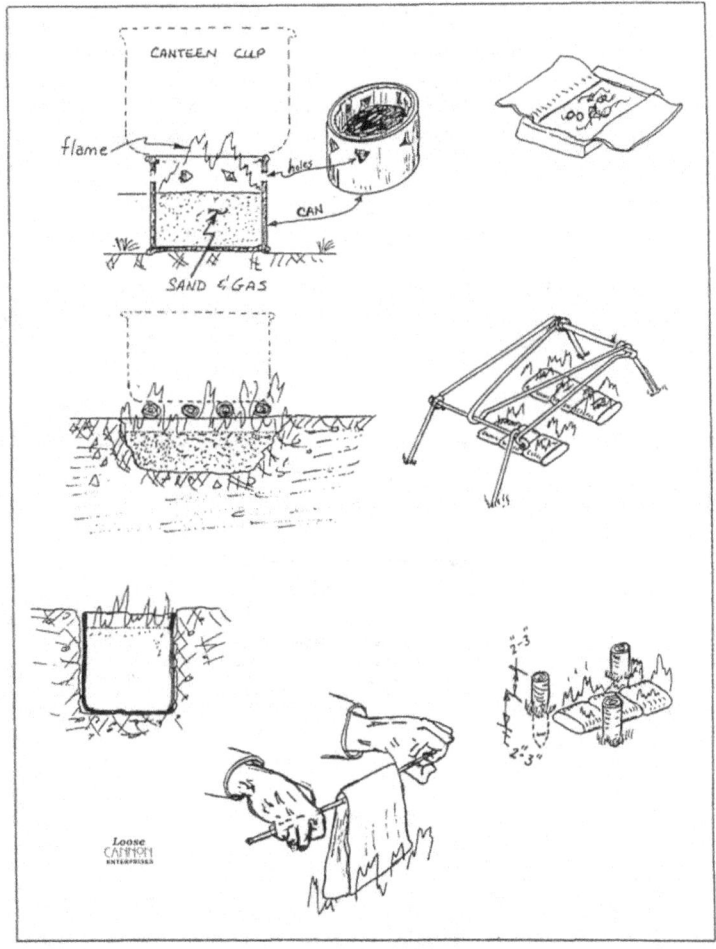

SOLDIER CRAFT

To use gasoline as a fuel source you'll need a small can and some dirt. Take a knife or other sharp instrument and punch several large holes around the top of the can, these holes are necessary to provide a source of air to the fuel which would otherwise be cut off when you set the canteen cup on the can itself. Then put several inches of "fine " dirt or sand in the can and add gasoline until the dirt sand is completely saturated and pour off any excess gasoline that does not get absorbed. And then light it, place your food or canteen cup on top of the can and your stove will burn for quite some time.

If you don't have a can, you can dig a small hole in hard soil then put the pulverized dirt back in and add the gasoline. Though this method will not bum as long for the same amount of gas, but it will work if a tin can is not available. A couple of half inch diameter green limbs laid across the fire will support the canteen cup while giving some space for some air to feed the flame.

If you heat a packet of food by boiling the unopened packet in a canteen cup of water, the army says not to drink the water because the food packet (dyes or whatever) may contaminate the water. A way around this is to carry a couple of plastic zip-lock bags. Place the food packet in the zip-lock bags and then place it into the canteen cup of water. This will isolate the water from the supposedly contaminated packets and allow you to use the hot water for coffee or chocolate powder or whatever.

A simple small cooking grill can be made from a heavy duty coat hanger by simply cutting, bending, and twisting the wire a certain way until it has four legs that fold up so that it can be packed away inside a rucksack or even a shirt pocket.

*The Complete Ranger Digest: Vol. IX*

# HOW TO TIE A FEW KNOTS

Attention **Readers**: If **you would've** ordered a "How To Tie **Knots**" reference **card** from one of the **many military, camping,** & outdoor supply **stores,** they would have charged you $S.QO for it. And if you photocopy these two pages and laminate both sides, it'll be just as **useful,** durable, & **waterproof as** the "Real McCoy."

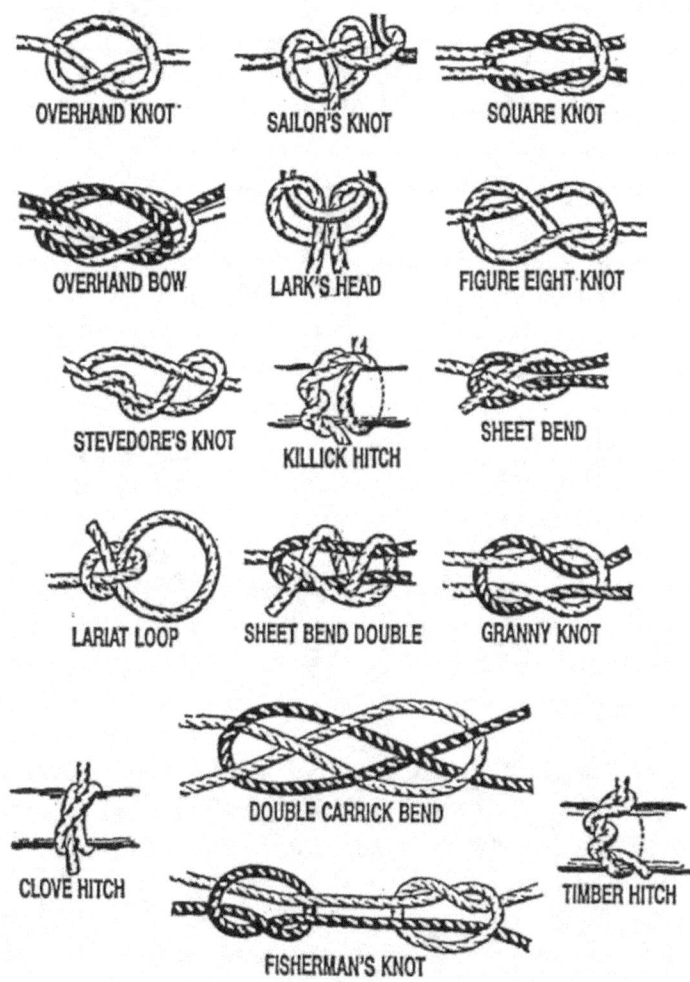

373

*Rick F. Tscherne*

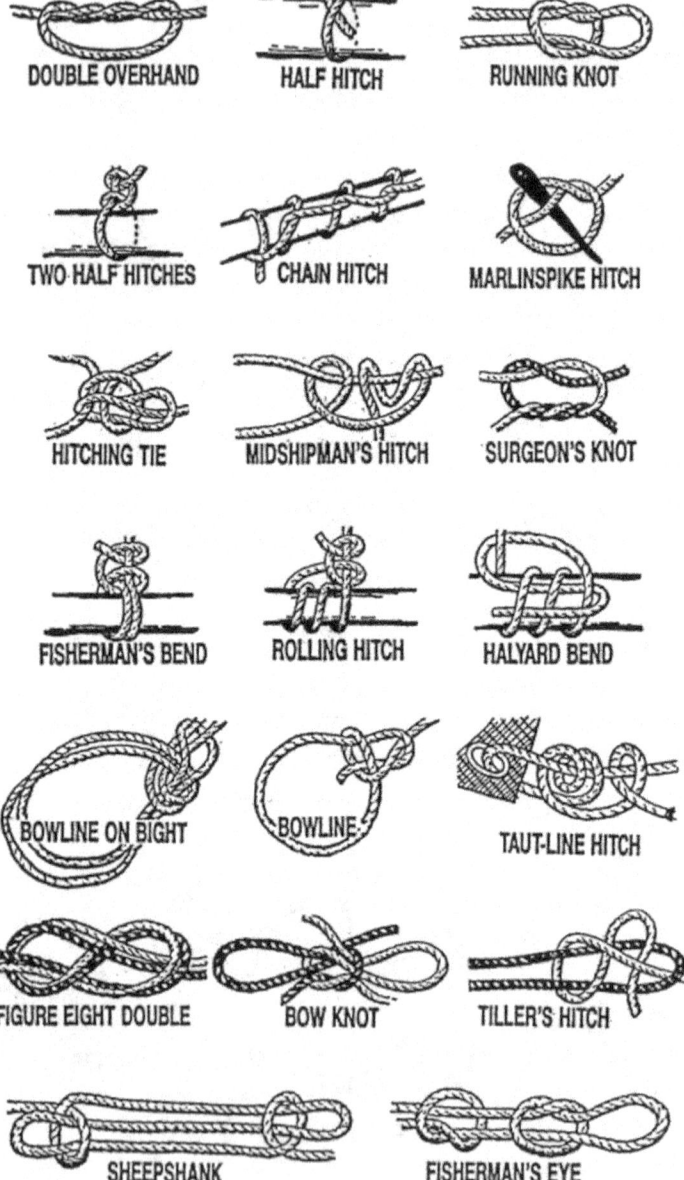

# OLD 45. CAL AMMO POUCHES AND WHAT YOU CAN DO WITH 'EM

Now I know you've seen these before because they use to hold two 7 round magazines for the old Gov't (1911) Colt 45. ACP pistols. Unfortunately when the military switched over to the new 9mm Beretta pistols there wasn't much anything else you could use these "web pouches" for. So Uncle Sam sold'em all to the Army & Navy Surplus Stores, and because not even the civilians could figure out what else to use'em for, you can buy'em today pretty cheap, a few $$$. Well, here's a few ideas that I've been using mine for.

**GI 45 MAG POUCH**

Government model Colt .45 ACP Magazine belt carrier holds two clips. Rare new surplus from the 1950's in heavy web canvas with Lift-o-Dot snaps and two belt keepers. Great for .45 shooters, or for pocket knife and doodads. Size: 5 3/8" x 3-1/2". Olive drab. [6oz/128gm]

Important: Before you can place a couple of plastic 35mm containers inside the pouch, you MUST stretch the pouch so they'll fit inside, and it can be easily done too. Just find yourself a thick round piece of wood (like a broom handle) slightly thicker and wider than the 35mm containers and repeatedly "shove it in & out" of the pouch until the film containers fit snuggly inside. Then remove the lid off one film container, slide it over the bottom of a second container, and place it inside the pouch.

## A WRISTWATCH GPS?

G- damn it! I knew as soon as I bought me one of those handheld GPS they were going to come up with one of these, a Wristwatch GPS. Though they're pretty damn expensive right now, I guarantee they'll come down in price just as soon as enough people start buying'em. And if ya all can't wait for the price to come down, then do me a favor and order yours now so the price will come down and I can afford to buy one. OK? Is it a deal? Here's the address on where you can order 'em.

This watch was released in 1999, but was since discontinued.
http://world.casio.com/corporate/history/chapter03/contents11/

*Rick F. Tscherne*

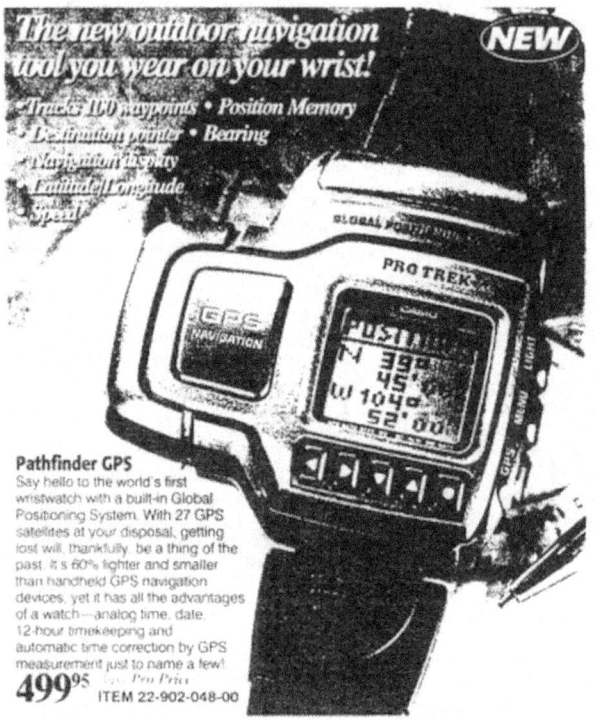

UPDATE: Several new GPS watches packed with some serious features have recently been released (2016...)

- Garmin tactix Bravo, Epix  (GPS & GLONASS)
- Suunto Traverse Alpha  (GPS & GLONASS)
- Timex Ironman GPS Watch  (only $99 !!)

For more info on Garmin models: http://www.strohmanenterprise.com/

## MILITARY HUMOR

One day a private was walking by the motor pool office when he heard the phone ring, there wasn't anyone else around so he decided to pick it up. He said, "Motor Pool!"

The voice on the other end asks, "Soldier, can you tell me how many vehicles there are in the motor pool? "

The private says, "Sure, I see six Abrams, twelve Bradleys, and fat ass Johnson's Humvee."

The voice on the other end asks, "Soldier, do you know who the hell you're talking to?"

The private says, "Nope, sure don't."

The voice screams, "This is Colonel Johnson - your commander!"

The private, stumbling for words asks, "Uh sir, do you know who you're talking to?"

The colonel says, "No, not yet, but... "

"Good," says the private, "So long Fat-Ass!"

*******************************

A colonel and a sergeant were both in a barbershop getting a haircut and a shave, when their barbers finished they were about to put some after shave lotion on them.

The colonel yells, "Hey, don't put that shit on me, my wife will think I've been in a whorehouse." The sergeant turns to his barber and says, "You can go ahead and put some on me, my wife has never been inside of a whorehouse and doesn't know what one smells like."

***************************

Thank you for calling the United States Army, we're sorry but all our units are currently deployed. At the tone of the beep, please leave a message with the name of your country, the nature of the crisis, and a telephone number where we can reach you. And as soon as we have sorted out our missions in Bosnia, Kosovo, Korea, Kuwait, Central America, and a few other places, we'll get back to you. But if your crisis is serious, please listen carefully to the following instructions:

If your crisis is small and near the sea or ocean, press 1 for the United States Marine Corp.

If your crisis is far inland and can be solved by one or two low risk, high altitude bombing runs, press 2 for the United States Air Force. But be advised this service is unavailable on the weekends and after 1630 hours during the weekdays.

If your crisis is in need of a good marching band, press 3 for the United States Navy.

If your crisis is not that urgent, press 4 for the United States Army Rapid Deployment Force.

If your crisis is urgent, press 5 and your call will be transferred to the United States Army Special Operation Command. And then please have ready your credit card number so we can bill you for all the meals, hotel rooms, & other classified expenses that are needed to solve your urgent crisis.

If you would like to join the Army, earn minimum wage, experience the different types of environments in far away exotic places and risk your life feeding, protecting, or killing people from other countries. Please hold and your call will be transferred to a bitter sweet, passed over for promotion Army Recruiter who's sitting on his fat ass doing nothing but waiting for your call.

Thank you for calling *the United States Army*, we hope we have been of some assistance to you and look forward to possibly resolving your crisis sometime in the near future when we have more personnel and less real world crisis and deployments on our hands. *Have a nice day!*

# ICE PINS

*Submitted By: Friedrich W. Eickelen*

A German friend of mine sent me this next idea, they're called Ice Pins. The Scandinavians are known to carry them whenever they have to cross frozen lakes and rivers. So what I did was make a pair of these for myself out of two pieces of wood and two large nails.

Now to be honest with you, I've never tried out my ice pins. But if I was gonna be around frozen water I'd sure as hell carry 'em, and wear 'em either around my neck or through the sleeves of my jacket. So if I accidentally fell through the ice I'd be able to pull myself out or at least keep my head above the water until help arrived. That is if I didn't freeze to death first.

Now there are several ways you can make 'em, with or without wooden handles. I prefer 'em with wooden handles so they'll float if accidentally dropped in the water. Or should you see someone fall through the ice, instead of trying to be a hero in rescuing them and possibly endangering your own life, you can toss 'em the ice pins without worrying about 'em sinking in the water. Makes sense, don't ya think? Sure!

When attaching some cord to your ice pins, make sure it's strong enough to hold the weight of an average person, like 550 para-cord. Check out these different types of ice pins that I made.

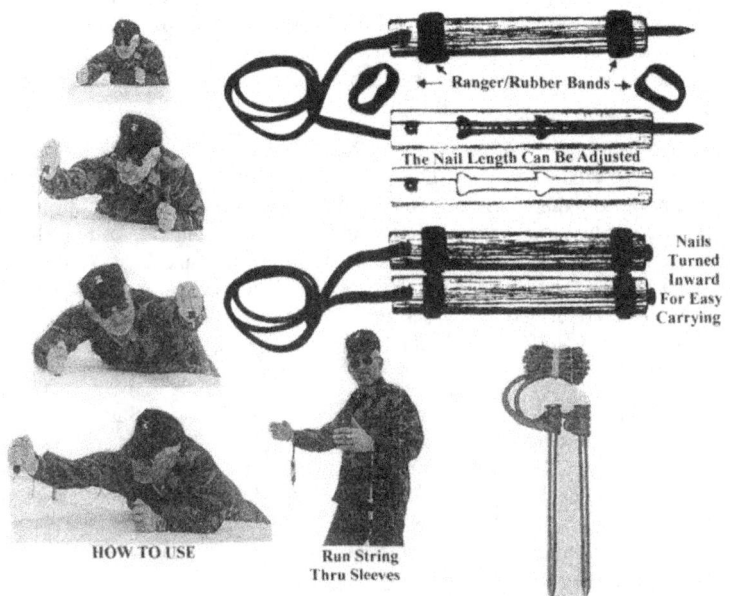

## CARRY AN ANTI-FROSTBITE ANTI-HYPOTHERMIA KIT

If you've been reading my books for awhile, you know I rarely repeat the same tip or trick twice unless there's a better way in doing it. And although I showed you in my *Ranger Digest I* (page 3) how to keep warm out in the field with a candle and a poncho/liner, this next tip is very similar.

I'm sure you've read or heard stories about lost skiers, hikers, and stranded motorists freezing to death due to prolong exposure to severe cold weather conditions. And or if they did survive their ordeal they had to have their hands/feet amputated due to severe frostbite.

Now with all these high-speed items on the market today, especially the easy-to-carry-in-your-pocket kind of stuff, it's hard to believe why some outdoor "winter enthusiasts" don't better prepare themselves before venturing out in the cold. I mean, "Hello, don't ya know it can get pretty darn cold outside?"

And whether you're only going away for a few hours or for the entire day, if you're gonna be off the main roads, trails and or away from civilization, you should at least carry in your pocket a few basic "winter survival items." Such as a small signal mirror, a small compass, a small florescent orange handkerchief, etc. And if nothing else, at least some matches, a candle and a compact pocket-size emergency space blanket. (See photos/drawings on how to use.)

Now one day while waiting in the checkout line at the PX, I just happen to see a small metal "booze flask" on a nearby shelf. And I said to myself, "'Hmmmm, now that could possibly come in handy for something out in the field, and not just for booze neither." And so 1 bought me one, filled it with kerosene, placed a homemade "cloth wick" inside if it and lit it. And guess what? It burned for approximately 10 hours and 30 minutes. Not bad! It's not only better than Sterno, but it's refillable, lightweight, and spill-proof too. And if you can't find any kerosene, you can always use lamp oil, citronella, or paraffin.

Then the other day I saw an advertisement in the American Survival Guide magazine about some survival candles that burn for 6 1/2 days each. Wow! That's even better. But the only problem with this is the price, though they only cost $3.95 each, you gotta buy 12 of them for $35.00 and pay another $10.75 for shipping and handling fees. Yipes! Yea, they're a little bit pricey, but if you can talk a few of your buddies into buying one or two candles, it'll be less expensive for you.

If you do buy a pocket-size emergency space blanket, wrap several layers of 100 mph (duck) tape around it so it'll stay tightly wrapped and you'll be able to reuse the tape as described in my Ranger Digest VI (page 87). And 1 encourage not only all outdoor winter enthusiasts to carry an "Anti-Frostbite/Hypothermia Kit" in their pocket, but field medics should also carry them in their aid bag too. What better way to defrost a frozen soldier out in the field during a military tactical operation when open fires are not permitted and a life is at stake. Think about it.

### 6 ½ Day Candles

Excellent Emergency Candles. One of these candles will burn for 6 ½ days straight. 12 candles to a case, 5 cases will provide over one year of continuous light. Buy quantity and save! #24454 Reg. $35.00 per cs.

**Y2K Preparedness Headquarters
The Survival Center ~**
America's Oldest Continually Operating Survival Center
Box 234 Dept. ASG McKenna, Wa 98579

**1-800-321-2900**

Open 10 am - 6 pm M-F (PT)
Internet site: http://survivalcenter.com
e-mail sales@survivalcenter.com

You Won't Believe How Nice & Warm It Is Until You Try It.

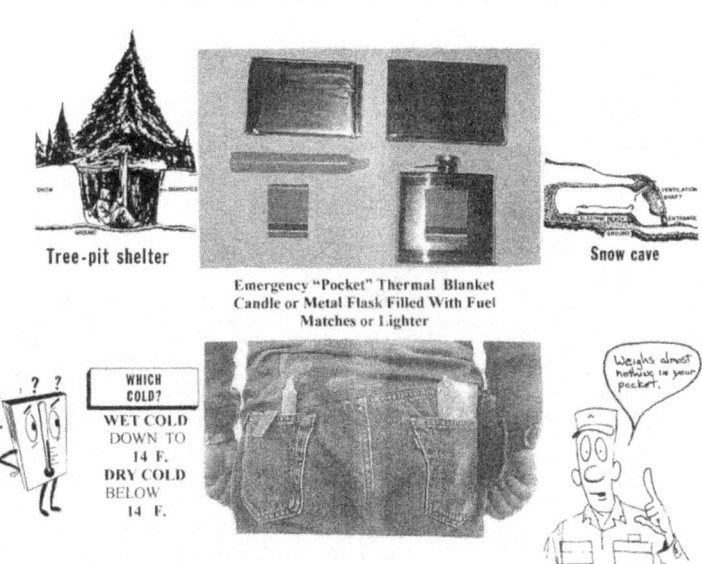

Tree-pit shelter

Emergency "Pocket" Thermal Blanket
Candle or Metal Flask Filled With Fuel
Matches or Lighter

Snow cave

WHICH COLD?
WET COLD DOWN TO 14 F.
DRY COLD BELOW 14 F.

# HOW TO MAKE FIELD EXPEDIENT SNOWSHOES

Have you ever walked in snow that was up to your ass? If you have, then you know it's physically exhausting and can dehydrate a body in no time due to rapidly burning up an enormous amount of body fluids and calories. And if you don't replenish the lost body fluids, it can lead to severe illness.

In a winter or cold weather environment when there's lots of snow on the ground, chances are you'll probably encounter some knee or ass deep snow and find yourself struggling at a slow creeping turtle pace. Unless of course, you're wearing a pair of snowshoes, which not many soldiers, campers, hikers, hunters, etc take along with them because it's just too much extra weight to carry.

But what if you desperately needed a pair, would you know how to improvise and make a set? Check out these drawings of some field expedient snowshoes that I found in a survival manual.

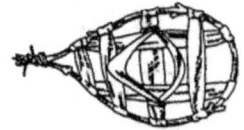 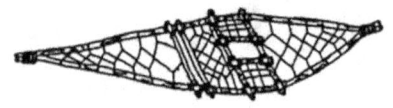

You think you could remember how to make a pair of these? Though they're not that very hard to make, provided you've got everything you need like a knife, saw, string, wood, etc. The most important part to remember is how they're suppose to be attached and worn on the feet, like this:

Now if you make 'em out of thick and heavy wood, it's going to be very difficult and hard to walk around with them. And if you use wood that's too dry and thin, they're not going to support your weight and instead break apart. Which is why it's important you select and cut down only live trees/branches rather than dead dry ones. The greener or liver they are, the more flexible and easier they'll be to shape into snowshoes, not to mention they won't crack or break apart neither.

Now I can go on and on and explain how to make 'em or I can show you some pictures in how to make 'em, which do you prefer? OK, I knew you'd want me to shut up. And so here are some photos & drawings of how I made a pair out of fiberglass tent poles, wood, and tree branches too. Check it out!

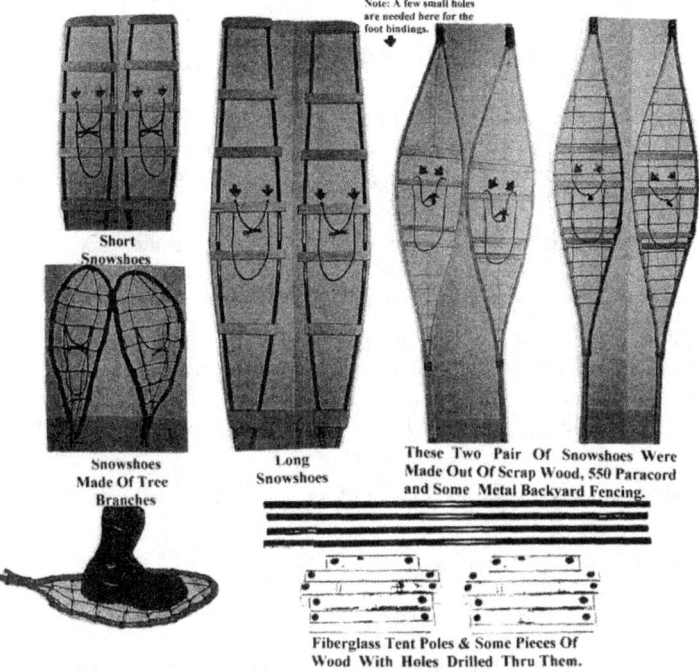

Construction details.

Yep, And As You Can See, They All Work Too.

*Rick F. Tscherne*

# HOW TO MAKE A CARGO SLED FROM A PAIR OF SKIS

You're gonna love this next invention of mine, especially if you're a winter camper, hunter, trapper, ice fisherman, or survivalist. I call it a *Ski-Ahkio*. It's a device you hook up to an ordinary pair of skis so you can haul cargo across the snow with it.

And it's not that difficult to make either, all you need is some wood, screws, and a pair of skis. And although you don't have to make your *Ski-Ahkio* as fancy as the one below or in the photos, here's how to make one.

1 - Take a pair of ski-boots and place them on top of a 2 1/2 x 12 inch piece of wood, trace it with a pencil and then cut 'em out. If you don't have any ski-boots for your skis, then you'll have to measure the distance between the front & back bindings to determine the size of the wood you'll need to cut out. (A)

2 - Take a 1 x 2 1/2 x 60 inch piece of wood and cut it into the following pieces:

    1 x2 1/2 x 12 inch – 4 pieces (B)
    1 x2 1/2 x 3 inch  - 2 pieces (C)
    1 x 2 1/2 x 5 inch - 2 pieces (D)

3 - Take a 1/2 x 12 x 50 inch piece of plywood and round off the comers. ( E )

4 - Now attach the two (A) 1 x 2 1/2 x 12 inch pieces of wood to the pieces that will go in the ski bindings and attach the two (C)1 x 2 1/2 x 3 inch piece of wood on top of them. Take the two (B) 1 x 2 1/2 x 12 inch pieces of wood and attach it to the top of the two ( C ) 1 x 2 1/2 x 3 inch pieces of wood. Lay your skis down on the ground, open up the ski-bindings, place the pieces of wood that go inside the bindings and lock'em in. Take the two (D) 1 x2 1/2 x 5 inch pieces of wood and the two ( B ) 1 x 2 1/2 x 12 inch pieces of wood and place them forward of the ski bindings. Place the ( E ) 1/2 x 12 x 50 inch piece of plywood on top and determine where you want it positioned, which should be somewhere in the middle of the skis. When you are satisfied where you want it, attach it to the wood.

Did I lose or confuse you somewhere along the way? Did I? Ok, don't worry, just look closely at the photos and drawings and you'll be able to figure out where everything goes, or at least be able to make a similar *Ski-Ahkio* based on this general design Good luck with it!

## Ski Ahkio

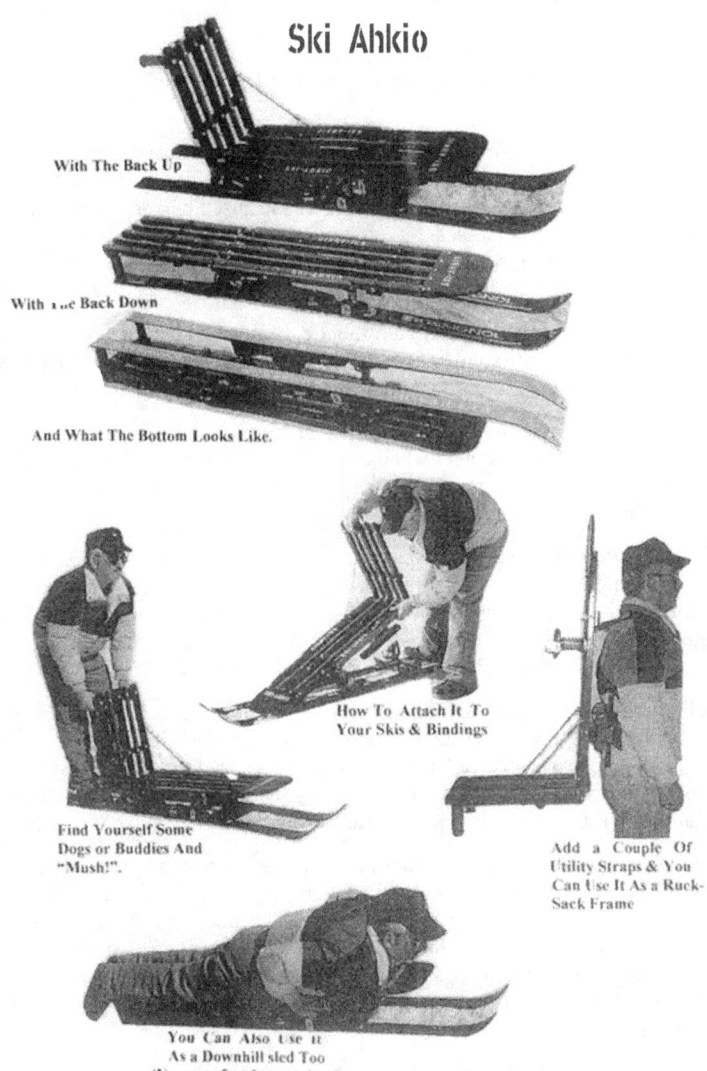

Rick F. Tscherne

# IMPROVISED INDIVIDUAL RADIATION DOSIMETER

Submitted By: Friedrich W. Eickelen

Here's something pretty interesting, a German friend of mine sent this info to me. He wrote:

As you know there's been a lot of studies on what caused the Gulf War Syndrome, and nobody really seems to know the answer to what caused this illness. But one theory is that it may have been caused by "depleted uranium ammunition," as there was a lot of it used during the Gulf War. Not to mention, there were other military things used that were known to emit low dosages of "Gamma Radiation."

But most armies don't spend enough money on equipment to detect radiation contamination. So if a soldier does become contaminated during a mission, it's difficult for him to know for sure if he is or isn't contaminated.

So between 1956 and the early 1960 the German Army started using dog tag size x-ray dosimeters to detect whether personnel have been contaminated by Gamma Radiation. It's something very similar to what medical personnel wear while working around X-ray machines.

This miniature dosimeter is nothing more than a small piece of film, almost the same type used in dental clinics for X-raying teeth and it's kept sealed in a "light tight" plastic or rubber coated pouch. And when an individual believes he or she may have been contaminated during a mission, all they have to do is turn in the film and wait for the results/evaluation.

Well to the best of my knowledge we don't have this in our Army, but it sure sounds like a damn good idea. And if you're in the military and you're concerned about possible radiation contamination, try to convince your local base dental clinic to give you one of their rubber coated "dental film negatives." And if they won't, no problem, you can always make your own.

How? All ya need is a roll of 35mm film, a piece of paper and some 100 mph tape. Then lock yourself in a dark closet, pull out about 3 inches of film and cut it off. Then cut off from the end of this film about a VA inch piece of film. Why? Because usually the first few inches of the film (the part that gets loaded inside the camera) has already been "exposed to light." So it's very important that you cut it from "the end" of the film and NOT from "the beginning."

After you have cut it off, cover it with a piece of paper so the tape won't stick to the film when it's removed for evaluation. Then seal it with some 100 mph tape to make sure it stays lightproof & waterproof. If you leave a short piece of tape hanging off one side you'll be able to attach it to your dog tags, lbe, helmet, or uniform. Then whenever you think you've been exposed to radiation just take the film to a hospital X-ray room to confirm your suspicions. And don't forget to write down your name, social security, and unit on the tape too.

If you think this German individual radiation dosimeter sounds like a good idea, make a photocopy of this page and send it to: **The Department of the Army, Research, Development & Acquisition, 103 Pentagon, Washington, D.C. 20310-0103.**

## MAGAZINE MAGPULS, "LOV 'EM or LEAV 'EM"

One of my readers sent me a couple of these MAGPULS and so I decided to test 'em out.

Well, I know I'm gonna piss off the developer, a former US Marine Corp Recon Sergeant by the name of Richard Fitzpatrick. But who cares, opinions are like assholes - everyone has one. So here's my asshole opinion on them, I don't like 'em, I'd rather use 100 mph tape or 550 paracord than these Magpuls.

Why? Because once you have attached the Magpuls to the bottom of the magazines, you have to place them back inside the ammo pouch the same way you pulled them out - "bottoms up." And if you try to place 'em back inside the pouch the other way around, you'll have difficulties closing the pouch. Which means you'll either waste time trying to force the pouch close or turning 'em back around so you can close it. Precious time you don't need to waste especially if you're in the middle of a firefight. Ya know what I mean there, Private Gomer Pyle?

I also had trouble extracting & inserting the magazines in and out of the ammo pouch because the rubber Magpuls made the magazines a bit too tight and snug to fit inside the nylon ammo pouch. But as Mr. Fitzpatrick points out in his very informative Magpul brochure...

*"While the Magpuls are compatible with almost all 5.56mm NATO magazines, it's usefulness will depend entirely on the type of pouch used,*

certain ammunition pouches may not function at all when the Magpuls are installed on the magazines."

READERS BEWARE: Before ordering a set of these Magpuls, buy just one to make sure you can close your ammo pouch with a Magpul attached to one of your magazines. And if you can't, well then you'll only have wasted your money on one instead of a set of them.

## RADIO LISTENERS

*Submitted By: 1 Lt. John Davis*

Dear Ranger Rick,
How many times have you gone to the field and everyone wanted to know what was being said on the radio? Haw many times did you wish a portable speaker came with your military radio so you didn't have to keep the damn handset glued to your ear all the time? A bunch of times, right?

Well, you might not be able to use this device while on the move, only in stationary positions, but it sure beats the hell out of everyone bugging you and asking, "What are they saying? Huh? What are they talking about? Huh? " It's called a Telephone Listener.

It's a device that hooks up to a telephone handset and amplifies your call so everyone can listen in on it. It weighs 6 oz, runs on a 9 volt battery, and Radio Shack sells them for about $10. And they 're easy to install too, you just attach a suction cup coil to the handset, turn on the power and adjust the volume.

Note: Because a military radio handset is not as smooth as a commercial telephone handset, you'll need to wrap a rubber band or some tape around the suction cup coil to keep it in place.

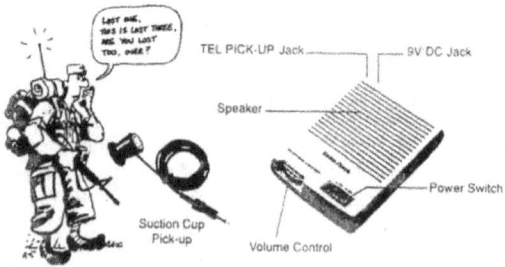

## MAN, WHAT I HATE TO SEE...

There's two things I hate to see soldiers wear on their lbe, and one of em is a damn upside down first aid pouch. Why? Well, if you think you can remove the field dressing faster and easier in this mode and it's gonna make a drastic difference in saving yours or someone's else life, you're a dumbshit. Because regardless of how you wear your first aid pouch, upside down, right side up, sideways or whatever, getting the bandage out of the pouch ain't the time consuming part. It's tearing the wrapper apart, unrolling the bandage and placing it properly on the wound to stop the bleeding. That is, if you still got it inside your pouch and you haven't lost it, yet.

I won't lie to you, back in the old days I use to wear my first aid pouch on my lbe upside down too. That is until the snap button on the pouch kept getting snagged & popping open on wait-a-minute vines. And before I knew it.... my field dressing was GONE. And this didn't just happen to me, no sir, it happened to other guys in my unit too. And wasn't long before our chain of command put out the word to either start wearing them "right side up" or place a rubber band around the pouch to prevent the bandages from accidentally falling out.

Now some guys in my unit replaced the snap button with some velcro, they claimed it was more secure and didn't pop open so easily. Maybe it was or wasn't, I don't know, I didn't go to this extreme. I just took it off my lbe suspenders and wore it on my web belt instead, "right side up." Why on my belt and not on my suspenders? Regardless of where you get wounded, the upper or lower part of the body, it's the halfway point where you or someone else can easily reach for it regardless of where they're standing over you. Go figure it out, bubba.

And the second thing that I hate to see soldiers wear on their lbe - is a shiny snap-link. Man, it seems like every time I pick up an Army Times or Soldier Magazine there's a photo of some soldier all decked out, camouflaged up with war paint on his face, vegetation in his helmet, etc and on the upper part of his lbe is a snap-link."

What's wrong with this? Nothing, nothing at all if you want the enemy to spot you coming. And if you think a shiny snaplink attached to the upper part of an lbe can't be seen a distance away, again you're a dumbshit. Don't think just because a snaplink is narrow and round that it can't reflect that much light, because it can, and at night too. If mere's a "full moon out" and the enemy has night vision capability, trust me the odds are in his favor he'll be able to see the moonlight glare coming off your snaplink. Try it out if you don't believe me.

The solution? Well, the best solution is to wear your snap-link somewhere else where it can't be seen or throw off a glare, like inside a pouch. Or spray paint 'em flat black or o.d. green "before, after, and every time" you come in out in the field, but I know this is going a little bit too extreme.

Now if they're the military issued "steel" type (10mm oval) and not the lightweight civilian aluminum ones, you can tone down the shine by heating 'em over a fire. Just grab the snaplink with a pair of pliers and hold it over an open flame for a few (but not more than 2-3) minutes and then dip it in cold water. If it's still kinda shiny, repeat this procedure again but not more than twice or you'll risk damaging the strength of the metal.

WARNING: Never heat snap-links till it turns "red hot" or you will definitely damage and weaken the strength of the metal by making it too brittle and too dangerous for rappelling.

UPDATE: A newer option is to use a polymer, low-IR signature Grimloc carabiner: http://www.itwmilitaryproducts.com/content/grimloc .

## FIELD EXPEDIENT EMERGENCY GLASSES

I discovered this trick by accident one day while deployed to Turkey on a training exercise. I was trying to peek through a "pin hole" on the side of my tent to see if I could see my buddy outside. I had my glasses on and couldn't see very well through it, and so 1 removed them.

"Hrnmm, this is interesting..." I said to myself. Without my glasses on I can't see things far away, but when I look through this pin hole, I could see pretty clearly. Not 100% clearly like with my glasses on, but clear enough to see things far away.

So I grabbed me a piece of paper, poked "2 x pin holes" in it about eyeball distance apart and held it up in front of me. Well I'll be dam, it worked! I could even read street signs that I normally couldn't read without my glasses on. Don't ask me how this works - it just works!

So what I did was attach this piece of paper onto my forehead just slightly above my eyes so I could walk without having to look through the pinholes all the time. And when I wanted to see something clearly a distance away, I just tilted my head downward so I could peek through them.

I know what you're thinking, big f—— deal, right? Well it is to people like me who wear glasses.

One time while out on night maneuvers I lost my glasses, and because no one can see very well in total darkness, it really didn't bother me much, at least not until the sun came up. And when it did, EVERYTHING WAS A BLURR. I couldn't even make out my buddies unless they were talking or standing right next to me. And if I had known this "pin hole in a piece of paper" trick, I would have used it. Because walking around in the boonies for days in total blurrness until we returned to garrison was a blinding experience that I never want to go through again.

Well if you or someone else should someday lose a pair of glasses while out in the boonies, or you find yourself a POW and your captors take away your glasses so you won't be able to see clear enough to escape, remember this pin hole trick. And if you ain't got any paper, try. Punching some pin holes into something else, a piece of cardboard, tree bark, or whatever. Better to see clearly "sporadically" than not to see clearly at all, ya know what I mean? You guys and gals who wear glasses know where I'm coming from.

## Dear Ranger Digest Readers,

Yep, another one of my articles made it into the "European Stars & Stripes Newspapers." Hooah! Now the only reason why I decided to write this article, is because I got tired of hearing about a couple of (active & retired) senior NCOs & officers here in US Army SETAF who are constantly bad mouthing me to their fellow co-workers and troops. I've heard they've told others they know me personally, which is a crock of shit. I've never shared a drink, meal, nor ever associated with them on or off duty. Hell, I don't even waste my breath to say "hello" to these scumbags when our paths cross. Why? Well even though we once served in the same battalion together, 1st Bn 509th ABCT. I can't stand NCOs or officers who are known liars, bullshiters, backstabbers, alcoholics, kiss asses, adulterous, and live only by "pocket values." (See Ranger Rick's Commentary for definition of "pocket values.") ATTENTION SETAF SOLDIERS: Believe nothing what you hear and only half of what you see. And if you want to know something about me or about something you heard about me, hell just stop me on the street and ask, I ain't gonna bite your head off. Unlike these lowlifes, I ain't got no "skeletons in my closet" nor do I have anything to hide, so why should I lie. .

*STARS AND STRIPES, SUNDAY ·*     Sunday, October 24, 1999

## Spreading rumors, gossip a 'jealous and envious' act

Not long ago, someone placed a handwritten message up on the post office/community bulletin board. It read:

"Rumors destroy lives. If you don't have anything nice to say about someone, then don't say it at all."

The handwriting on the paper made it obvious the writer was very upset. Either they were a victim of some rumors or someone they really cared about was a victim and they were expressing their anger to the community.

I know how that person feels because I've been a victim of gossip and rumors myself. And if I did half the things that some of these rumors say I did, the MPs at the gate should either roll out the red carpet or pull out their pistols when they see me coming. It's unbelievable how fast gossip and rumors can spread on a small installation like Caserma Ederle in Vicenza, Italy.

I have to admit, some of these rumors did bother me. But not anymore, not after listening to some human behavior experts (or whatever they call themselves) on the Oprah, Leeza and Ricky Lake shows.

**Be Our Guest**

Now according to these experts, more than 50 percent of what people hear about others is false, misleading or exaggerated information. And less than 50 percent is true. (Hmmm, now you know why the tabloids are always getting sued.)

And there are two types of gossipers, the initiator and the spreader. The initiator is someone who starts a rumor by talking bad about another individual to another person. And then this other person becomes a spreader when he or she tells another person, and then he tells another person, and he tells another and so on. And as each individual tells another person, the information becomes more and more distorted and exaggerated. (Hmmm, sounds like a virus.)

What type of person spreads gossip and rumors about another person? The most common type is someone who is "jealous and envious" of another person's accomplishments. For what reason? To destroy that person's reputation and credibility. Maybe not directly and intentionally, but indirectly and discretely.

Yep. I know how that person who put the message on the bulletin board feels. Because ever since I've written and published several military training handbooks, I, too, have been a victim of gossip and rumors. And they weren't started by people who couldn't care less about me or my accomplishments, but rather by some "jealous and envious" active and retired military members in our community.

Rumors "can" and "do" destroy lives, but only if you let them.

The next time you hear someone talking bad about another person, ask if they know that person personally or if they have ever worked alongside or socialized with them. And if the answer is "no," then tell them to "stop spreading gossip and rumors about a person they know nothing about."

Or, I'm sure they'll deny that's what they're doing. But I guarantee they'll feel "lower than whale poop" when you suddenly turn and walk away without saying another word.

When it comes to gossip and rumors, "believe nothing you hear and only half of what you see." Some good advice, wouldn't you agree?

*Rick F. Tscherne is retired from the Army and lives in Bardolino, Italy.*

## HOME MADE FACE VEILS

(OR HOW TO AVOID PUTTING ON CAMMIE PAINT)

I know I'm not the only one who hates to put on camouflage paint, but if there's a time of the year or type of climate that I really hate to put it on, and that's during cold weather conditions. And not just because the camouflage stick is hard and the paint won't smear on so easily, but because it blocks the tiny pores in the skin and makes your face colder. And according to the Encyclopedia:

The most important care that can be given to the skin is to keep it clean- By keeping it clean it prevents the tiny pores, or mouths of glands in the skin from becoming clogged, hinders the spread of infections and helps regulate the body temperature too.

You didn't know this, did you? Now don't let your alligator mouth overload your hummingbird ass and tell your chain of command "Hey Ranger Rick says we shouldn't put on camouflage paint..:' That's NOT what I'm saying.

But what I am going to suggest depends on your unit's tactical situation, alert status, and chain of command. And if you're in a defensive position, such as in a dug out fighting position and it's pretty dam cold out, your chain of command shouldn't mind you doing this. And that's to make a camouflage face veil out of a camouflage handkerchief.

The key to making one of these is being able to see & hear with it on without reducing you field of vision and hearing capabilities. Which will depend on how you cut out the holes for your eyes and ears. If you make them too small - you'll drastically reduce your field of vision and hearing. And if you make them too large, you'll expose too much skin which will defeat the purpose of making a camouflage face veil.

When making one, make the holes big enough so you can hear clearly, tie some string between the eye so it will give you a much greater field of vision, and wear it securely around your head so it doesn't shift or move around Try it and you'll see it works pretty good - but only during cold weather conditions and in stationery positions, of course.

*Rick F. Tscherne*

# HOW TO MAKE A LOW-COST KNIFE SHEATH

*Submitted By: Sgt William Johnvin*

Dear Ranger Rick,
Sometimes knife sheaths wear out faster than knives do, especially if they're made of some cheap leather or plastic material. And to find a store that sells sheaths separately...forget it, impossible! So when mine wore out I went through my "military junk box " and made my own. All you need is some 100mph/ducktape, 550 paracord, an empty MRE box, and an old belt or rifle sling.

Take an **MRE** cardboard box and cut out a piece of cardboard about twice the width **of** your knife and **just slightly longer** than the blade.

Bend the cardboard in half the long way and then tape it in place with some 100 **mph/duck** tape.

Measure the length of your knife blade, multiply **it by X** 6, **find an old belt** or rifle sling **and** cut this amount off

Fold it in half, place the ends together **side-by-side** and run it from the front to the rear of the cardboard sheath and tape it

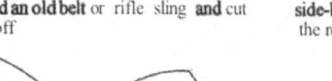

Wrap a 24 inch piece of 550 **paracord** around the lower portion of the sheath and tie it in place.

Starting at the top or **bottom**, take the 550 paracord and begin wrapping it around the cardboard sheath the same way shown in **RD** VII in how to make a knife para-grip.

When the 550 paracord is securely wrapped around the cardboard **sheath**, add either a thick rubber band or short piece of 550 paracord around **the** belt loop for securing the knife handle **to** the sheath to prevent it from falling out

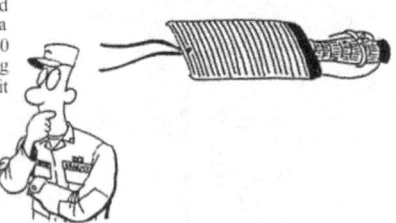

## HOW TO MAKE A BELT EXTENDER FOR YOUR KNIFE SHEATH

*Submitted By: Joseph Richer*

Dear Ranger Rick,
I'm a cadet at Marion Military Institute and I have enclosed is a tip on how to make a "belt extender" for your knife sheath.

1 - Determine how far down your leg you want your knife sheath and then measure the distance between this point on your leg and the belt on your pants

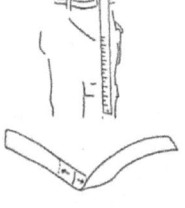

2 - Multiply this length 2 X 2, add +2 inches for the belt loop in the knife sheath and then add +6 inches for the two belt loops you'll need to make for your knife sheath extender.

3 - Take an old belt, strap, or rifle sling and cut off this amount of material and then fold it in half. Then measure 1 inch to the left and right of this center point and then "fold and sew" it in place.

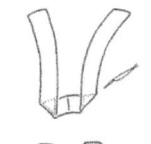

4 - Take the two running ends, measure off about 3 inches, fold it over and then sew these in place. Note: If you so desire, you can add velcro to these two running ends so you can adjust the belt extender to any length you want.

5 - Now all you have to do is attach your knife sheath to your belt extender and slide it on your belt.

## FIELD EXPEDIENT LITTERS

*Submitted By: Capt Sheran L. Benerth*

Dear Ranger Rick,
A handy and surprisingly sturdy field expedient litter can be made out of tree saplings when there's no wooden or metal poles available to use.

Cut 18-20 tree saplings about the thickness of your thumb and tapering to about the thickness of your index finger and no longer than 8 feet in length. Then lay them down side-by-side on the ground alternating "large and thin ends" and then tie six 10 foot lengths of paracord equally spaced out all along one of the outside poles. (See drawings.)

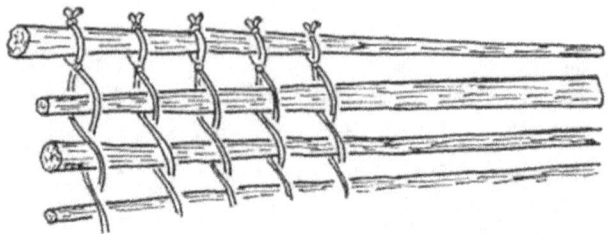

Then take one cord at a time and weave it "under & over" all the tree saplings and then do the same thing with the other 5 cords. Then tie all the cords off with a double square knot or surgeon's knot and cover the stretcher with a sleeping pad, jacket, blankets, cut-up cardboard boxes, etc for added support and comfort.

# MAKING A SIGNAL PANEL OUT OF A HANDKERCHIEF

Whether you're in the military or not, a hunter, hiker, camper, survivalist, cross country skier, mountain biker, or other type of outdoor enthusiast. You can never go wrong in carrying two things - a signal mirror & signal panel. And if they're small, compact, and lightweight, you should get in the habit of carrying 'em in your shirt or pants pocket before departing on one of your outdoor adventures, because they just might save your life.

Now when I was in die military I always carried my signal mirror inside my first aid pouch right between my 2 x first aid field dressings so it wouldn't get broke. But later on when I found some bright "orange cloth" to use as an improvised signal panel, I removed one of my field dressings and stuffed it inside instead. Which I then called this pouch, "my ass saving pouch." Go figure.

Well, then I discovered something that works a lot better, something more compact, lightweight, and useful than a US Army VS-17 Signal Panel. A spray pained "fluorescent orange" white handkerchief. And all you need to do is:

Buy a $ .99 "white handkerchief," a can of "fluorescent orange" spray paint and some 550 parachute cord.

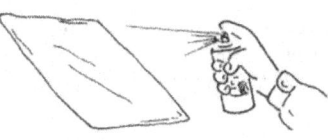

Then all ya gotta do is spraypaint both sides of the handkerchief,

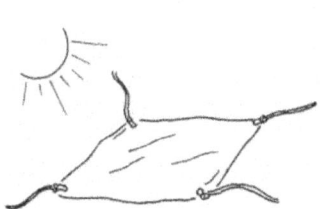

Wait until it dries and then sew some 550 paracord onto the corners.

Then you'll be able to use it for signaling aircraft, personnel, or for marking LZ/PZ.

Attention Non-Military Personnel: If you want, print in big, black, bold letters "HELP" on one side of it so others will know you need assistance and you're not just waving it to say "Hello."

## COFFEE FILTER USES

Have ya ever filled your canteens from a stream, creek, or river? If ya have, then you know it's very important to filter the water before adding water purification tablets to it. Ya don't wanna swallow some pesky little critters or pieces of dirt, do ya? No, of course not.

But how do you normally filter your water? Like everyone else, right? You place your handkerchief or T-shirt over the opening of the canteen and either submerge it or pour the water inside of it with a cup, right? Ain't this a hassle? And does it really, really filter everything out? No, I don't think so neither.

Well, instead of getting your handkerchief, T-shirt, or sock wet, why don't you just carry a couple paper "coffee filters" in each of your canteen pouches next to your water purification tablets. Then when you need to fill your canteens from a creek or stream, just pull out your coffee/water filters, place it over the opening of the canteen and either submerge it or pour the water through it.

NOTE: Water filtering means removing "visual" particles and matter from the water. And water purification means removing "invisible" bacteria and harmful elements from the water to make it safe for consumption.

Here's something else you can use a coffee filter for. When your weapon cleaning solvent becomes filthy, dirty, and grimy, don't dump it out or throw it away, just filter it with a coffee filter and you'll be able to reuse it over and over and over again. But remember not to get it mixed up with your water filter, throw it away after every use, (Auugh!)

## THERE ONCE WAS A FIRST SERGEANT AND A 2d LIEUTENANT...

A First Sergeant and a 2d Lieutenant in two separate cars collided on Fort Bragg.

The First Sergeant, seeing that the Lieutenant was a pretty shaken up, helps him out of his car and offers him a drink from his hip flask.

The Lieutenant drinks from it and then hands it back to the First Sergeant who then closes it up and puts it away.

The Lieutenant asks, "Aren't you going to drink any yourself, 1SG ?

The First Sergeant says, "Sure am, but only after the MPs leave. "

### THERE ONCE WAS A SERGEANT MAJOR AND A 2d LIEUTENANT...

A Sergeant Major and a 2d Lieutenant were walking together through the woods when they came upon a big old vicious grizzly bear.

The Sergeant Major stops, opens up his rucksack, pulls out his tennis shoes and quickly starts to put them on.

The Lieutenant looks down at him and says, " Sergeant Major, are you crazy? You 'll never be able to outrun that bear? "

The Sergeant Major looks up at the lieutenant and says, "I don't have to, sir, I only have to outrun you."

### THE HIGHWAY CODE WAS NEVER LIKE THIS

*Rick F. Tscherne*

## AMERICAN SURVIVAL GUIDE MAGAZINE

Now here's a magazine I recommend you get every month, it's called The American Survival Guide. It's a source of information for hunters, campers, hikers, survivalist and other outdoor enthusiasts who want to learn how to survive and thrive in the outdoors. It's excellent!

The editorial focuses on the philosophy of how to be ready to meet the challenges posed by threatening situations and man's relationship with nature. It also contains a mixture of information on technology, hardware, gear, supplies, and basically - how to become more independent and self-reliant living in the outdoors.

When it comes to learning how to survive outdoors, what the military forgot or failed to teach you, this magazine will fill in the gap. And if you want to keep up with the latest outdoor techniques and learn some new tips, tricks, and ideas on survival, then this magazine is for you. Trust me, you won't be disappointed and I guarantee what you learn from it can be utilized in the military too.

UPDATE: This magazine is no longer in print. Another magazine to check out is Survival Quarterly Magazine, created by the late Ron Hood: www.survivalQ.com

## U.S. Army Retirees - A Valuable Asset

*Written By: SGM George & Kulas, USA Ret.*

A few years ago someone e-mailed this article to me, I don't remember who sent it or where they got it from, but I found it very interesting. So Sgm George S. Kulas, I don't know who you are or where you live, but I hope you don't mind me sharing your article with my readers, because I agree with it 100%. Hooah!

\*\*\*\*\*\*\*\*\*\*\*\*\*\*\*\*\*\*\*\*\*\*\*\*\*\*\*\*

## Active Retirees and the Reserves

The U.S. Army has available "volunteers" who are highly skilled, trained, motivated and proven soldiers who could contribute substantially to it's combat readiness. These soldiers, U.S. Army active duty retirees, are an asset the Army could use in it's ready reserve. These retirees spent 20 or more years in the active Army. They knew their jobs well. Most of them were in leadership positions. Many trained junior soldiers during much of their careers. Many are combat veterans.

There are over 500,000 Army retirees, this more than the entire Army active duty force. Though it is likely that only a relatively small number are fully qualified, many of these retirees are prepared to continue serving in the Army on a part-time basis as members of the reserve components. Members of the selected reserve attend paid weekend drills and at least two weeks of active duty for training each year. Ready reserve soldiers are the most likely to be called to active duty during a war or national emergency.

Under current law, section 269d of Title 10 of the U.S. Code, military retirees are not allowed to join a ready reserve unit unless the Secretary of the Army "makes a special finding that the member's services in the ready reserve are indispensable." Additionally, according to Section 684 of the Title 10 of the U.S. Code, a soldier cannot receive both retired pay and reserve pay concurrently. Should a retiree be allowed to participate in the ready reserve under the provision of the statute, he or she will either have to decline reserve pay or forfeit retired pay for the number of days duty is performed.

Retirees Subject to Recall

The same Title 10 of the U.S. Code that virtually bars Army retirees from being in the ready reserve states, in Section 688, that an Army retiree can be ordered to active duty at any time. In fact, the Army's "Policies and Procedures for Pre-assigning and Recalling Retired Army Personnel During a War or National Emergency" states that retirees under age 60 are subject to be recalled to active duty within seven days of being notified. These retirees, if fully qualified, can also be assigned to deploying units that will fight the war. Active duty Army retirees are issued pre-assignment orders directing them where to report when fully mobilization is declared. A retiree who fails to comply with the orders may be considered Absent Without Leave (AWOL) and could be subject to disciplinary action by the Army, including suspension of retired pay.

On one hand, retirees are told they cannot join a ready reserve unit to train and maintain proficiency in basic soldier skills and their military occupational specialty. At the same time they are told that they are subject to being called back to active duty. Many of these retirees were senior soldiers and may be required to lead and or manage in areas that have undergone major technological changes since they retired from active duty. It would make sense to keep otherwise qualified retirees that may be called up trained to standards. This can be done through their participation in the reserve components.

Representative Thomas Petri (R-WI) agrees. In a letter dated September 16, 1993, Mr.Petri states that making it easier for an active duty retiree to join a Guard or Reserve unit seems to make sense. He says it is certainly possible for retirees who have been retired for a while to not be qualified physically or current in tactics and technology. Mr.Petri has taken this up with the chairman of the House Armed Services Committee and says, "I will be watching his response but action on this issue does not seem warranted."

Impact On Involuntarily Separated Soldiers

Francis M. Rush, Jr., the principal director for manpower and personnel in the Office of the Assistant Secretary of Defense for Reserve Affairs, is concerned about the impact or retired volunteers on newly separated active duty soldiers. In a letter dated October 5, 1993, he cites Section 269d of Title 10, U.S. Code as the law which precludes retired members from serving in the ready reserve. He goes on to say that the Department of Defense, however, is also committed to giving priority for membership in the Guard and Reserve to those individuals who were involuntarily separated from the service during the drawdown and who have not yet qualified for retirement.

There is no doubt whom we want to lead America's reserve soldiers into combat: The Army will need outstanding, motivated, experienced leaders who have excelled in soldiering and now want to continue serving the Army and the country. Otherwise qualified soldiers who were involuntarily separated from active service and now serve in the Guard or Reserve are invaluable. But there are also qualified soldiers who have voluntarily retired. The Army should take the best of die best from both groups. The law which says it can't be done needs to be changed so ti can be done. If Congress is serious about having the best quality Army with limited manpower, they should consider changing the law to give the Army flexibility in tapping personnel resources to assure the nation of a quality Army.

Impact on Reserve Promotions

There is another argument against allowing active duty retirees to serve in the ready reserves. It is believed that retirees would create a promotion stagnation and take up many of the higher grade positions and leave no upward mobility for junior leaders.

But this promotion stagnation could partly be overcome as long as active duty retirees agree to serve in positions far lower than which they retired at. And whether or not they are career reservists should not be a factor as long as they are physically fit and highly qualified volunteers.

Experienced active duty retirees could help reserve units to achieve and maintain combat readiness which many have had difficulties in doing. During Operation Desert Storm some Army National Guard units were called up and discovered not to be fully combat ready. Consequently these same units had to spend valuable time training and were not deploy to the Gulf. Perhaps these units would have been more prepared and ready to fight if they had experienced active duty retirees assigned to them.

# A FEW MINUTES WITH RANGER RICK

*A Ranger Rick Commentary*

The Army has "seven core values" that every soldier from private through general are suppose to live by and uphold. Which are:

- DUTY
- LOYALTY
- SELFLESS SERVICE
- HONOR
- COURAGE
- RESPECT
- INTEGRITY

Now even though I've been retired from active duty since 1993, I still try to live by these values today. And if you ask me which one I think is the most important, it would have to be INTEGRITY. Why? Because it would best describe myself - a person who lives by a set of values and moral codes.

Unfortunately there are some leaders today who claim they live by these values when they really don't. But to me they're not leaders, they're just sergeants and officers with "pocket values," meaning they pull 'em out and uphold 'em only when they want and then they put 'em away until they need 'em for another day.

A fine example of a leader with "pocket values" is someone who tells you don't drink and drive, don't bounce checks, don't commit adultery and you find out later on they got caught doing it. It falls in the same category as double standards and don't do what I do -just do what I say.

Now I've known quite a few NCOs who had pocket values, and I'm sorry to say some of them made it to the top three enlisted ranks (E7, E8, & E9) via their social "drinking & ass kissing" skills. Not to mention, bad mouthing and putting down other leaders in their unit to make themselves look better.

But what's interesting is while these lowlifes are busy sucking up to their company 1SG, battalion Command Sergeant Major and other members in their chain of command, the troops they lead know what kind of leaders they are. And it's just too damn bad there's nothing they can really do about it but hope they never have to serve under them again in the future.

Now one time I worked for a company called Military Professional Resources Incorporated (MPRI) in Bosnia under the US State Department approved "Train & Equip Program." And basically what this company does is hire ex-service members to help train foreign armies who are friendly towards the United States. And most, if not all are former or retired US Army Rangers, Special Forces, Drill Instructors or leaders who have exceptional teaching, training, and leadership skills. Or I should say "they're suppose to have these qualifications and skills."

And because most, if not all of them were former senior NCOs & officers with twenty or more years in service, you'd think they would all have high standards, morals, and live by these seven core values. And most of them did, but there were a few former military officers (an O4/Major, O5/Lt.Colonel, & a O6/Colonel) in my training brigade who had only "pocket values."

Now because they were making big bucks and wanted to hold onto their jobs as long as possible, they were willing to "sell their integrity" and "bark & wag their tails" when their masters pulled on their chain. And if we were having some training problems or something was not going according to plans, they would cover it up and tell the big boss, *"..everything's going fine and morale is high!"* (Sound familiar?)

For example:

One time about a hundred Bosnian (Muslim) NCOs and officers were given a two day class on how to call in artillery fire, and during the testing phase I was tasked to evaluate them, and they all failed my station miserably.

My supervisor, Major "Phil" was worried about our boss getting upset when he saw how many Bosnians failed the test, and so he asked me to grade them on "a curve" so it wouldn't look so bad on his report. I blew my stack and told him, *"No way, either they can do it or they can't, how would you like one of these guys to call in a fire mission for your unit?"* He says, "Well, no, not really, but... " And I told him, *"Forget it, I'm not passing any of them."* And a few days later I find out he told our boss it was my fault they failed and that I was giving him a hard time.

Another time we spent several weeks training a Bosnian battalion on how to fire & maneuver, and we were suppose to teach it by the book too. First you teach 'em how to fire & maneuver as a buddy team, then as a fire team, a squad and finally as a platoon. But while I was away on leave my boss, Lt. Colonel "Frank" instructed my fellow co-workers to skip the squad level. And when I came back from leave I was tasked to evaluate the unit on platoon level fire & maneuver, but they couldn't even do it as squads, and then later I found out why.

Well, when we had a meeting that day to discuss training, Lt. Colonel Frank says, *"Well I guess maybe we should have taught them squad level fire & maneuver before progressing to platoon level."* And I jumped on him and said, *"No shit, Sherlock..."* And the next thing I know I'm being blamed for this training and tagged a "loose cannon."

But what really upset me is when one of my fellow co-workers came to me and said, *"Rick, I gotta hand it to you buddy, you got a pair of balls. And if I didn't need this job, I'd be right there beside you telling 'em how f—— they are. But because I need the cash, if they want me to teach something differently, even though I know it's wrong, I'm gonna do what they tell me to do."*

And this my friends, is an example of leaders who live by "pocket values," they pull 'em out and uphold 'em only when they want and then they put 'em away until they need 'em for another day.

Till next time kiddies...

US Army, Retired (E-7 ½)

*Rick F. Tscherne*

# The Complete RANGER DIGEST

## Revised Edition - Misc. Updates

**Ranger School Tips:**
Training tips: http://72deuce.blogspot.com/2006/12/ranger-school-tips-train-up.html
A packing list: http://72deuce.blogspot.com/2006/12/ranger-school-tips-packing-list.html

**Military / Wilderness medical gear:**
http://www.chinookmed.com/

**Wilderness Medical Training**
http://wildernessmedicine.com/

**Equipment / Gear blogs:**
http://www.soldiersystems.net/
http://kitup.military.com/
http://www.militarymorons.com/

**ALICE pack Manual:**
http://www.georgia-outfitters.com/_alice/alicemanual.htm

**ALICE/Molle "Hellcat" Hybrid mod by Rod Teague** :
http://libertytreeblogs.blogspot.com/2011/04/building-hellcat-hybrid-ruck-from-us.html
or
.http://www.survivalistboards.com/showthread.php?t=47472&page=2

**ALICE Add-a-Pouch:**
http://shop.vtarmynavy.com/alice-pack-add-a-pouch-p4945.aspx

**Homemade backpacking gear plans:**
http://www.backpacking.net/makegear.html

**Ghillie Suits / Sniper head cloaks:** http://www.tacticalconcealment.com/
 Ghillie/camo material supplies: http://www.tacticalconcealment.com/cat-accessories.cfm
Camo Burlap: http://shop.vtarmynavy.com/camo-burlap-c636.aspx

## Lighting:
**Helmet Light:**
Princeton Tech MPLS: http://www.princetontec.com/?q=mpls

**Light stick alternatives:**
*Krill light*
The Krill Light electronic electroluminescent (EL) lightstick originated form the need for a less wasteful, more efficient replacement to chemical lightsticks. Krill Lights were put into production and the U.S. Military immediately saw the value in these marking and lighting tools. Today, the Krill Light is available in three sizes with various lighting options and colors including infra-red and strobing versions.
http://www.kriana.com/how-to-choose

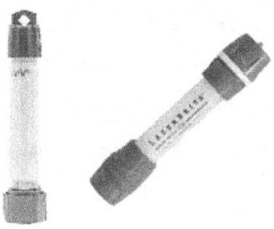

Left: Krill , Right: Lazerbrite

*Lazerbrite MultiLux:*
Similar to the Krill light above, this light stick replacement can change colors/convert to a flashlight of bright spot/and even breakdown into two lights , in case your buddy needs one. They have an iris shield to aid in light discipline and also sell LZ marking kits. Awesomely smart design, made in the USA and not too expensive.
https://www.lazerbrite.com/

Reusable glow light: http://www.uvpaqlite.com/tooblite.html
PaqLite: http://www.uvpaqlite.com/
Keychain glowtube:
http://www.jsburlys.com/index.php?app=ecom&ns=prodshow&ref=texaccessoriesa3bglow
Glow in dark Ranger eyes:
http://www.tripleaughtdesign.com/Apparel/Apparel-Accessories/FrontSight-Ranger-Eye

Superbright glow-in-the-dark paint: http://glonation.com/pigmented-glow-paint.html
Luminous Tape: http://www.amazon.com/Rothco-Luminous-Tape-Yds/dp/B0036VLEA0

## Weapon stuff:
MAG Pod: https://www.facebook.com/MagPod

Rail integrated weapon light: Inforce XML
http://www.inforce-mil.com/

M4 Cleaning tool: Combat Applications Tool

http://www.optactical.com/catm4tool.html

**Tools:**
Leatherman MUT: Great review and pics  http://www.2centtac.com/?p=633

Battery caddy:
http://www.adventuresurvivalequipment.com/survival-accessories/powerpax-battery-caddy.html
http://www.tripleaughtdesign.com/Equipment/Equipment-Accessories/Battery-Case
http://www.gggaz.com/preparedness-gear/batteries-and-battery-caddies.html

Tear-Aid repair heavy-duty tape:
http://www.kk.org/cooltools/archives/006215.php

Velcro adhesive tape, including Multicam velcro:
http://www.supplycaptain.com/index.cfm?fuseaction=category.display&category_ID=14

**Water/Hydration:**
Collapable water bladders:
  Evernew bladders: http://www.evernewamerica.com/EBY205208.htm
  Platypus: http://www.amazon.com/exec/obidos/ASIN/B002LSS68C
  Vapur (US Made): http://vapur.us/

Inexpensive water filter straw:
http://www.cheaperthandirt.com/CAMP174-1.html

New UV water purifier system, Camelbak's "All Clear" :
http://www.camelbak.com/Sports-Recreation/Purification/Intro.aspx

**More Survival kits in a tin**
http://www.adventuresurvivalequipment.com/military-pocket-survival-tin.html
http://www.adventuresurvivalequipment.com/ultimate-adventurer-survial-kit.html

**Survival Gill Net:**
http://www.adventuresurvivalequipment.com/survival-gill-net-best-glide.html

**Rail-mounted Spare parts/Survival kit:**
http://soldiersystems.net/2011/10/21/e-pak-from-ares-defense/

**Electronics:**
Head / Weapon Cams, rail mountable HD cameras
  GoPro: http://gopro.com/
  Contour: http://contour.com/
  ReplayXD: http://replayxd.com/
  The Replay is available from *UStacticalsupply.com*

**Military Smartphone Apps:**
http://www.networkworld.com/news/2011/041411-military-iphone-apps.html

http://articles.latimes.com/2011/sep/25/business/la-fi-isoldiers-20110926

http://www.navytimes.com/offduty/technology/offduty-technology-hot-military-apps-practical-103111w/

Extended battery pack iPhone case:
http://www.mophie.com/juice-pack-PRO-Rugged-case-for-iPhone-4-4S-p/2120_jppr-ip4-blk.htm

# Are You Really Survival Ready? Anytime? Anywhere? Are You?

## Announcing Some of the "Most Advanced & Jealously Guarded 169+ Survival Secrets."

Hi, My name is Joseph A. Laydon Jr. and I'm a retired US Army Special Forces Instructor. And like Ranger Rick, I also publish my own line of handbooks too, under the name of Intensive Research Information Services and Products. And to name just a few of the Most Advanced & Jealously Guarded Survival Secrets you won't find in any military or civilian survival book.... (If you don't believe me, try looking 'em up.)

Where to get plenty of water where none is seen, no digging and no hard work neither

Catch fish lea and right with this simple item that can be found in almost any household.

Navigate using simple 3rd grade math instead of wandering around lost like an idiot.

How to hypnotize snakes and other tasty critters and be in total control of them.

Build a special type of shelter that "energizes" your tired body, science is baffled!

How to forecast weather using insects, pine cones, birds, the sky, moon, sun...

How to make a home-made compass that you can walk and run with it too, cost you nothing.

How to build a field expedient "unsinkable raft" that 12 men can't even submerge, cheap too.

Got a knife but no compass? How you can use it to navigate without guessing.

How to cocoon yourself & stay toasty warm - it's cheap, waterproof & critter proof too.

How to construct 21 international traps & snares for snakes to deer that are illegal in most states and countries except in real "life or death" survival situations.

Did you find any? I'll bet you didn't! And these are nothing compared to my XXX-Rated Anytime Anywhere Survival Program which contains 12 packages of information that will insure a lifetime of safe outdoor adventure for you, your family and unit members too. And if you're not completely

100% satisfied - you have my No-Risk Unconditional Money Back Guarantee. For a Free Special Report and more info, remove this page, fill it out and send it to:

**IRISAP**
**http://survivalexpertbooks.com/**
1157 Dawn Drive, Dept RRFT-1
Belleville, IL 62220-3325

NAME:_____ DOB: ____
STREET      ADDRESS:_____
CITY:_____ STATE:_____ ZIP:_____
PHONE_____FAX_____E-MAIL:_____

IMPORTANT: To receive special bonuses, discounts, & other information, you must remove and send in this page or make a photocopy of it, please do not send any letters or postcards.

ANNOUNCING
THE NEWEST MEMBER OF
# RANGER RICK'S FAMILY
# THE NEXT FUTURE US ARMY

### ALLEN E. LOVE
Born: October 27, 1999

To The Parents Of:
Dana & Jennifer (Tscherne) Love

PS - This is a photo of my "Little Ranger Buddy" catching some Z's before his next night mission - Operation: Keep Mom & Pop Up All Night  Hooah

---

Also available:
# THE COMPLETE RANGER DIGEST
## VOLUMES 1 – 5

In both print and ebook editions
at AMAZON.com

# ABOUT THE AUTHOR...

**Richard F. Tscherne** (nicknamed "Ranger Rick") was a member of the United States Army who successfully graduated from the U.S. Army Ranger School, the French Army commando School, and the Belgium Army Commando School.

His awards include the U.S. Army Ranger Tab, Master Parachutist Wings, Drill Instructor Badge, Expert Rifleman Badge, Jungle Expert Patch, 5 AAM, 1 ARCOM, 3 MSM, 4 Overseas Ribbons and an assortment of other U.S. military medals. His foreign awards include the French Army Commando Badge, the Belgium Army Commando Badge, the Italian Army Parachutist Wings and the German Army Weapon's Qualification Badge.

Ranger Rick served over 13 of 21 years overseas in Italy, Germany, and Korea. His vast experiences there included duties as a Rifleman, Machine Gunner, RTO, Recon Scout, Armor, Cold Weather Instructor, Drill Instructor, Recon Gun Jeep Section Leader, Anti-Tank Squad Leader, Airborne/Ranger Platoon Sergeant, and Asst. Bn Operation Sergeant. He has served in the following units;

| | |
|---|---|
| 1st Bn 87th Inf. (Mech) Germany | 1st Bn 31st Inf. (Mech) Korea |
| 1st Bn 509th ABCT (Abn) Italy | 3rd Bn 325th Inf. (Ft. Bragg) |
| 1st Bn 75th Rangers (Ft. Stewart) | Drill Instr. A-4-3 (Ft. Dix) |

In September 1992, he was selected by DA for advancement to Master Sergeant/E-8, but he refused the promotion and on January 1st, 1993, he retired from the United States Army to live in Italy.

# Be sure to check out Rick's new website:

## www.SurvivalOutdoorSkills.com

- Outdoor Survival Tips and Info
- SOS Survival Kits (developed by "Rick")
- Survival Training

### Loose Cannon Military History DVDs
- The Generals/ Great Battles series
- Historical footage / battlefield newsreels
- Hard-to-find vintage training films
- Hybrid discs include Field and Technical Manuals

www.loose-cannon.com/military

Other Websites:

AnySoldier.com - Send a care package to an overseas soldier

**"Simple Survival**
A Family Outdoors Guide" is more than a book —it is an outdoor resource bible that every family should have a copy of. This is one of those books that you should have in your camping bag along with the tent and other equipment. However, reading it at home before you go off on some outdoor adventure would be a great help when potential situations happen.
   Available at Amazon and other online bookstores

**Impending Disasters** - This helpful and comprehensive book covers most major disasters and how to stay safe if you decide to evacuate or stay. It has a section on prolonged survival, which will assist keeping you alive after the natural disaster has done its damage. Many people die following natural disasters, from one mishap or another, but you can learn to survive.
Learn to deal with Tornadoes, ice storms, hurricane, flooding, blackouts, riots, and much more. Contains easy to understand information, and critical gear/equipment lists you will need.
   Available at Amazon and other online bookstores

On a trip to the Lake Clark area of the Alaskan bush, a sudden arctic weather system forces down the small plane of Dr. Jim Wade, and his son David. Both have survived the crash, but not unscathed. Food, fire and shelter are all a priority. Following the death of his father, now it is up to David to figure out what to do next, and how to survive, on a remote Alaskan mountain—in winter!

This is a fictional story of survival, resilience and of the spirit to live. It is both authentic and accurate, having been written by a former Air Force life support survival instructor. For ages 10 and up

**Both are available at Amazon and other online bookstores**

Set adrift, a family of three are cast out to sea in a rubber raft, where they must find a way to conquer one terrifying tragedy after another or die in the process.

In this gripping story of survival everyone will be tested to their limits. Christian faith and hope are hallmarks of this tale that will touch your heart..

www.ingramcontent.com/pod-product-compliance
Lightning Source LLC
Chambersburg PA
CBHW071553080526
44588CB00010B/900